Computational Thermodynamics of Materials

A unique and comprehensive introduction, offering an unrivalled and in-depth understanding of the computational-based thermodynamic approach and how it can be used to guide the design of materials for robust performances. This valuable resource integrates basic fundamental concepts with experimental techniques and practical industrial applications to provide readers with a thorough grounding in the subject. Topics covered range from the underlying thermodynamic principles to the theory and methodology of thermodynamic data collection, analysis, modeling, and verification, with details on free energy, phase equilibrium, phase diagrams, chemical reactions, and electrochemistry. In thermodynamic modeling, the authors focus on the CALPHAD method and first-principles calculations. They also provide guidance for the use of YPHON, a mixed-space phonon code, developed by the authors for polar materials, based on the supercell approach. Including worked examples, case studies, and end-of-chapter problems, this is an essential resource for students, researchers, and practitioners in materials science.

Dr. Zi-Kui Liu is a professor at the Department of Materials Science and Engineering at The Pennsylvania State University. He has been the Editor-in-Chief of CALPHAD journal since 2001 and the President of CALPHAD, Inc. since 2013. Dr. Liu is a Fellow and a member of the Board of Trustees of ASM International and was a member of the TMS Board of Directors. His awards include the ASM J. Willard Gibbs Phase Equilibria Award, the TMS Brimacombe Medalist Award, the ACers Spriggs Phase Equilibria Award, the Wilson Award for Excellence in Research from The Pennsylvania State University, the Chang Jiang Chair Professorship of the Chinese Ministry of Education, and the Lee Hsun Lecture Award of the Institute of Metal Research, Chinese Academy of Sciences.

Dr. Yi Wang is a senior research associate of Materials Science and Engineering at The Pennsylvania State University. He has been working on method development and computerized simulation of material properties, using a range of disciplines including condensed matter theory, quantum chemistry, thermodynamics, high-temperature and high-pressure reactions, elastic/plastic mechanics, molecular dynamics, and all first-principles calculation related subjects.

Computational Thermodynamics of Materials

ZI-KUI LIU
The Pennsylvania State University

YI WANG
The Pennsylvania State University

CAMBRIDGE
UNIVERSITY PRESS

University Printing House, Cambridge CB2 8BS, United Kingdom

Cambridge University Press is part of the University of Cambridge.

It furthers the University's mission by disseminating knowledge in the pursuit of education, learning and research at the highest international levels of excellence.

www.cambridge.org
Information on this title: www.cambridge.org/9780521198967

First published 2016

Printed in the United States of America by Sheridan Books, Inc.

A catalog record for this publication is available from the British Library

Library of Congress Cataloging-in-Publication Data
Names: Liu, Zi-Kui. | Wang, Yi. (Writer on thermodynamics)
Title: Computational thermodynamics of materials / Zi-Kui Liu, Yi Wang.
Description: Cambridge : Cambridge University Press, 2016. | Includes bibliographical references and index.
Identifiers: LCCN 2015040101 | ISBN 9780521198967 (Hardback : alk. paper)
Subjects: LCSH: Materials–Thermal properties. | Thermodynamics. | Heat–Transmission–Mathematical models.
Classification: LCC TA418.52 .L58 2016 | DDC 620.1/1296–dc23 LC record available at http://lccn.loc.gov/2015040101

ISBN 978-0-521-19896-7 Hardback

Additional resources for this publication at www.cambridge.org/9780521198967

Contents

1 Laws of thermodynamics

1.1 First and second laws of thermodynamics

Thermodynamics is a science concerning the state of a system when interacting with the surroundings; it is based on two laws of nature, the first and second laws of thermodynamics. The interactions can involve exchanges of any combinations of heat, work, and mass between the system and the surroundings, dictated by the boundary conditions between the system and the surroundings. The first law of thermodynamics describes those interactions, while the second law of thermodynamics governs the evolution of the state inside the system. Consequently, the combination of the first and second laws of thermodynamics provides an integration of the external and internal parts of a system.

A system typically consists of many chemical components. The first law of thermodynamics states that the exchanges of heat, work, and individual components with the surroundings must obey the law of conservation of energy. In the domain of materials science and engineering, the energy of interest is at the atomic and molecular levels. The energies at higher and lower levels such as nuclear energy and the kinetic and potential energies of a rigid body are usually excluded from the discussion of the thermodynamics of materials.

Let us consider a system receiving an amount of heat, dQ, an amount of work, dW, and an amount of each independent component i, dN_i, from the surroundings. Such a system is called an open system in contrast to a closed system when $dN_i = 0$ for all components, i.e. there is no exchange of mass between the system and the surroundings. Other types of systems commonly defined in thermodynamics include adiabatic systems, those without exchange of heat, i.e. $dQ = 0$, and isolated systems, those without exchange of any kind, i.e. $dQ = dW = dN_i = 0$.

The corresponding change of energy in the system, i.e. the internal energy change, dU, is formulated in terms of the first law of thermodynamics as follows,

$$dU = dQ + dW + \sum H_i dN_i \qquad 1.1$$

where H_i is the unit energy of component i in the surroundings, and the summation is for all components in the system which can be controlled independently from the surroundings, i.e. the independent components of the system.

It is self-evident that the left-hand side of Eq. 1.1 refers to the change inside the system, while its right-hand side is for the contributions from the surroundings to the

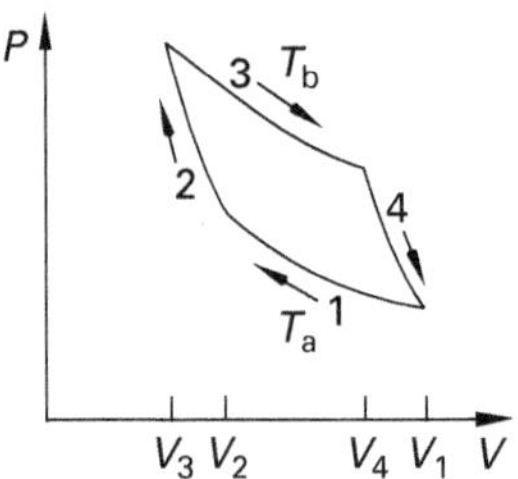

Figure 1.1 *Schematic diagram of the Carnot cycle, from [1] with permission from Cambridge University Press.*

system. In principle, no matter how the heat and mass are added, or how work is done to the system, as long as their summation is the same the change of the internal energy from the first law of thermodynamics will be the same, indicating that for a closed system the system always reaches the same state. The internal energy is thus a state function in a closed system as it does not depend on how the state is reached.

On the other hand, for the purpose of easy mathematical treatment, a reversible process can be considered for a closed system, in which the initial state of the system can be restored reversibly without any net change in the surroundings. Therefore, the heat transferred and the work done to the system are identical to the heat and work lost by the surroundings and vice versa. The classic example of reversible processes is the Carnot cycle, which is shown in Figure 1.1. It consists of four reversible processes for a closed system. The four reversible processes are compression at constant temperature T_1 (isothermal), compression without heat exchange (adiabatic) ending at T_2, isothermal expansion at T_2, and adiabatic expansion ending at T_1.

The Carnot cycle involves a simple type of mechanical work, either hydrostatic expansion or compression, with the work that the surroundings does to the system represented by

$$dW = -PdV \tag{1.2}$$

with P being the external pressure that the surroundings exerts on the system and V the volume of the system. It is now necessary to differentiate the external and internal variables for further discussion, with the former representing variables in the surroundings and the latter representing variables in the system. For the isothermal processes in the Carnot cycle, the entropy change of the system, dS, can be defined as the heat exchange divided by temperature:

$$dS = \frac{dQ}{T} \tag{1.3}$$

In addition to processes involving heat, work, and mass exchanges between the system and the surroundings, there can be internal processes taking place inside the system. As the system cannot do work to itself, the criterion for whether an internal process can occur spontaneously must be related to the heat exchange, which is related to the entropy change as shown by Eq. 1.3.

It is a known fact that heat will spontaneously transfer from a higher temperature (T_2) region to a lower temperature (T_1) region inside a system if heat conduction is allowed, and this process is irreversible because heat cannot be conducted from a low temperature region to a high temperature region spontaneously. Equation 1.3 indicates that for the same amount of heat change, the entropy change at T_1 is higher than that at T_2, and the heat conduction thus results in a positive entropy change in the system, i.e.

$$\Delta S = -\frac{dQ}{T_2} + \frac{dQ}{T_1} = \frac{dQ}{T_2 T_1}(T_2 - T_1) > 0 \qquad 1.4$$

Consequently, the second law of thermodynamics is obtained, which states that for an internal process to take place spontaneously, or irreversibly, this internal process (*ip*) must have positive entropy production, which can be written in differential form as follows:

$$d_{ip}S > 0 \qquad 1.5$$

From the definition of entropy change shown by Eq. 1.3, the amount of heat produced by this irreversible internal process can be calculated as follows:

$$d_{ip}Q = Td_{ip}S \qquad 1.6$$

Let us represent this internal process by $d\xi$ and define the driving force for the internal process by D. The work done by this internal process is thus $Dd\xi$, which is released as heat, i.e.

$$Dd\xi = d_{ip}Q = Td_{ip}S \qquad 1.7$$

An irreversible process thus must have a positive driving force in order for it to take place spontaneously.

1.2 Combined law of thermodynamics and equilibrium conditions

For a system with an irreversible internal process taking place, the entropy change in the system consists of three parts: the heat exchange with the surroundings, defined by Eq. 1.3, the entropy production due to the internal process, represented by Eq. 1.5, and the entropy of mass exchange with the surroundings. The total entropy change of the system can thus be written as follows:

$$dS = \frac{dQ}{T} + d_{ip}S + \sum S_i dN_i \qquad 1.8$$

where S_i is the unit entropy of component i in the surroundings, often called the partial entropy of component i, which will be further discussed in Chapter 2.

Combining Eq. 1.7 and Eq. 1.8 and re-arranging, one obtains

$$dQ = TdS - Dd\xi - \sum TS_i dN_i \qquad 1.9$$

Inserting Eq. 1.2 and Eq. 1.9 into Eq. 1.1 yields the combined law of thermodynamics from the first and second laws of thermodynamics,

$$dU = TdS - PdV + \sum (H_i - TS_i)dN_i - Dd\xi \qquad 1.10$$

The internal energy of the system is thus a function of the variables S, V, N_i and ξ of the system, which are called natural variables of the internal energy, i.e. $U(S,V,N_i,\xi)$. The other variables are dependent variables and can be represented by partial derivatives of the internal energy with respect to their respective natural variables with other natural variables kept constant, as shown below:

$$T = \left(\frac{\partial U}{\partial S}\right)_{V,N_i,\xi} \qquad 1.11$$

$$-P = \left(\frac{\partial U}{\partial V}\right)_{S,N_i,\xi} \qquad 1.12$$

$$\mu_i = H_i - TS_i = \left(\frac{\partial U}{\partial N_i}\right)_{S,V,N_{j\neq i},\xi} = U_i \qquad 1.13$$

$$-D = \left(\frac{\partial U}{\partial \xi}\right)_{S,V,N_i} \qquad 1.14$$

In Eq. 1.13, a new variable, μ_i, is introduced. This is called the chemical potential and is defined as the internal energy change with respect to the addition of the component i when the entropy, volume, and the amount of other components of the system are kept constant. It may be worth pointing out that for a system at equilibrium, i.e. $d_{ip}S = 0$, and with constant entropy, $dS = 0$, if the system exchanges mass with the surroundings, $dN_i \neq 0$, then it must also exchange heat with the surroundings at the same time in order to keep the entropy invariant as demonstrated by Eq. 1.8.

The pairs of natural variables and their corresponding partial derivatives are called conjugate variables, i.e. S and T, V and $-P$, N_i and μ_i, and ξ and $-D$. There are minus signs in front of P and D as the increase of volume and the progress of the internal process decrease the internal energy of the system. The importance of this conjugate relation will be evident when various forms of combined thermodynamic laws and various types of phase diagrams are introduced.

The last pair of conjugate variables, ξ and $-D$, is worthy of further discussion. Based on the second law of thermodynamics, i.e. Eq. 1.5, no internal processes take place spontaneously if there is no entropy production, i.e. $D \leq 0$ or $d\xi = 0$ and $D > 0$. With $D \leq 0$, there is no driver for any internal processes, and the system is in a full equilibrium state. The last term in Eq. 1.10 drops off, and ξ becomes a dependent variable of the system and can be calculated from the equilibrium conditions. With $d\xi = 0$ and $D > 0$, the system is under a constrained equilibrium or freezing-in condition when the internal process is constrained not to take place, and ξ remains an independent variable of the system.

These two cases represent the two branches of thermodynamics: equilibrium, i.e. reversible, thermodynamics and irreversible thermodynamics. It is clear from the above discussions that these two branches are identical if the internal energy is not only a function of S, V, and N_i, but is also a function of any internal process variable ξ. This means that one should be able to evaluate the internal energy of a system for any freezing-in equilibrium conditions in addition to the full equilibrium conditions. In the rest of the book, freezing-in equilibrium and full equilibrium are not differentiated unless specified.

As the mechanical work under hydrostatic pressure is very important in experiments, let us define a new quantity called the enthalpy as follows:

$$H = U + PV \tag{1.15}$$

Its differential form can be obtained from Eq. 1.1 as

$$dH = dU + d(PV) = dQ + VdP + \sum H_i dN_i \tag{1.16}$$

There are two significant consequences of the above equation. First, for a closed system under constant pressure, i.e. $dN_i = dP = 0$, one has $dH = dQ$. This implies that the enthalpy change in a system is equal to the heat exchange between the system and the surroundings of the system, which is why enthalpy and heat are often used exchangeably in the literature. Second, for an adiabatic system under constant pressure, i.e. $dQ = dP = 0$, Eq. 1.16 can be re-arranged to the following equation:

$$H_i = \left(\frac{\partial H}{\partial N_i}\right)_{N_{j \neq i}, dQ=dP=0} \tag{1.17}$$

The quantity H_i is thus the partial enthalpy of component i and will be further discussed in Chapter 2. The chemical potential of component i defined in Eq. 1.13 is thus related to the partial enthalpy and partial entropy of the component.

To further define the equilibrium conditions of a system, consider a homogeneous system in a state of internal equilibrium, i.e. no spontaneous internal processes are possible with $Dd\xi = 0$ and Eq. 1.10 becomes

$$dU = TdS - PdV + \sum \mu_i dN_i = \sum Y_i dX_i \tag{1.18}$$

where X represents S, V, N_i, and Y represents their conjugate variables T, $-P$, μ_i. The state of the system with c independent components is completely determined by $c + 2$ variables, i.e. S, V, and N_i with i ranging from 1 to c.

To simplify the situation, let us limit the discussion to an isolated equilibrium system, i.e. $dU = 0$, and conduct a virtual internal experiment inside the system by moving an infinitesimal amount of X_i, dX_i, with other X_j kept constant, from one region of the system to another region of the system as schematically shown in Figure 1.2.

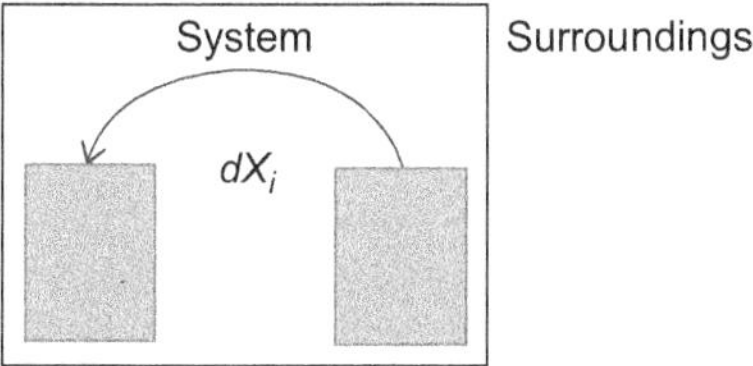

Figure 1.2 *Virtual experiment for a system at equilibrium.*

As the system is homogeneous and at equilibrium, $-dX'_i = dX''_i = dX_i$. The total change of the internal energy for this internal process is the combination of the changes in the two regions, i.e.

$$dU = dU' + dU'' = Y'_i dX'_i + Y''_i dX''_i = \left(-Y'_i + Y''_i\right) dX_i = 0 \qquad 1.19$$

Therefore, $Y'_i = Y''_i$ for T, $-P$, and μ_i, indicating that T, $-P$, and μ_i are homogeneous in the system, respectively, and are thus named as potentials of the system. Furthermore, these potentials are independent of the size of the system and are often referred to as intensive variables in the literature. On the other hand, all X, i.e. S, V, and N_i, are proportional to the size of the system and can be normalized with respect to the size of the system, usually in terms of the total number of moles,

$$N = \sum N_i \qquad 1.20$$

They are thus called molar quantities and are often referred to as extensive variables, and the respective normalized variables are molar entropy, molar volume, and mole fractions, defined as follows:

$$S_m = \frac{S}{N} \qquad 1.21$$

$$V_m = \frac{V}{N} \qquad 1.22$$

$$x_i = \frac{N_i}{N} \qquad 1.23$$

Consider a small subsystem in this homogeneous system at equilibrium and let the subsystem grow in size. The entropy, volume, and mass enclosed in the subsystem increase as follows:

$$dS = S_m dN \qquad 1.24$$

$$dV = V_m dN \qquad 1.25$$

$$dN_i = x_i dN \qquad 1.26$$

The corresponding change in the internal energy of the subsystem becomes

$$dU = TdS - PdV + \sum \mu_i dN_i = \left(TS_m - PV_m + \sum \mu_i x_i\right) dN = U_m dN \qquad 1.27$$

By integration one obtains the integral form of the internal energy as

$$U = \left(TS_m - PV_m + \sum \mu_i x_i\right) N = U_m N = TS - PV + \sum \mu_i N_i \qquad 1.28$$

Similarly, the molar enthalpy can be defined as follows:

$$H = U + PV = U_m N + PV_m N = (U_m + PV_m) N = H_m N \qquad 1.29$$

In the case when a potential is not homogeneous in a system, the system will not be in a state of equilibrium. Let us consider the same virtual experiment as shown in Figure 1.2 for an isolated system that is not in equilibrium, i.e. by moving an infinitesimal amount of X_i, dX_i, with other X_j kept constant, from one region of the system to another region of the system with the two regions having different potentials. The total internal energy change is equal to zero as the virtual experiment has $dU = 0$. Similarly, each region can be considered to be homogeneous by itself, and one has $-dX_i' = dX_i'' = dX_i$. The total internal energy change in the system is thus the sum of that for these two regions plus the entropy production due to the internal process with $d\xi = dX_i$, i.e.

$$dU = dU' + dU'' + Dd\xi = Y_i'dX_i' + Y_i''dX_i'' + Dd\xi = \left(-Y_i' + Y_i''\right)dX_i + Dd\xi = 0 \qquad 1.30$$

Consequently, one obtains the following:

$$D = Y_i' - Y_i'' \qquad 1.31$$

The driving force thus represents the difference of the potential at the two regions, and the internal process acts to eliminate inhomogeneity of the potential by means of heat transfer from high temperature regions to low temperature regions, or volume shrinkage of low pressure regions (high $-P$) and volume expansion of high pressure regions (low $-P$), and/or the transport of components from high chemical potential regions to low chemical potential regions.

1.3 Stability at equilibrium and property anomaly

As shown by Eq. 1.19, potentials are homogenous for a homogeneous system in a state of internal equilibrium. To study the stability of the equilibrium state, one considers the entropy production due to a fluctuation of a molar quantity as an internal process. Based on the second law of thermodynamics, the driving force, as the first derivative of the entropy production with respect to the internal process, is zero for such a fluctuation at equilibrium, i.e. $D = 0$, and the entropy of production thus depends on the second derivative. It can be written as follows in terms of Taylor expansion:

$$Td_{ip}S = \frac{\partial_{ip}S}{\partial\xi}d\xi + \frac{1}{2}\frac{\partial_{ip}^2 S}{\partial\xi^2}(d\xi)^2 = Dd\xi - \frac{1}{2}D_2(d\xi)^2 \qquad 1.32$$

with $D_2 = -\partial_{ip}^2 S/\partial\xi^2$. When $\partial_{ip}^2 S/\partial\xi^2 < 0$ or $D_2 > 0$ along with $D = 0$, the fluctuation does not produce positive entropy of production and thus cannot develop further. The equilibrium state of the system is therefore stable against the fluctuation. On the other hand, when $\partial_{ip}^2 S/\partial\xi^2 > 0$ or $D_2 < 0$ along with $D = 0$, the fluctuation creates positive entropy of production and can continue to grow. The equilibrium state of the system is therefore unstable against the fluctuation. In connection with Eq. 1.8, one can realize that for a system at stable equilibrium without heat and mass

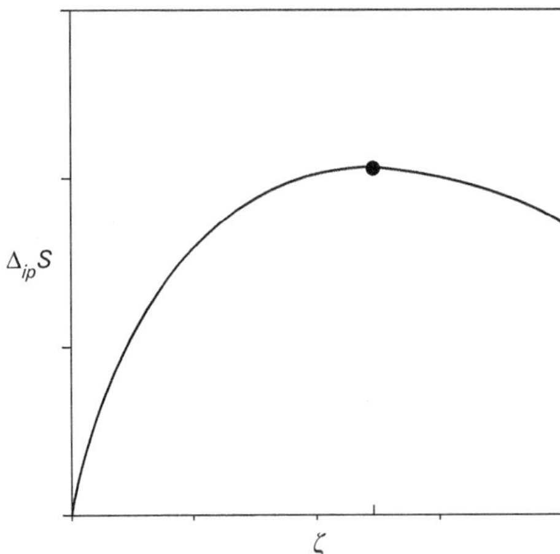

Figure 1.3 *Schematic diagram showing maximum entropy, from [1] with permission from Cambridge University Press.*

exchange with the surroundings, its entropy is at its maximum and there are no other internal processes which could produce any more entropy. This is schematically shown in Figure 1.3.

Using Eq. 1.10, Eq. 1.18, and Eq. 1.32, the combined law of thermodynamics can be written as

$$dU = \sum Y_i dX_i - D d\xi + \frac{1}{2} D_2 (d\xi)^2 \qquad 1.33$$

Let us carry out the same virtual internal experiment as in Section 1.2, i.e. moving an infinitesimal amount of X_i in an isolated homogenous system with the other X_j kept constant, i.e. $dU = 0$ and $D = 0$. The internal energy change due to this internal process is

$$dU = \frac{1}{2} D_2 \left\{ \left(dX_i'\right)^2 + \left(dX_i''\right)^2 \right\} \qquad 1.34$$

For a homogeneous system in a state of stable equilibrium with $\left(dX_i'\right)^2 = \left(dX_i''\right)^2 = \left(dX_i\right)^2$, this internal process must result in an increase of internal energy, $dU > 0$, and thus gives

$$D_2 = 2\left(\frac{\partial^2 U}{\partial X_i^2}\right)_{X_j} = 2\left(\frac{\partial Y_i}{\partial X_i}\right)_{X_j} > 0 \qquad 1.35$$

Equation 1.35 shows that for a system to be stable, any pair of conjugate variables must change in the same direction when other independent molar quantities are kept constant. For the conjugate variables discussed so far, this means that for a stable system, the addition of entropy increases with temperature if $\partial T/\partial S > 0$, the volume decreases with pressure or increases with the negative of pressure if $\partial(-P)/\partial V > 0$, and the chemical

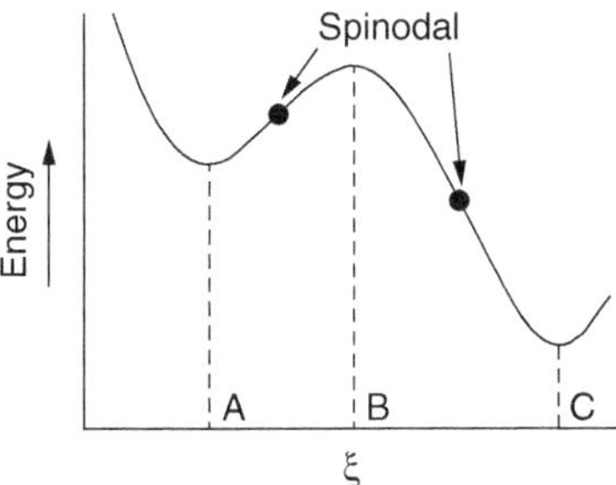

Figure 1.4 *Schematic diagram showing the metastable (A), unstable (B), and stable (C) equilibrium states.*

potential of a component increases with the amount of the component, i.e. $\partial\mu_i/\partial N_i > 0$, where the derivatives are taken with all other molar quantities kept constant. The limit of stability is reached when Eq. 1.35 becomes zero, i.e.

$$D_2 = 2\left(\frac{\partial Y_i}{\partial X_i}\right)_{X_j} = 0 \tag{1.36}$$

Figure 1.4 shows schematically the energy as a function of configuration including three states: unstable, stable, and metastable. Both the stable and metastable states have positive curvatures due to $D_2 > 0$, while the unstable state has a negative curvature due to $D_2 < 0$. There is an inflection point, at which $D_2 = 0$, for a state between a stable or metastable state with $D_2 > 0$ and an unstable state with $D_2 < 0$. These two inflection points, called spinodal, represent the limit of stability. The states between the two inflection points are unstable, and the other states are either stable or metastable. The two inflection points can move apart from or close to each other depending on the change of external conditions, i.e. the natural variables. One extreme situation is when these two inflection points merge into one point, and the instability occurs only at this particular point. It is evident that then all three states, stable, metastable, and unstable, also merge into one point. This point is called the critical or consolute point, beyond which the instability no longer exists.

To mathematically define the consolute point, the third derivative needs to be added to Eq. 1.32 because both D and D_2 vanish at this point, i.e.

$$Td_{ip}S = \frac{\partial_{ip}S}{\partial\xi}d\xi + \frac{1}{2}\frac{\partial_{ip}^2 S}{\partial\xi^2}(d\xi)^2 + \frac{1}{6}\frac{\partial_{ip}^3 S}{\partial\xi^3}(d\xi)^3 = Dd\xi - \frac{1}{2}D_2(d\xi)^2 + \frac{1}{6}D_3(d\xi)^3 \tag{1.37}$$

$$dU = \sum Y_i dX_i - Dd\xi + \frac{1}{2}D_2(d\xi)^2 - \frac{1}{6}D_3(d\xi)^3 \tag{1.38}$$

At the consolute point, the third derivative also becomes zero, i.e.

$$D_3 = \frac{\partial_{ip}^3 S}{\partial\xi^3} = 0 \tag{1.39}$$

Let us further discuss the properties of the system in relation to the critical point. By taking the inverse of the equation for the limit of stability, Eq. 1.36, one obtains

$$\left(\frac{\partial X_i}{\partial Y_i}\right)_{X_j} = +\infty \tag{1.40}$$

i.e. all X_i quantities diverge at the critical point. Therefore, when a system approaches the critical point from its stable region, the change of a molar quantity with respect to its conjugate potential varies dramatically and becomes infinite at the critical point, resulting in property anomalies in the system. In the unstable region, the system will thus separate into stable subsystems and become heterogeneous, and the X_i will change discontinuously between subsystems. In the stable region, the change of a molar quantity with respect to its conjugate potential decreases as the system moves away from the critical point and remains positive due to the stability criterion denoted by Eq. 1.35.

However, it is not clear how a molar quantity changes with respect to a non-conjugate potential at the critical point. From the Maxwell relation, one has

$$\left(\frac{\partial Y_i}{\partial X_j}\right)_{X_{k\neq j}} = \frac{\partial^2 U}{\partial X_i \partial X_j} = \left(\frac{\partial Y_j}{\partial X_i}\right)_{X_{k\neq i}} \tag{1.41}$$

$$\left(\frac{\partial X_j}{\partial Y_i}\right)_{X_{k\neq j}} = \left(\frac{\partial X_i}{\partial Y_j}\right)_{X_{k\neq i}} \tag{1.42}$$

Since all the X_i diverge at the critical point, both derivatives in Eq. 1.42 should also go to infinity at the critical point. To investigate their signs, let us carry out a virtual experiment similar to that used to derive the stability condition (Eq. 1.34 and Eq. 1.35). In this case, two internal processes are needed for moving two molar quantities simultaneously in an isolated system, i.e.

$$dU = -D_{\xi_1} d\xi_1 - D_{\xi_2} d\xi_2 + D_{\xi_1\xi_2} d\xi_1 d\xi_2 + \frac{1}{2} D_{2\xi_1} (d\xi_1)^2 + \frac{1}{2} D_{2\xi_2} (d\xi_2)^2 \tag{1.43}$$

Based on the above discussions, in a stable system at equilibrium with $D_{\xi_1} = D_{\xi_2} = 0$, $D_{2\xi_1} > 0$, and $D_{2\xi_2} > 0$, the sign of $D_{\xi_1\xi_2}$ cannot be unambiguously determined when keeping the change of internal energy positive, i.e. $dU > 0$. This indicates that the quantities in Eq. 1.41 can be either positive or negative in the stable region and become zero at the critical point. By the same token, the quantities in Eq. 1.42 can be either positive or negative and become either positive or negative infinity at the critical point.

A profound conclusion from this analysis is that in a stable system, even though a molar quantity always changes in the same direction as its conjugate potential, the same molar quantity may change in the opposite direction to a non-conjugate potential, resulting in additional anomalies represented by Eq. 1.40. One example of Eq. 1.42 is the thermal expansion in a closed system, i.e. $dN_i = 0$, as follows

$$\left(\frac{\partial V}{\partial T}\right)_S = \left(\frac{\partial S}{\partial(-P)}\right)_V \tag{1.44}$$

The left-hand side of Eq. 1.44 can be understood as follows: with the increase of temperature, the system regulates its pressure in order to keep the entropy from increasing, which results in a volume change of the system. The behavior of the system depends on whether the pressure decreases or increases in order to maintain the entropy of the system constant. If the pressure decreases to maintain the entropy of the system constant, the volume will increase with the increase of temperature, i.e. the left-hand side of the equation has a positive sign, which is also shown by the right-hand side of the equation as the changes of S and $-P$ have the same sign. That the volume increases with temperature is the normal scenario. On the other hand, if the pressure increases to maintain the entropy of the system constant, the volume will decrease with the increase of temperature, resulting in a negative sign for the left-hand side of the equation. This decrease of volume with the increase of temperature is usually considered to be anomalous, originating from the increase of entropy by the decrease of $-P$, i.e. the increase of pressure. More discussions on entropy will follow in Section 5.2.5 and Chapter 9.

1.4 Gibbs–Duhem equation

In experiments, it is difficult to control the variables S and V of a system in comparison with their conjugate variables T and $-P$. It is thus desirable to construct new functions to represent the system with T and $-P$ as natural variables of the functions. One of them is the enthalpy, defined in Eq. 1.15, and the other two can be defined as follows:

$$F = U - TS = -PV + \sum \mu_i N_i \qquad 1.45$$

$$G = U - TS + PV = \sum \mu_i N_i = H - TS = F + PV \qquad 1.46$$

where F and G are the Helmholtz energy and Gibbs energy, respectively. The middle part of Eq. 1.46 is obtained using U from Eq. 1.28. The corresponding combined law of thermodynamics in terms of H, F, and G can be obtained through the Legendre transformation of Eq. 1.10 as

$$dH = TdS - Vd(-P) + \sum \mu_i dN_i - Dd\xi \qquad 1.47$$

$$dF = -SdT - PdV + \sum \mu_i dN_i - Dd\xi \qquad 1.48$$

$$dG = -SdT - Vd(-P) + \sum \mu_i dN_i - Dd\xi \qquad 1.49$$

The independent variables in each of the above forms are regarded as the natural variables of the corresponding function. The integral forms of all the functions can thus be written as follows with their natural variables listed within parentheses:

$$U = U(S, V, N_i, \xi) \qquad 1.50$$

$$H = H(S, -P, N_i, \xi) \qquad 1.51$$

$$F = F(T, V, N_i, \xi) \qquad 1.52$$

$$G = G(T, -P, N_i, \xi) \qquad 1.53$$

By differentiating Eq. 1.46, one obtains

$$dG = \sum \mu_i dN_i + \sum N_i d\mu_i = -SdT - Vd(-P) + \sum \mu_i dN_i - Dd\xi \qquad 1.54$$

For a system at equilibrium, $Dd\xi = 0$, re-arranging Eq. 1.54 gives the Gibbs–Duhem equation

$$0 = -SdT - Vd(-P) - \sum N_i d\mu_i \qquad 1.55$$

This equation indicates that for a homogeneous system with c independent components at equilibrium, there is a direct relation among all the $c+2$ potentials, which are the c chemical potentials (μ_i), temperature, and pressure. Consequently, only $c+1$ potentials can change independently, and the remaining potential is dependent on the other potentials. As discussed in connection with Eq. 1.18, there are $c+2$ independent variables for an equilibrium system with c independent components, all of which are molar quantities.

With the relationships between potentials and molar quantities defined by Eq. 1.11 to Eq. 1.13, one can switch between potentials and molar quantities as natural variables of the system. For example, one can define a new energy function Φ when the chemical potential of one component is controlled by the surroundings instead of its content and obtain the following combined law of thermodynamics:

$$\Phi = G - \mu_1 N_1 = \sum_{i=2}^{c} \mu_i N_i \qquad 1.56$$

$$d\Phi = -SdT - Vd(-P) - N_1 d\mu_1 + \sum_{i=2}^{c} \mu_i dN_i - Dd\xi \qquad 1.57$$

However, even though the $c+2$ molar quantities are independent of each other, Eq. 1.55 indicates that not all the $c+2$ potentials are independent, i.e., if chemical potentials of all components are changed to natural variables, one would obtain Eq. 1.55. Therefore, among the $c+2$ independent variables used to define the system, the maximum number of independent potentials is $c+1$, and at least one of the $c+2$ independent variables must be a molar quantity. This variable is usually chosen to be the size of the system or of the major element in the system. The Gibbs–Duhem equation is used to derive the Gibbs phase rule in heterogeneous systems, which is discussed in Section 3.2.

Exercises

1. Consider a closed system with a spontaneous internal process under such conditions that there is no exchange of heat or work with the surroundings.
   a. Calculate the equilibrium value of ξ if the internal entropy production has the form $\Delta_{ip}S = -R[\xi \ln\xi + (1-\xi)\ln(1-\xi)]$, where ξ is a measure of the progress of the internal process.

(cont.)

b. Estimate the change of internal energy.

c. If one would like to keep the temperature constant, should one put heat into the system or extract heat from the system? How much?

2. In thermodynamics, state functions such as internal energy, enthalpy, and entropy are path independent. Consider the two differentials

$$df_1 = y(3x^2 + y^2)dx + x(x^2 + 2y^2)dy$$

$$df_2 = y(3x^2 + y)dx + x(x^2 + 2y)dy$$

Integrate them between the points $x = 0$, $y = 0$ and $x = 1$, $y = 1$, along two different paths, $y = x$ and $y = x^2$. Are the functions f_1 and f_2 state functions?

3. The combination of the first and second thermodynamic laws of a closed system yields the following combined law of thermodynamics

$$dU = dQ + dW = TdS - PdV \tag{1}$$

Other characteristic state functions are defined as $F = U - TS$, $H = U + PV$, $G = U - TS + PV$. Answer the following questions

$$\left(\frac{\partial S}{\partial V}\right)_T = \left(\frac{\partial P}{\partial T}\right)_V \tag{2}$$

a. Derive the equations for dF, dH, and dG.

b. List the natural variables of U, H, F, and G.

c. What natural variables and characteristic state function should one use in order to prove equation 2?

d. Prove equation 2.

4. Through virtual experiments, show in detail, by means of the combined law of thermodynamics, that each potential has the same value everywhere in a homogeneous system at equilibrium.

5. Use a Maxwell relation to check whether the two functions in Exercise 2 are state functions or not.

6. Entropy change and heat exchange of a system are two quantities closely related to each other. Discuss in detail whether it is possible to have a system with a spontaneous process such that dS and δQ have different signs under isothermal conditions. If your answer is yes, find such a system.

7. Enthalpy and partial enthalpy can be expressed by $H = G + TS$ and $H_j = G_j + TS_j$, respectively. Derive an expression for H_j in terms of the partial derivatives of the molar Gibbs energy, G_m, with respect to the natural variables of G_m.

8. Consider a system that can exchange energy with the environment through magnetic work in addition to the usual heat flow and volume work. The

(cont.)

magnetic work can be represented by HdM, where M is the magnetization and H the applied field.

a. Prove that:

$$\left(\frac{\partial V}{\partial M}\right)_{T,P} = \left(\frac{\partial H}{\partial P}\right)_{H,T}$$

b. Derive one more Maxwell relation with either H or M in it.

9. If a rubber band is stretched, the reversible work is given by $\delta W = \tau dL$ where τ is the tension on the band and L is the length.

a. If the stretching is carried out at constant pressure and the volume of the band also remains constant during expansion/contraction, derive a thermodynamic function (G) which is a function of L and T.

b. Show that

$$\left(\frac{\partial \tau}{\partial T}\right)_L = -\left(\frac{\partial S}{\partial L}\right)_T$$

c. Derive the following from thermodynamic principles:

$$\left(\frac{\partial U}{\partial L}\right)_T = \tau + T\left(\frac{\partial S}{\partial L}\right)_T = \tau - T\left(\frac{\partial \tau}{\partial T}\right)_L$$

d. For an ideal gas, U is a function of T only, and as a result it can be shown that

$$\frac{1}{P}\left(\frac{\partial P}{\partial T}\right)_V = \frac{1}{T}$$

Show that a corresponding equation exists for an "ideal" rubber band.

10. Prove the identity

$$T\left(\frac{\partial^2 P}{\partial T^2}\right)_V = \left(\frac{\partial C_V}{\partial V}\right)_T$$

11. Derive the relation

$$H_j = \left(\frac{\partial(\mu_i/T)}{\partial(1/T)}\right)_{P,N_i}$$

from

$$H = \left(\frac{\partial(G/T)}{\partial(1/T)}\right)_{P,N_i}$$

2 Gibbs energy function

As shown in Eq. 1.50 through Eq. 1.53, all functions have N_i and ξ as natural variables while they differ in the other two natural variables. In typical materials-related experiments, temperature and pressure are the two variables that are controlled. They are also the natural variables of the Gibbs energy. Consequently, the Gibbs energy is the most widely used function in the thermodynamics of materials science. The rest of this book focuses on the Gibbs energy for this reason. In this chapter, the mathematical formulas for the Gibbs energy of phases with fixed and variable compositions are discussed. These are needed for quantitative calculations of the Gibbs energy under given values of its natural variables.

From Eq. 1.46, the molar Gibbs energy can be defined as

$$G_m(T, P, x_i, \xi) = \frac{G}{N} = \sum \mu_i x_i \quad 2.1$$

The molar entropy, molar volume, chemical potential, and the driving force can be obtained from Eq. 1.49 as

$$S_m = \frac{S}{N} = -\frac{1}{N}\left(\frac{\partial G}{\partial T}\right)_{P,N_i,\xi} = -\left(\frac{\partial G_m}{\partial T}\right)_{P,x_i,\xi} \quad 2.2$$

$$V_m = \frac{V}{N} = \frac{1}{N}\left(\frac{\partial G}{\partial P}\right)_{T,N_i,\xi} = \left(\frac{\partial G_m}{\partial P}\right)_{T,x_i,\xi} \quad 2.3$$

$$\mu_i = \left(\frac{\partial G}{\partial N_i}\right)_{T,P,N_{j\neq i},\xi} \quad 2.4$$

$$-D = \left(\frac{\partial G}{\partial \xi}\right)_{T,P,N_i} \quad 2.5$$

Based on Eq. 1.46, the molar enthalpy is written as

$$H_m = G_m + TS_m \quad 2.6$$

Other physical properties of the system can also be represented by the partial derivatives of Gibbs energy, such as the heat capacity, C_P, volume thermal expansivity, α_V, isothermal compressibility, κ_T, under constant pressure or temperature:

$$C_P = \left(\frac{\partial Q}{\partial T}\right)_P = \left(\frac{\partial H}{\partial T}\right)_P = T\left(\frac{\partial (G+TS)}{\partial T}\right)_P = T\left(\frac{\partial S}{\partial T}\right)_P = -T\left(\frac{\partial^2 G}{\partial T^2}\right)_P \quad 2.7$$

$$\alpha_V = \left(\frac{\partial V}{\partial T}\right)_P \Big/ V = \left(\frac{\partial\left(\frac{\partial G}{\partial(-P)}\right)_T}{\partial T}\right)_P \Big/ \left(\frac{\partial G}{\partial(-P)}\right)_T = \frac{\partial^2 G}{\partial T \partial(-P)} \Big/ \left(\frac{\partial G}{\partial(-P)}\right)_T \quad 2.8$$

$$\kappa_T = \left(\frac{\partial V}{\partial(-P)}\right)_T \Big/ V = \left(\frac{\partial\left(\frac{\partial G}{\partial(-P)}\right)_T}{\partial(-P)}\right)_T \Big/ \left(\frac{\partial G}{\partial(-P)}\right)_T$$
$$= \frac{\partial^2 G}{\partial(-P)^2} \Big/ \left(\frac{\partial G}{\partial(-P)}\right)_T = \frac{1}{B} \quad 2.9$$

where the N_i and ξ are kept constant for all partial derivatives, and B is the bulk modulus.

In Eq. 2.4, G cannot be directly replaced by G_m because N also depends on N_i. The thermodynamic quantities under such conditions, i.e. variations in the amount of a component at constant temperature and pressure, are called partial quantities; these were introduced in Eq. 1.8 for partial entropy and Eq. 1.17 for partial enthalpy. This type of definition can be extended to all molar quantities such as partial volume and partial Gibbs energy. Partial quantities of a molar quantity, A, can thus be defined in general as

$$A_i = \left(\frac{\partial A}{\partial N_i}\right)_{T,P,N_{j\neq i},\xi} \quad 2.10$$

The general differential form of a molar quantity for a system at equilibrium can be represented by its partial quantities as

$$dA = \left(\frac{\partial A}{\partial T}\right)dT + \left(\frac{\partial A}{\partial P}\right)dP + \sum\left(\frac{\partial A}{\partial N_i}\right)dN_i \quad 2.11$$

where the subscripts representing variables that are kept constant, i.e. the remaining natural variables of Gibbs energy not in the denominator, are omitted for simplicity. This will be done throughout the book unless specified otherwise.

Using the following relations, $A = NA_m$, $N = \sum N_j$, $x_i = N_i/N$, $\partial x_i/\partial N_i = (1-x_i)/N$, and $\partial x_k/\partial N_i = -x_k/N$, Eq. 2.10 can be expressed as, under constant T and P,

$$A_i = A_m + N\sum_{j=1}^{c}\frac{\partial A_m}{\partial x_j}\frac{\partial x_j}{\partial N_i} = A_m + \frac{\partial A_m}{\partial x_i} - \sum_{j=1}^{c} x_j\frac{\partial A_m}{\partial x_j} \quad 2.12$$

where the summation is for all c components and the partial derivatives are taken with other mole fractions kept constant. However, mole fractions are not independent, but follow the relation $\sum x_i = 1$. Taking $x_1 = 1 - \sum_{j=2}^{c} x_j$ as the dependent mole fraction, Eq. 2.12 can be rewritten as

$$A_i = A_m + \left(\frac{\partial A_m}{\partial x_i} - \frac{\partial A_m}{\partial x_1}\right) - \sum_{j=2}^{c} x_j\left(\frac{\partial A_m}{\partial x_j} - \frac{\partial A_m}{\partial x_1}\right) \quad 2.13$$

Applying Eq. 2.12 and Eq. 2.13 to the Gibbs energy, the partial Gibbs energy or chemical potential of component i is obtained as

$$\mu_i = G_i = G_m + \frac{\partial G_m}{\partial x_i} - \sum_{j=1}^{c} x_j \frac{\partial G_m}{\partial x_j} = G_m + \left(\frac{\partial G_m}{\partial x_i} - \frac{\partial G_m}{\partial x_1}\right) - \sum_{j=2}^{c} x_j \left(\frac{\partial G_m}{\partial x_j} - \frac{\partial G_m}{\partial x_1}\right) \quad 2.14$$

The derivatives in the stability equation, Eq. 1.35, are defined with the molar quantities kept constant. On the other hand, the Gibbs energy has two potentials, temperature and pressure, as natural variables instead. One would thus need to compare the stability conditions when a variable kept fixed is changed from a molar quantity to its conjugate potential. This can be carried out through the use of Jacobians to change the independent variables:

$$\frac{\partial(Y_i, Y_j)}{\partial(X_i, X_j)} = \left(\frac{\partial Y_i}{\partial X_i}\right)_{Y_j}\left(\frac{\partial Y_j}{\partial X_j}\right)_{X_i} = \left(\frac{\partial Y_i}{\partial X_i}\right)_{X_j}\left(\frac{\partial Y_j}{\partial X_j}\right)_{X_i} - \left(\frac{\partial Y_i}{\partial X_j}\right)_{X_i}\left(\frac{\partial Y_j}{\partial X_i}\right)_{X_j} \quad 2.15$$

For a stable system, both $(\partial Y_i/\partial X_i)_{X_j}$ and $(\partial Y_j/\partial X_j)_{X_i}$ are positive based on Eq. 1.35. Using the Maxwell relation shown by Eq. 1.41, one thus obtains

$$\left(\frac{\partial Y_i}{\partial X_i}\right)_{X_j} - \left(\frac{\partial Y_i}{\partial X_i}\right)_{Y_j} = \left(\frac{\partial Y_i}{\partial X_j}\right)_{X_i}\left(\frac{\partial Y_j}{\partial X_i}\right)_{X_j} \Big/ \left(\frac{\partial Y_i}{\partial X_i}\right)_{X_j} \geq 0 \quad 2.16$$

This means that $(\partial Y_i/\partial X_i)_{Y_j}$ will go to zero before $(\partial Y_i/\partial X_i)_{X_j}$ does. It indicates that the stability condition becomes more restrictive when potentials are kept constant in place of their conjugate molar quantities. Based on the Gibbs–Duhem equation, Eq. 1.55, the maximum number of independent potentials is $c + 1$, and the last potential is dependent, i.e.

$$\left(\frac{\partial Y_{c+2}}{\partial X_{c+2}}\right)_{Y_{j\leq c+1}} = 0 \quad 2.17$$

Therefore, the limit of stability is determined when the derivative becomes zero with one molar quantity kept constant, for example

$$\left(\frac{\partial Y_{c+1}}{\partial X_{c+1}}\right)_{Y_{j<c+1}, X_{c+2}} = 0 \quad 2.18$$

The reason is that this derivative reaches zero faster than any other derivatives with more molar quantities kept constant. Equation 2.18 shows that all molar quantities diverge at the limit of stability. The consolute point is obtained with c additional conditions as follows, based on Eq. 1.39:

$$\left(\frac{\partial^2 Y_i}{\partial (X_i)^2}\right)_{Y_{j\leq c+1, j\neq i}, X_{c+2}} = 0 \quad 2.19$$

Together with Eq. 2.18, all $c + 1$ independent potentials at the consolute point can be determined. It is evident that the consolute point is a zero-dimensional point in a two-dimensional space of independent potentials in a one-component system. With the addition of a second component to form a binary system, this consolute point in the one-component system extends into a one-dimensional line. This line represents the limit of stability of the binary system, and a consolute point is located at the end of this line. It is thus evident that in a system with c independent components, the limit of stability is a $(c - 1)$-dimensional hypersurface in a space of $c + 1$ independent potentials, while the consolute point is a zero-dimensional point in all systems, and may be called the invariant critical point.

2.1 Phases with fixed compositions

The homogeneous system discussed so far has only one phase in the system, i.e. it is a single-phase system. A phase with a fixed composition can be a pure element or a stoichiometric compound. There is thus only one independent component in the system. A stoichiometric compound contains more than one element, but the relative amounts of each element are fixed by the stoichiometry and cannot vary independently, i.e., $dN_i = x_i dN$. The combined law of thermodynamics becomes

$$dG = -SdT - Vd(-P) + \left(\sum x_i\mu_i\right)dN - Dd\xi = -SdT - Vd(-P) + G_m dN - Dd\xi \quad 2.20$$

Here G_m is the molar Gibbs energy of the stoichiometric compound and can be regarded as the chemical potential of the stoichiometric phase α:

$$G_m = \mu^\alpha = \sum x_i\mu_i \quad 2.21$$

The chemical potential of an individual component in the phase cannot be defined because the amount of each component cannot be varied independently. For a stoichiometric phase of N moles of atoms at equilibrium with $dG = Nd\mu^\alpha + \mu^\alpha dN$, Eq. 2.20 reduces to

$$0 = -SdT - Vd(-P) - Nd\mu^\alpha \quad 2.22$$

which is the Gibbs–Duhem equation, Eq. 1.55, applied to a stoichiometric phase. It can be represented graphically by a surface in a three-dimensional space defined by μ^α, T, and $-P$. A direction on the surface is represented by the three partial directives between any two of μ^α, T, and $-P$ with the third one kept constant, i.e.

$$\left(\frac{\partial\mu^\alpha}{\partial T}\right)_P = -\frac{S}{N} = -S_m \quad 2.23$$

$$\left(\frac{\partial\mu^\alpha}{\partial(-P)}\right)_T = -\frac{V}{N} = -V_m \quad 2.24$$

$$\left(\frac{\partial(-P)}{\partial T}\right)_{\mu^\alpha} = -\frac{S}{V} = -\frac{S_m}{V_m} \quad 2.25$$

Based on Nernst's heat theorem, the entropy difference between two crystals approaches zero when the temperature approaches absolute zero. It is thus a common practice to put $S = 0$ for a crystal at 0 K. This is usually referred to as the third law of thermodynamics. From the definition of entropy change in Eq. 1.3, S or S_m is always positive at finite temperatures as the system or the crystal absorbs heat from the surroundings to increase its temperature. The volume V or V_m of a phase is a well-defined physical quantity, and its absolute value can be given and is always positive. The above three equations can be written in a general form as

$$\left(\frac{\partial Y_i}{\partial Y_j}\right)_{Y_k} = -\frac{X_j}{X_i} < 0 \qquad 2.26$$

The surface thus has negative slopes in all directions. The curvature of the surface can be derived from Eq. 2.26:

$$\begin{aligned} \left(\frac{\partial^2 Y_i}{\partial (Y_j)^2}\right)_{Y_k} &= -\left(\frac{\partial\left(\frac{X_j}{X_i}\right)}{\partial Y_j}\right)_{Y_k} = -\frac{1}{X_i}\left(\frac{\partial X_j}{\partial Y_j}\right)_{Y_k} + \frac{X_j}{(X_i)^2}\left(\frac{\partial X_i}{\partial Y_j}\right)_{Y_k} \\ &= -\frac{1}{X_i}\left[\left(\frac{\partial X_j}{\partial Y_j}\right)_{Y_k} - \frac{X_j}{X_i}\left(\frac{\partial X_i}{\partial Y_i}\right)\left(\frac{\partial Y_i}{\partial Y_j}\right)_{Y_k}\right] \\ &= -\frac{1}{X_i}\left[\left(\frac{\partial X_j}{\partial Y_j}\right)_{Y_k} + \left(\frac{X_j}{X_i}\right)^2\left(\frac{\partial X_i}{\partial Y_i}\right)_{Y_k}\right] < 0 \end{aligned} \qquad 2.27$$

Both terms inside the last bracket are positive for a system in a state of stable internal equilibrium, and the surface thus has a negative curvature and is convex everywhere as shown in Figure 2.1.

From experimental observations, it is known that $S_m^{vapor} \gg S_m^{liquid} > S_m^{solid}$. The curves of G_m or μ^α plotted with respect to T at constant P would thus have the most negative slope for a vapor phase, followed by those for its liquid and solid phases. As an example, Figure 2.2 shows the Gibbs energy of Zn in its solid, liquid, and vapor forms as a function of T at constant $P = 1$ atmospheric pressure.

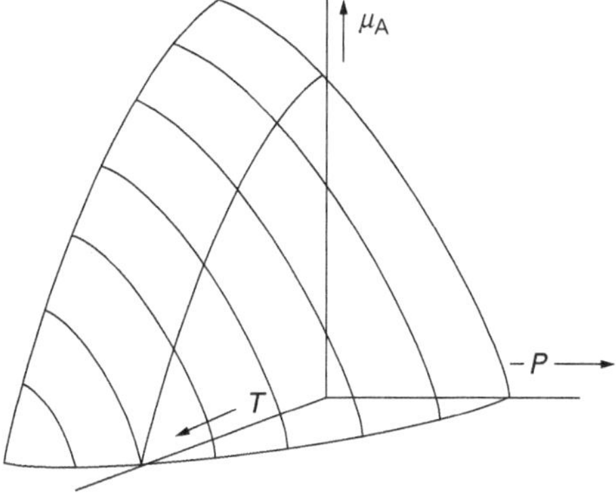

Figure 2.1 *Gibbs energy of a one-component phase as a function of temperature and negative pressure, showing the convex shape, from [1] with permission from Cambridge University Press.*

Similarly it is well known that $V_m^{vapor} \gg V_m^{liquid} > V_m^{solid}$, and the curves of G_m or μ^{α} plotted with respect to P at constant T would thus have the most positive slope for a vapor phase, followed by those for its liquid and solid phases, though there are cases for which $V_m^{liquid} < V_m^{solid}$ such as that of water and ice. As an example, Figure 2.3 shows the

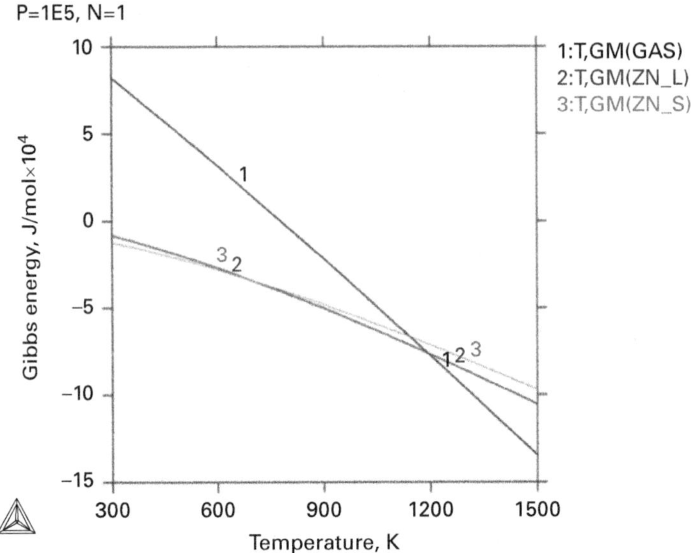

Figure 2.2 *Molar Gibbs energy of Zn as a function of T at constant P.*

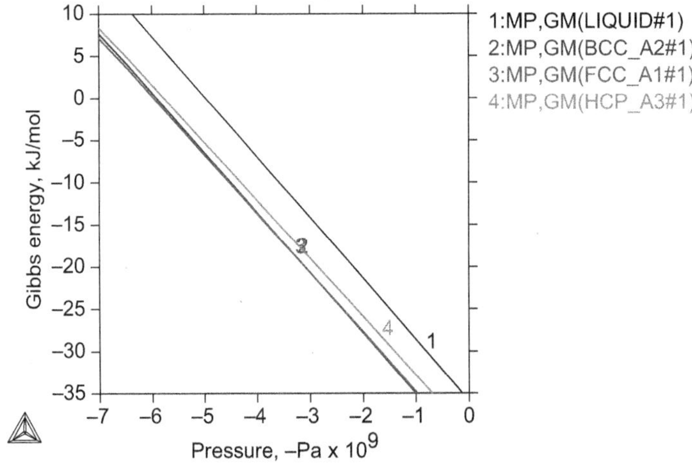

Figure 2.3 *Molar Gibbs energy of Fe as a function of P at constant T = 100 K.*

Gibbs energy of Fe in its three solid (fcc, hcp and bcc), and liquid forms as a function of P at constant $T = 1000$ K. The gas phase is out of the chart.

The quantities measurable by experiments typically include temperature, pressure, volume, composition, and amount of heat flow, in the combined law of thermodynamics discussed so far. By measuring the heat needed to increase the temperature of a phase, the heat capacity of the phase is obtained, as shown by Eq. 2.7. A typical heat capacity curve as a function of temperature is shown in Figure 2.4 for fcc-Al, hcp-Mg, and an intermetallic phase, $Al_{12}Mg_{17}$.

There are various theoretical models, to be discussed in Chapter 5, for the heat capacity under constant volume, which is defined as

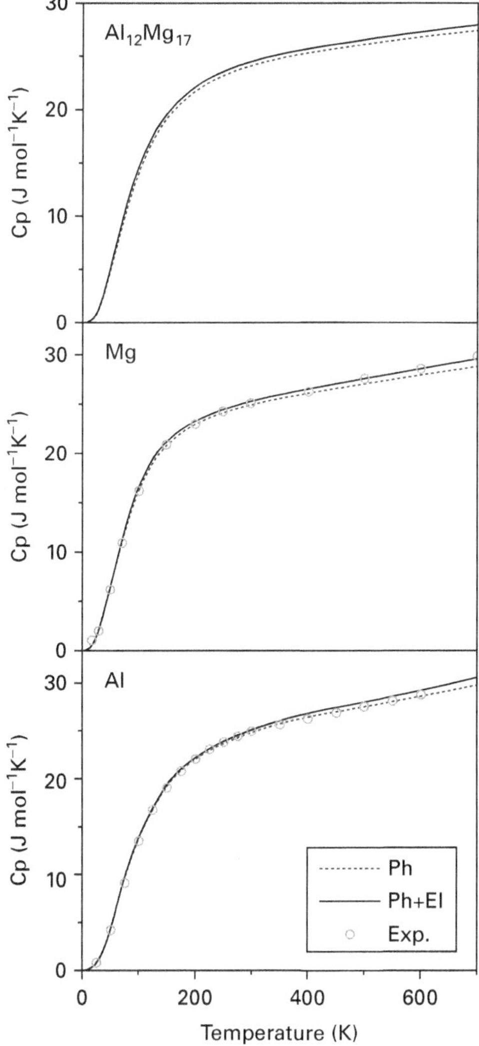

Figure 2.4 *Heat capacities of fcc-Al, hcp-Mg, and $Al_{12}Mg_{17}$ as a function of temperature.*

$$C_V = \left(\frac{\partial U}{\partial T}\right)_V = T\left(\frac{\partial (F+TS)}{\partial T}\right)_V = T\left(\frac{\partial S}{\partial T}\right)_V = -T\left(\frac{\partial^2 F}{\partial T^2}\right)_V \qquad 2.28$$

To establish the relationship between C_P, defined by Eq. 2.7, and C_V, U needs to be represented as a function of T and V in terms of G and its derivatives with respect to the natural variables of the Gibbs energy, T and P. It can be done as follows:

$$dV = \frac{\partial V}{\partial T}dT + \frac{\partial V}{\partial(-P)}d(-P) = -\frac{\partial^2 G}{\partial T\partial(-P)}dT - \frac{\partial^2 G}{\partial(-P)^2}d(-P) \qquad 2.29$$

$$dU = \frac{\partial(G+TS-PV)}{\partial T}dT + \frac{\partial(G+TS-PV)}{\partial(-P)}d(-P) = -\left(T\frac{\partial^2 G}{\partial T^2} - P\frac{\partial^2 G}{\partial T\partial(-P)}\right)dT$$
$$-\left(T\frac{\partial^2 G}{\partial T\partial(-P)} + P\frac{\partial^2 G}{\partial(-P)^2}\right)\left(-1\Big/\frac{\partial^2 G}{\partial(-P)^2}dV + \frac{\partial^2 G}{\partial T\partial(-P)}\Big/\frac{\partial^2 G}{\partial(-P)^2}dT\right)$$
$$= -\left[T\frac{\partial^2 G}{\partial T^2} - T\left(\frac{\partial^2 G}{\partial T\partial(-P)}\right)^2\Big/\frac{\partial^2 G}{\partial(-P)^2}\right]dT + \left(-T\frac{\partial^2 G}{\partial T\partial(-P)}\Big/\frac{\partial^2 G}{\partial(-P)^2} + P\right)dV \qquad 2.30$$

$$C_V = C_P + T\left(\frac{\partial^2 G}{\partial T\partial(-P)}\right)^2\Big/\frac{\partial^2 G}{\partial(-P)^2} = C_P - \frac{\alpha_V^2 VT}{\kappa_T} = C_P - \alpha_V^2 BVT \qquad 2.31$$

where the thermal expansion, α_V, and the compressibility κ_T or bulk modulus B, are defined by Eq. 2.8 and Eq. 2.9, respectively. From the heat capacity, the enthalpy and entropy can be obtained by integration of Eq. 2.7 at constant pressure:

$$S = S_0 + \int_0^T \frac{C_P}{T}dT = S_0 + \int_0^{298.15}\frac{C_P}{T}dT + \int_{298.15}^{T}\frac{C_P}{T}dT = S_{298.15} + \int_{298.15}^{T}\frac{C_P}{T}dT \qquad 2.32$$

$$H = H_0 + \int_0^T C_P dT = H_0 + \int_0^{298.15} C_P dT + \int_{298.15}^{T} C_P dT = H_{298.15} + \int_{298.15}^{T} C_P dT \qquad 2.33$$

In the above equations, two temperature ranges of integration are chosen for practical applications, as most processing procedures in the field of materials science and engineering take place at temperatures above room temperature. Based on the third law of thermodynamics, $S_0 = 0$, $S_{298.15}$ can be obtained by integration. On the other hand, for $H_0 = U_0 + PV$ one does not know the absolute value of the internal energy and thus one has to select a reference state for H. In principle, the reference state can be arbitrarily chosen. A widely used reference state in thermodynamic modeling practice is that for which $H_{298.15}^{SER} = 0$ at ambient pressure for pure elements in their respective stable structures at room temperature; it is called the stable element reference (SER) state, with

$$G_{298.15}^{SER} = H_{298.15}^{SER} - TS_{298.15}^{SER} = -TS_{298.15}^{SER} \qquad 2.34$$

It is further noted that after defining $S_{298.15}$ and $H_{298.15}$, one only needs the heat capacity at higher temperatures. This makes the mathematical representation of heat capacity simpler due to the relatively simple temperature dependence of heat capacity at higher temperatures in comparison with the variation at lower temperatures. One common expression for heat capacity at high temperatures and ambient pressure is as follows:

$$C_P = c + dT + \frac{e}{T^2} + fT^2 \quad 2.35$$

where c, d, e, and f are parameters fitted to experimental or theoretical data and compiled in various handbooks.

Corresponding expressions for S, H, and G are obtained as

$$S = b' + c\ln T + dT - \frac{e}{2T^2} + \frac{f}{2}T^2 \quad 2.36$$

$$H = a + cT + \frac{d}{2}T^2 - \frac{e}{T} + \frac{f}{3}T^3 \quad 2.37$$

$$G = H - TS = a - bT - cT\ln T - \frac{d}{2}T^2 - \frac{e}{2T} - \frac{f}{6}T^3 \quad 2.38$$

with $b = b' - c$. The integration constants b' and a are evaluated from $S_{298.15}$ and $H_{298.15}$, respectively. As an example, the enthalpy and entropy of Zn in solid (hcp), liquid, and gas forms are plotted in Figure 2.5 and Figure 2.6, respectively. The distances between any two curves in Figure 2.5 and Figure 2.6 represent the enthalpy or entropy differences between the two phases. It can be seen that the gas has much higher enthalpy and entropy than the solid and liquid.

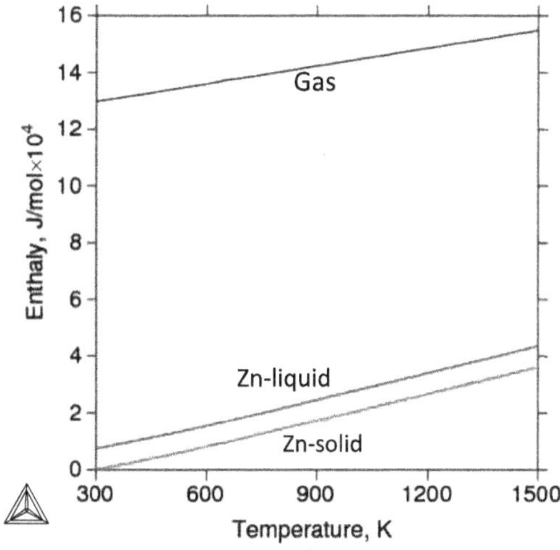

Figure 2.5 *Enthalpy of Zn as a function of temperature at one atmospheric pressure.*

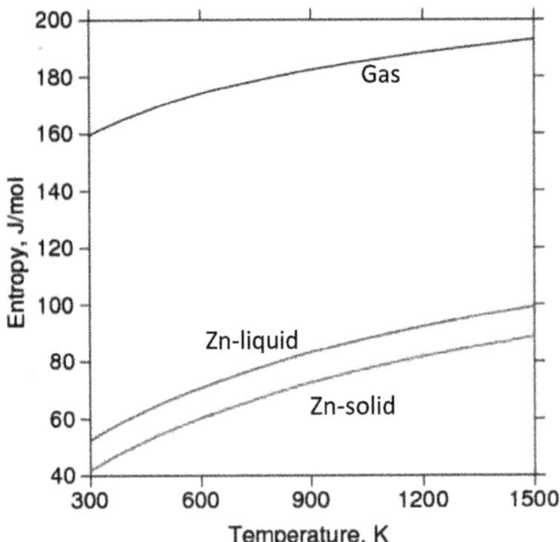

Figure 2.6 *Entropy of Zn as a function of temperature at one atmospheric pressure.*

Similarly, one can add the pressure dependence into the Gibbs energy function, obtaining for example

$$G = a - bT - cT\ln\mathrm{T} - \frac{d}{2}T^2 - \frac{e}{2T} - \frac{f}{6}T^3 + gP + hTP + mP^2 \qquad 2.39$$

where g, h, and m are parameters fitted to experimental or theoretical data and compiled in various handbooks.

The expression for V can be derived as

$$V = g + hT + 2mP \qquad 2.40$$

The Helmholtz energy can be expressed as a function of its natural variables by solving P from Eq. 2.40:

$$F = G - PV = a - bT - cT\ln T - \frac{d}{2}T^2 - \frac{e}{2T} - \frac{f}{6}T^3 - \frac{(g + hT - V)^2}{4m} \qquad 2.41$$

In the literature there are many models available to represent the temperature and pressure dependences of thermodynamic properties. The Gibbs energy difference between a stoichiometric compound and the components of which the compound is composed at their reference states, ${}^0G_i^{ref}$, is termed the Gibbs energy of formation, i.e.

$$\Delta_f G = G - \sum N_i {}^0G_i^{ref} \qquad 2.42$$

with N_i being the stoichiometry of the compound. Similarly, the enthalpy of formation, entropy of formation, and heat capacity of formation with respect to components in their reference states, ${}^0H_i^{ref}$, ${}^0S_i^{ref}$, and ${}^0C_{Pi}^{ref}$, can be defined as

$$\Delta_f H = H - \sum N_i{}^0 H_i^{ref} \tag{2.43}$$

$$\Delta_f S = S - \sum N_i{}^0 S_i^{ref} \tag{2.44}$$

$$\Delta_f C_P = C_P - \sum N_i{}^0 C_{Pi}^{ref} \tag{2.45}$$

It should be mentioned that one mole of a compound usually refers to one mole of formula the stoichiometric of the compound. With a formula like $A_a B_b C_c$, the compound is composed of a total $(a + b + c)$ moles of components. One should thus be very careful when dealing with numerical values to be sure whether the data are given per mole of formula or per mole of components. At the same time the reference states must be clearly defined. When the SER state defined in Eq. 2.34 is selected as the reference state, the above formation quantities are called standard formation quantities, such as the standard enthalpy of formation.

Since there are only two independent potentials in a one-component system, its limit of stability can be evaluated with one potential kept constant, i.e. either T or P. Consequently, either the Helmholtz energy or enthalpy is to be used in deriving the limit of stability of a homogeneous system. For practical purposes, let us use the Helmholtz energy because its natural variables T and V are measurable quantities in typical experiments, while one of the natural variables of enthalpy, the entropy, is not. From Eq. 1.48 and Eq. 2.16, the limit of stability for a one-component system at constant temperature can be written as

$$\left(\frac{\partial(-P)}{\partial V}\right)_{T,N} = F_{VV} = \frac{1}{V\kappa_T} = \frac{B}{V} = 0 \tag{2.46}$$

where the isothermal compressibility and bulk modulus, κ_T and B, are defined in Eq. 2.9. The limit of stability is thus determined when the isothermal compressibility diverges or the bulk modulus becomes zero, because V has a finite value at any temperature. It is evident that the Helmholtz energy must have higher order dependence on volume than in Eq. 2.41 for a system with instability because F_{VV} as derived from Eq. 2.41 is constant.

From Eq. 2.19, the consolute point is defined by

$$F_{VVV} = \left(\frac{\partial^2(-P)}{\partial V^2}\right)_{T,N} = \frac{\partial(1/V\kappa_T)}{\partial V} = -\frac{1 + (V/\kappa_T)(\partial\kappa_T/\partial V)}{\kappa_T V^2} = 0 \tag{2.47}$$

Since κ_T becomes infinite at the limit of stability, $\partial\kappa_T/\partial V$ approaches negative infinity when the critical/consolute point is approached, so that $(V/\kappa_T)(\partial\kappa_T/\partial V) = -1$ and $F_{VVV} = 0$.

2.2 Phases with variable compositions: random solutions

The combined law of thermodynamics and the Gibbs–Duhem equation for a solution phase with variable composition are shown by Eq. 1.49 and Eq. 1.55, respectively.

A phase can be represented by a $(c+1)$-dimensional surface in a $(c+2)$-dimensional space of potentials based on the Gibbs–Duhem equation. The directions and curvature of the surface are represented by the partial derivatives shown in Eq. 2.26 and second derivatives shown in Eq. 2.27, both being negative for a stable phase. To develop a mathematical formula for the Gibbs energy of a phase with variable compositions, one can consider a phase as a mixture of independent components that make up the phase. Its Gibbs energy function can be postulated as the sum of the Gibbs energies of the independent components of a solution with the same structure, ${}^0G_i$, plus the contribution due to the mixing, ${}^{mixing}G$ or ${}^MG$:

$$G = \sum N_i {}^0G_i + {}^MG \tag{2.48}$$

Since the system size is usually not important in thermodynamics, properties are typically normalized to one mole with its composition represented by molar fractions of components. The molar Gibbs energy is obtained as shown below with the molar Gibbs energy of mixing separated into two parts: the ideal Gibbs energy of mixing assuming no chemical interaction among components, ${}^{ideal}G_m$ or ${}^IG_m$, and the excess Gibbs energy of mixing due to chemical reaction among components, ${}^{excess}G_m$ or ${}^EG_m$:

$$G_m = \sum x_i {}^0G_i + {}^MG_m = \sum x_i {}^0G_i + {}^IG_m + {}^EG_m \tag{2.49}$$

From Eq. 2.14, the chemical potential of a component is thus

$$\mu_i = {}^0G_i + {}^IG_m + {}^EG_m + \frac{\partial({}^IG_m + {}^EG_m)}{\partial x_i} - \sum_j^c x_j \frac{\partial({}^IG_m + {}^EG_m)}{\partial x_j} \tag{2.50}$$

One can define the chemical activity of component i, a_i, as follows:

$$\begin{aligned} RT\ln a_i &= \mu_i - {}^0G_i = {}^IG_m + \frac{\partial^I G_m}{\partial x_i} - \sum_{j=1}^{c} x_j \frac{\partial^I G_m}{\partial x_j} \\ &+ {}^EG_m + \frac{\partial^E G_m}{\partial x_i} - \sum_{j=1}^{c} x_j \frac{\partial^E G_m}{\partial x_j} \end{aligned} \tag{2.51}$$

In this definition, the chemical activity or simply activity is calculated with respect to the pure elements in the structure of the solution for practical reasons, as one would like to understand the chemical potential difference of components both in the solution and by itself, with the same structure. It should be noted that this reference state for chemical activity is usually different from the SER reference state defined in Eq. 2.34 as the solution may have a different structure from that of pure components in their SER states. On the other hand, the activity under the SER reference state can be easily obtained by replacing ${}^0G_i$ with ${}^0G_i^{SER}$ from Eq. 2.34. In principle, one may choose any structure as the reference state in order for the activity to be useful for practical applications, i.e.

$$RT\ln a_i^{ref} = \mu_i - {}^0G_i^{ref} \tag{2.52}$$

For example, the activity of a component in a liquid solution is defined with respect to the pure component in its liquid form from Eq. 2.51, but can also be referred to its SER state which is solid using Eq. 2.52. The following sections will discuss in more detail how components mix when they are brought together, including concepts such as random mixing, short-range ordering, and long-range ordering.

The limit of stability of a solution with respect to composition fluctuation under constant T, P, and N_i can be derived as follows from Eq. 2.16 and Eq. 2.18:

$$\left(\frac{\partial\mu_i}{\partial N_i}\right)_{T,P,N_{j\neq i},i>1} > \left(\frac{\partial_i}{\partial N_i}\right)_{T,P,N_1,\mu_2,N_{j\neq i},i,j>2} > \ldots > \left(\frac{\partial\mu_c}{\partial N_c}\right)_{T,P,N_1,\mu_2..\mu_{c-1}} = 0 \quad 2.53$$

The first term can be derived from Eq. 2.14 as follows:

$$\begin{aligned}\left(\frac{\partial\mu_i}{\partial N_i}\right)_{T,P,N_{j\neq i},i>1} &= \sum_{j=1}^{c}\frac{\partial^2 G_m}{\partial x_i\partial x_j}\frac{\partial x_j}{\partial N_i} - \sum_{j=1}^{c}x_j\sum_{k=1}^{c}\frac{\partial^2 G_m}{\partial x_j\partial x_k}\frac{\partial x_k}{\partial N_i}\\ &= \frac{1}{N}\left(\frac{\partial^2 G_m}{\partial x_i^2} - \sum_{j=1}^{c}x_j\frac{\partial^2 G_m}{\partial x_j^2} - \sum_{j=1}^{c}x_j\frac{\partial^2 G_m}{\partial x_i\partial x_j} + \sum_{j=1}^{c}\sum_{k=1}^{c}x_jx_k\frac{\partial^2 G_m}{\partial x_j\partial x_k}\right)\end{aligned} \quad 2.54$$

Denoting $G_{ij} = (\partial\mu_i/\partial N_j)_{T,P,N_{k\neq j}}$ and using Equation 2.15 to change the variables kept constant from molar quantities to potentials one by one (see [1]), the limit of stability can be obtained as

$$\left(\frac{\partial\mu_c}{\partial N_c}\right)_{T,P,N_1,\mu_2,\ldots,\mu_{c-1}} = \frac{\det\left(G_{ij} : 2 \leq i,j \leq c\right)}{\det\left(G_{ij} : 2 \leq i,j \leq c-1\right)} = 0 \quad 2.55$$

where det stands for determinant. Equation 2.55 indicates that $\det\left(G_{ij} : 2 \leq i,j \leq c\right) = 0$ at the limit of stability. Considering $x_1 = 1 - \sum_{j\neq 1} x_j$, let us introduce

$$g_i = \mu_i - \mu_1 = \left(\frac{\partial G_m}{\partial x_i}\right)_{x_{k\neq i}} - \left(\frac{\partial G_m}{\partial x_1}\right)_{x_{k\neq 1}} \quad 2.56$$

and

$$g_{ij} = \frac{\partial g_i}{\partial x_j} = \frac{\partial(\mu_i - \mu_1)}{\partial x_j} = \frac{\partial^2 G_m}{\partial x_i\partial x_j} - \frac{\partial^2 G_m}{\partial x_1\partial x_j} - \frac{\partial^2 G_m}{\partial x_i\partial x_1} + \frac{\partial^2 G_m}{\partial(x_1)^2} \quad 2.57$$

The limit of stability can be rewritten as

$$\left(\frac{\partial(\mu_c - \mu_1)}{\partial x_c}\right)_{T,P,N,\mu_2-\mu_1,\ldots,\mu_{c-1}-\mu_1} = \frac{\det\left(g_{ij} : 2 \leq i,j \leq c\right)}{\det\left(g_{ij} : 2 \leq i,j \leq c-1\right)} = 0 \quad 2.58$$

i.e. $\det\left(g_{ij} : 2 \leq i,j \leq c\right) = 0$. The consolute point can be defined using Eq. 2.19:

$$\left(\frac{\partial^2\mu_c}{\partial(N_c)^2}\right)_{T,P,N_1,\mu_2,\ldots,\mu_{c-1}} = \left(\frac{\partial^2(\mu_c - \mu_1)}{\partial(x_c)^2}\right)_{T,P,N,\mu_2-\mu_1,\ldots,\mu_{c-1}-\mu_1} = 0 \quad 2.59$$

No closed mathematic form for it has been published in the literature.

2.2.1 Random solutions

The ideal Gibbs energy of mixing corresponds to an ideal solution in which all sites are equivalent and the distributions of components on the sites are completely random. The number of different configurations for arranging all components is

$$w = \frac{N!}{\prod(N_i!)} \quad 2.60$$

Based on Boltzmann's relation from statistical thermodynamics, when all configurations have the same probability of being observed, the ideal configurational molar entropy of mixing for the distribution is

$$^{ideal}S_m = {}^I S_m = \frac{R\ln w}{N} = R\frac{\ln N! - \sum \ln(N_i!)}{N} \cong R\frac{N\ln N - \sum N_i \ln N_i}{N} = -R\sum x_i \ln x_i \quad 2.61$$

where R is the gas constant. Equation 2.61 represents the entropy difference between that of the ideal solution and that of the individual components, i.e. it gives the entropy due to the mechanical mixing of the components. As x_i is smaller than unity, the entropy production on forming an ideal solution from pure components is thus positive, indicating that it is a spontaneous process. In such an ideal solution, it is assumed that there are no interactions between components, and the enthalpy of mixing is thus zero as the internal energy and the volume of the system do not change. The ideal Gibbs energy of mixing is written as

$$^I G = -T\,^I S_m = RT\sum x_i \ln x_i \quad 2.62$$

The Gibbs energy of real solutions, i.e. Eq. 2.49, becomes

$$G_m = \sum x_i {}^0 G_i + RT\sum x_i \ln x_i + {}^E G_m \quad 2.63$$

From Eq. 2.50, the chemical potential is obtained as

$$\mu_i = G_i = {}^0 G_i + RT\ln x_i + {}^E G_m + \frac{\partial^E G_m}{\partial x_i} - \sum_{j=1}^{c} x_j \frac{\partial^E G_m}{\partial x_j} \quad 2.64$$

From the chemical activity in Eq. 2.51, the activity coefficient, γ_i, can be defined as follows:

$$\gamma_i = \frac{a_i}{x_i} = \frac{1}{x_i}\exp\frac{G_i - {}^0 G_i}{RT} \quad 2.65$$

The solution is an ideal solution if $\gamma_i = 1$, and is said to be positively or negatively deviating from an ideal solution if $\gamma_i > 1$ or $\gamma_i < 1$, respectively. The chemical potential is related to the activity and activity coefficient by the following equation:

$$\mu_i = {}^0 G_i + RT\ln a_i = {}^0 G_i + RT\ln \gamma_i x_i = {}^0 G_i + RT\ln x_i + RT\ln \gamma_i \quad 2.66$$

Let us examine Eq. 2.14 in more detail in order to better understand the relation between G_m and μ_i. The partial derivatives in Eq. 2.14 represent the directions of the molar Gibbs energy in the composition space, i.e. the tangents of the molar Gibbs energy with respect to mole fractions of independent components. Collectively, they define the multi-dimensional tangent plane of the molar Gibbs energy at the given composition, x_i^0. The mathematical representation of this tangent plane, z_{G_m}, is defined by its directional derivatives and the distance from the point where the derivatives are taken,

$$z_{G_m} = G_m(x_i^0) + \sum_{j=1}^{c} \left(\frac{\partial G_m}{\partial x_i}\right)_{x_i^0} (x_i - x_i^0) \tag{2.67}$$

The intercept of this tangent plane at each pure component axis, i.e. $x_i = 1$ and $x_{j \neq i} = 0$, is obtained as

$$z_{G_m, x_i=1} = G_m(x_i^0) + \left(\frac{\partial G_m}{\partial x_i}\right)_{x_i=x_i^0} - \sum_{j=1}^{c} x_j^0 \left(\frac{\partial G_m}{\partial x_j}\right)_{x_i=x_i^0} \tag{2.68}$$

This is identical to Eq. 2.14 at the point x_i^0. It is thus shown that the chemical potential of a component in a solution is represented by the intercept of the tangent plane of the Gibbs energy of the solution and the G_m axis of the component. The distance between the intercept and the Gibbs energy of the pure component in the same solution structure is related to the chemical activity of the component as defined by Eq. 2.51. On the other hand, it is evident that one can choose any other structure of the pure element to define the chemical activity in order to compare the chemical potentials of the components as shown by Eq. 2.52.

The stability of a solution is evaluated using Eq. 2.55, and the derivatives of chemical potential with respect to the numbers of moles, i.e. the elements in the determinant, are obtained as follows from Eq. 2.12 and Eq. 2.66,

$$\frac{N}{RT}\frac{\partial \mu_i}{\partial N_i} = \frac{N}{RT} G_{ii} = \frac{1 - x_i}{x_i} + \frac{1}{\gamma_i}\left(\frac{\partial \gamma_i}{\partial x_i} - \sum_{j=1}^{c} x_j \frac{\partial \gamma_i}{\partial x_j}\right) \tag{2.69}$$

$$\frac{N}{RT}\frac{\partial \mu_i}{\partial N_k} = \frac{N}{RT} G_{ik} = -\frac{x_k}{x_i} + \frac{1}{\gamma_i}\left(1 - \sum_{j=1}^{c} x_j \frac{\partial \gamma_i}{\partial x_j}\right) \tag{2.70}$$

To study further the Gibbs energy of solution phases, let us discuss the details of the excess Gibbs energy of mixing. At this point, one can start with lower-order systems with fewer components, i.e. two-component and three-component systems, noting that the Gibbs energy of phases with one component was already presented in Section 2.1.

2.2.2 Binary random solutions

From Eq. 1.55, the Gibbs–Duhem equation of a binary system consisting of components A and B is written as

$$0 = -SdT - Vd(-P) - N_A d\mu_A - N_B d\mu_B \tag{2.71}$$

This equation represents a three-dimensional surface in a four-dimensional space. It is self-evident that both Eq. 2.26 and Eq. 2.27 hold for stable binary solutions too, i.e. the directions and the curvature of the surface are all negative. To visualize the three-dimensional surface in three-dimensional space, one needs to fix one of the four potentials. As T and P are the natural variables of the Gibbs energy, they are usually chosen to be kept constant. One can typically investigate behaviors of systems consisting of condensed phases by varying the temperature at constant pressure. Equation 2.71 at constant pressure thus becomes

$$0 = -SdT - N_A d\mu_A - N_B d\mu_B \qquad 2.72$$

Similarly to Eq. 2.22 and Figure 2.1, the property of a phase can be represented by a two-dimensional surface in the three-dimensional space composed of T, μ_A, and μ_B under constant P, keeping in mind the following:

$$G_m = x_A\mu_A + x_B\mu_B = x_A{}^0G_A + x_B{}^0G_B + RT(x_A\ln x_A + x_B\ln x_B) + {}^EG_m \qquad 2.73$$

Since ${}^EG_m$ must be zero for pure components A and B, it needs to be in the following form:

$${}^EG_m = x_A x_B L_{AB} \qquad 2.74$$

with L_{AB} a parameter denoting the interaction between components A and B, called the interaction parameter. When $L_{AB} = 0$, the solution is an ideal solution. When L_{AB} is a non-zero constant independent of temperature and composition, the solution is called a regular solution. Its excess entropy and excess enthalpy of mixing are obtained as

$${}^ES_m = \frac{\partial^E G_m}{\partial T} = 0 \qquad 2.75$$

$${}^EH_m = {}^EG_m - T^ES_m = x_A x_B L_{AB} \qquad 2.76$$

The chemical potential of component A or B in a binary regular solution can be derived as

$$\mu_i = {}^0G_i + RT\ln x_i + (1 - x_i)^2 L_{AB} \qquad 2.77$$

In a dilute solution with $x_i \to 0$, one has

$$RT\ln\gamma_i = (1 - x_i)^2 L_{AB} \approx L_{AB} \qquad 2.78$$

$$\gamma_i = e^{L_{AB}/RT} \qquad 2.79$$

The activity is thus proportional to its mole fraction, which is called Henry's law. By the same token, for the solvent, i.e. for $x_i \to 1$,

$$RT\ln\gamma_i = (1 - x_i)^2 L_{AB} \approx 0 \qquad 2.80$$

which gives $\gamma_i \approx 1$, and its activity approaches its mole fraction. This is called Raoult's law.

The stability of a binary regular solution is derived from Eq. 2.69 as

$$\left(\frac{\partial \mu_A}{\partial N_A}\right)_{T,P,N_B} = \left[\frac{RT}{x_A} - 2(1 - x_A)L_{AB}\right]\frac{1 - x_A}{N} \qquad 2.81$$

$$\left(\frac{\partial \mu_B}{\partial N_B}\right)_{T,P,N_A} = \left[\frac{RT}{x_B} - 2(1 - x_B)L_{AB}\right]\frac{1 - x_B}{N} \qquad 2.82$$

It should be noted that the two chemical potentials in a binary system at constant temperature and pressure are dependent on each other due to the Gibbs–Duhem equation shown in Eq. 2.72, i.e.

$$0 = -N_A d\mu_A - N_B d\mu_B \qquad 2.83$$

and the two chemical potentials depend on each other according to the following relation,

$$\left(\frac{\partial \mu_A}{\partial \mu_B}\right)_{T,P} = -\frac{N_B}{N_A} \qquad 2.84$$

Therefore, at the limit of stability, both Eq. 2.81 and Eq. 2.82 go to zero at the same time, which is obtained when

$$RT = 2x_A x_B L_{AB} \qquad 2.85$$

As the absolute temperature cannot be negative, Eq. 2.85 has no solution for a solution phase with $L_{AB} < 0$, i.e. the solution phase is stable with respect to the composition fluctuation. For a solution with $L_{AB} > 0$, its limit of stability is represented by Eq. 2.85.

A schematic of the molar Gibbs energy of a solution with $L_{AB} < 0$ at constant temperature and pressure is shown in Figure 2.7 along with the ideal and excess Gibbs energies of mixing. A tangent line is drawn on the molar Gibbs energy of the solution, and its two intercepts at $x_B = 0$ and $x_B = 1$ give the chemical potentials of components A and B, μ_A and μ_B, respectively by Eq. 2.68. It is evident that μ_A and μ_B are not independent of each other as they are two points on the same straight line. This is a graphic representation of the Gibbs–Duhem equation! of Eq. 2.83. The chemical activity of component B is also depicted, the reference state being pure B with the same structure. As shown in Eq. 2.52, other structures of pure B can be selected as the reference state of the chemical activity of component B, resulting in different distances from its chemical potential in the solution, and thus different values of its chemical activities. It is clear that this change of reference state for chemical activity does not affect the chemical potential of the component in the solution.

When $L_{AB} > 0$, Eq. 2.85 represents a parabola in the $T - x_i$ two-dimensional coordinate plane that is symmetric with respect to x_A and x_B, as shown in Figure 2.8; it is the spinodal

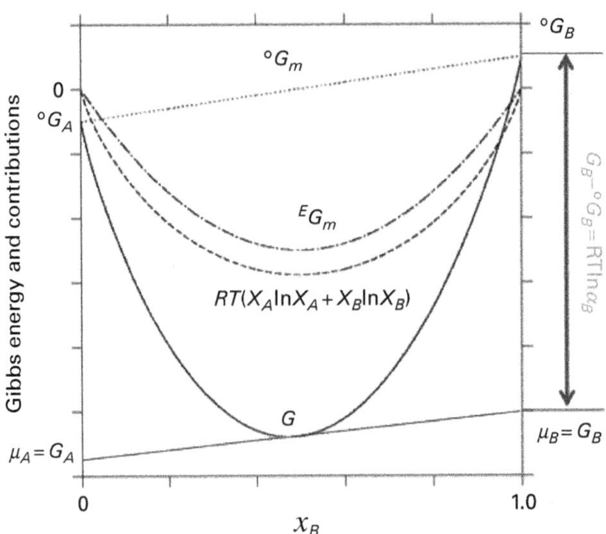

Figure 2.7 *Schematic molar Gibbs energy diagram with* $L_{AB} < 0$.

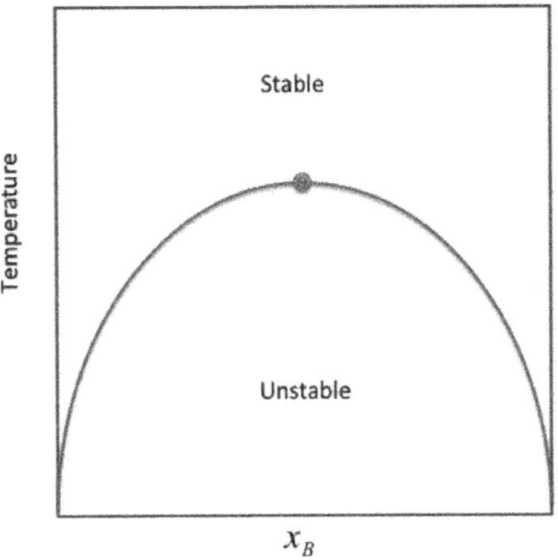

Figure 2.8 *A spinodal curve with* $L_{AB} > 0$.

of the solution. The consolute point is obtained by applying Eq. 2.59 to Eq. 2.81 and setting Eq. 2.81 equal to zero at the consolute point:

$$\left(\frac{\partial^2 \mu_A}{\partial N_A^2}\right)_{T,P,N_B} = \left[-\frac{RT}{x_A^2} + 2L_{AB}\right]\left(\frac{1-x_A}{N}\right)^2 = 0 \qquad 2.86$$

which gives

$$T_{cons} = 2x_A^2 L_{AB} \qquad 2.87$$

Solving Eq. 2.85 and Eq. 2.87, one obtains $x_A = x_B = 0.5$ and

$$T_{cons} = \frac{L_{AB}}{2R} \qquad 2.88$$

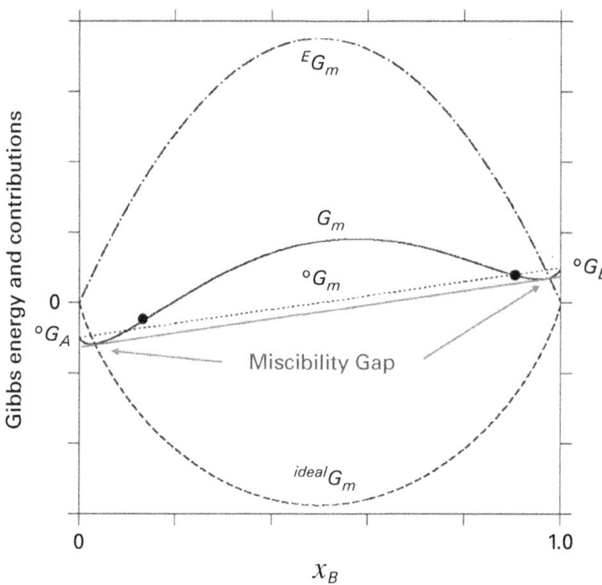

Figure 2.9 *Schematic molar Gibbs energy diagram with $L_{AB} > 0$.*

A schematic molar Gibbs energy diagram at temperatures below the consolute point is shown in Figure 2.9. It can be seen that part of the molar Gibbs energy has negative curvature, and the solution becomes unstable. The chemical potential thus does not change monotonically with respect to composition and its derivative changes sign at the inflexion point, as denoted by the two filled circles in Fig. 2.9, also called spinodal.

For more complex solutions, L_{AB} can be a function of temperature, pressure, and composition. In principle, the temperature and pressure dependences can be treated by means of a formula similar to Eq. 2.39. There are various approaches in the literature for considering the composition dependence of L_{AB}. The empirical Redlich–Kister polynomial stands out as the approach most widely used because it can be extrapolated to ternary and multi-component systems consistently; this will be discussed in Chapter 6.

2.2.3 Ternary random solutions

From Eq. 1.55, the Gibbs–Duhem equation of a ternary system consisting of components A, B, and C is written as

$$0 = -SdT - Vd(-P) - N_A d\mu_A - N_B d\mu_B - N_C d\mu_C \quad 2.89$$

This equation represents a four-dimensional surface in a five-dimensional space. It can be visualized in a three-dimensional space with two of the five potentials fixed. Usually T and P are kept constant as they are the natural variables of G, and Eq. 2.89 reduces to

$$0 = -N_A d\mu_A - N_B d\mu_B - N_C d\mu_C \quad 2.90$$

A phase can thus be represented by a surface in the three-dimensional space of μ_A, μ_B, and μ_C at constant T and P with similar geometric appearance of Figure 2.1.

From Eq. 2.63, the Gibbs energy of a ternary solution is written as

$$G_m = x_A{}^0G_A + x_B{}^0G_B + x_C{}^0G_C + RT(x_A\ln x_A + x_B\ln x_B + x_C\ln x_C) + {}^EG_m \qquad 2.91$$

When the mole fraction of one component approaches zero, ${}^EG_m$ reduces to the excess Gibbs energy of mixing of the binary systems of the remaining two components, represented by Eq. 2.74. However, for a given composition of a ternary solution, there is no unique way to assign the contributions from the ${}^EG_m$ of each binary to the ${}^EG_m$ of the ternary solution because the ${}^EG_m$ of the ternary solution contains information on both binary and ternary interactions. A variety of models is available in the literature (see [1]). One intuitive approach would be to use the same formula as that in the binary system, i.e. Eq. 2.74, with the mole fractions substituted by the values in the ternary system, and ${}^EG_m$ for a ternary solution may thus be defined as the following, by including the ternary interaction involving all three components,

$$^EG_m = x_Ax_BL_{AB} + x_Ax_CL_{AC} + x_Bx_CL_{BC} + x_Ax_Bx_CL_{ABC} \qquad 2.92$$

The chemical potential of a component is represented by Eq. 2.64. When all interaction parameters in Eq. 2.92 are constant, i.e. we have a ternary regular solution, the chemical potential of component A can be derived as

$$\begin{aligned}\mu_A = G_A &= {}^0G_A + RT\ln x_A + x_BL_{AB} + x_CL_{AC} - {}^EG_m \\ &= {}^0G_A + RT\ln x_A + x_B(1-x_A)L_{AB} + x_C(1-x_A)L_{AC} - x_Bx_CL_{BC} + x_Bx_C(1-2x_A)L_{ABC} \\ &= {}^0G_A + RT\ln x_B + x_B^2L_{AB} + x_C^2L_{AC} + x_Bx_C(L_{AB}+L_{AC}-L_{BC}) + x_Bx_C(1-2x_A)L_{ABC}\end{aligned} \qquad 2.93$$

Similar equations can be derived for components B and C with $L_{AB} = L_{BA}$, $L_{AC} = L_{CA}$, and $L_{BC} = L_{CB}$. A schematic molar Gibbs energy diagram at constant temperature and pressure is shown in Figure 2.10 with all three binary systems having $L_{ij} < 0$ of similar value.

To evaluate the stability of a ternary solution, one needs to calculate the elements in the determinants shown in Eq. 2.55. Using the moles of component C as the independent molar quantity, the limit of stability is expressed as

$$G_{AA}G_{BB} - G_{AB}G_{BA} = 0 \qquad 2.94$$

As an example, G_{AA} is shown in the following equation, which must be positive for the solution to be stable:

$$\begin{aligned}N\left(\frac{\partial\mu_A}{\partial N_A}\right)_{T,P,N_B,N_C} = NG_{AA} &= \frac{RT(1-x_A)}{x_A} - 2x_B^2L_{AB} - 2x_C^2L_{AC} - 2x_Bx_C \\ &\times(L_{AB}+L_{AC}-L_{BC}) - 2x_Bx_C(2-3x_A)L_{ABC}\end{aligned} \qquad 2.95$$

It is evident that any instability in binary systems with positive interaction parameters extends into the ternary system. It can also be seen that even if all binary interaction parameters are negative, i.e. there is no instability in the binary systems, it is possible for Eq. 2.95 to become negative for some combinations of the binary interaction parameters such that $\Delta L = L_{AB} + L_{AC} - L_{BC}$ becomes very positive and overshadows the

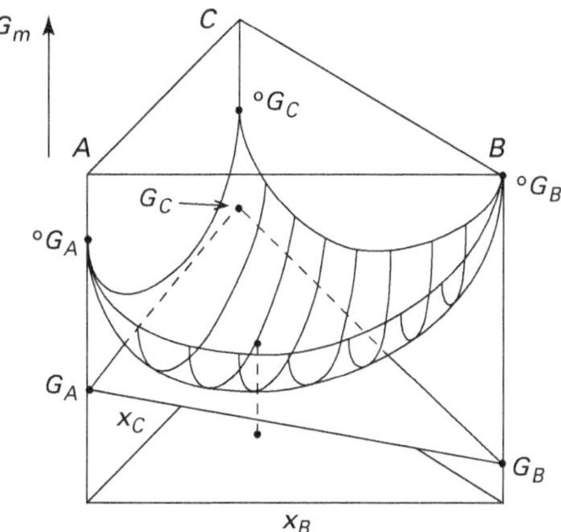

Figure 2.10 *Schematic ternary molar Gibbs energy diagram as a function of composition for given temperature and pressure, from [1] with permission from Cambridge University Press.*

contributions due to L_{AB} and L_{AC}, i.e. L_{BC} is more negative than L_{AB} and L_{AC} combined. In an extreme case with $L_{AB} = L_{AC} = L_{ABC} = 0$ and $L_{BC} < 0$, i.e. when we have ideal solutions for the $A - B$ and $A - C$ binary systems, a stable solution in the $B - C$ binary system, and no additional ternary interaction, Eq. 2.95 reduces to

$$N\left(\frac{\partial\mu_A}{\partial N_A}\right)_{T,P,N_B,N_C} = \frac{RT(1-x_A)}{x_A} + 2x_Bx_CL_{BC} \tag{2.96}$$

Setting $(\partial\mu_A/\partial N_A)_{T,P,N_B,N_C} = 0$, one obtains

$$-\frac{RT}{2L_{BC}} = \frac{x_Ax_Bx_C}{1-x_A} = \frac{(1-x_B-x_C)x_Bx_C}{x_B+x_C} \tag{2.97}$$

With $-RT/2L_{BC}$ being positive due to $L_{BC} < 0$, there is a parabola-shaped composition area in which the solution is unstable at constant temperature and pressure. This is reasonable because the system tends to maximize the number of *B–C* bonds in order to lower its energy, but this competes with the entropy of mixing among the three elements and results in the segregation of *B–C* bonds, and thus a miscibility gap at low temperatures.

To evaluate the ternary consolute point, the second derivatives for components *A* and *B* are obtained as

$$N\left(\frac{\partial^2\mu_A}{\partial N_A^2}\right)_{T,P,N_B,N_C} = \frac{RT(1-x_A)}{x_A^2} + 4x_B^2L_{AB} + 4x_C^2L_{AC} + 4x_Bx_C \times(L_{AB}+L_{AC}-L_{BC}) + 2x_Bx_C(7-9x_A)L_{ABC} = 0 \tag{2.98}$$

$$N\left(\frac{\partial^2\mu_B}{\partial N_B^2}\right)_{T,P,N_A,N_C} = \frac{RT(1-x_B)}{x_B^2} + 4x_A^2L_{AB} + 4x_C^2L_{BC} + 4x_Ax_C \times(L_{AB}+L_{BC}-L_{AC}) + 2x_Ax_C(7-9x_B)L_{ABC} = 0 \tag{2.99}$$

The consolute point can then be obtained using Eq. 2.94, Eq. 2.98 and Eq. 2.99.

It may be observed in Eq. 2.93 that $(1-2x_A)L_{ABC}=0$ at $x_A=0.5$, i.e. the ternary interaction parameter does not contribute to the chemical potential of A. It may also be observed in Eq. 2.95 that the contribution from the ternary interaction parameter changes sign at $x_i=2/3$ since $(2-3x_A)L_{ABC}=0$.

2.2.4 Multi-component random solutions

Similarly to that for a ternary solution, the excess Gibbs energy of mixing of a multi-component solution can be written as

$$ {}^{E}G_m^{'} = \sum_i \sum_j x_i x_j L_{ij} + \sum_i \sum_j \sum_k x_i x_j x_k L_{ijk} \qquad 2.100 $$

In principle, one can add interaction parameters for quaternary and higher order systems, but their contributions to the Gibbs energy are relatively minor; the major contributions have already been taken into account by the binary and ternary interactions. It is anticipated that not only are the interaction parameters of four or more components small, but also the mole fractions multiplying the interaction parameters diminish their contribution to the Gibbs energy even further.

Under the condition that all interaction parameters are constant, the chemical potential of a component in a multi-component system with binary and ternary interaction parameters can be extended from Eq. 2.93 as

$$ \mu_i = {}^{0}G_i + RT\ln x_i + \sum_{j\neq i} x_j^2 L_{ij} + \sum_{k>j} \sum_{j\neq i} x_j x_k \left[L_{ij} + L_{ik} - L_{jk} + (1-2x_i)L_{ijk}\right] \qquad 2.101 $$

The stability of the solution can also be extended from Eq. 2.95 as

$$ N\left(\frac{\partial \mu_i}{\partial N_i}\right)_{T,P,N_{j\neq i}} = NG_{ii} = \frac{RT(1-x_i)}{x_i} - 2\sum_{j\neq i} x_j^2 L_{ij} - 2\sum_{k>j} \sum_{j\neq i} x_j x_k \times \left[L_{ij} + L_{ik} - L_{jk} + (2-3x_i)L_{ijk}\right] \qquad 2.102 $$

The limit of stability of a multi-component random solution can be represented by Eq. 2.55 or Eq. 2.58.

2.3 Phases with variable compositions: solutions with ordering

2.3.1 Solutions with short-range ordering

The order in a system can be measured by correlation functions which describe how the various components are correlated in space. For simplicity, let us consider only the pairs of nearest neighbors, with the correlation function represented by the pair probability of nearest neighbor bonds between two components. In a random solution, the probability of finding nearest neighbor bonds between two components i and j is

$$ p_{ij} = x_i x_j \qquad 2.103 $$

When $p_{ij} \neq x_i x_j$, the nearest neighbors of component i are not occupied randomly by component j; rather, certain components are favored, resulting in short-range ordering or local clustering in the solution. When short-range ordering develops throughout the solution, long-range ordering takes place, and each component has its own primary sites in the solution, as discussed in Section 2.3.2. There are relations between bond probabilities and mole fractions of components due to the mass balance, as follows, with the assumption $p_{ij} = p_{ji}$:

$$\sum_i \sum_j p_{ij} = 1 \tag{2.104}$$

$$x_i = \sum_j p_{ij} \tag{2.105}$$

For small deviations from a random solution, one can consider the formation of $i-j$ bonds from $i-i$ and $j-j$ bonds and the ideal mixing of the three types of bonds, similarly to a typical Ising model. The bond reaction can be written as

$$(i-i)\text{ bonds} + (j-j)\text{ bonds} = 2\ (i-j)\text{ bonds} \tag{2.106}$$

with Gibbs energy

$$\Delta G_{ij} = 2G_{ij} - \left(G_{ii} + G_{jj}\right) \tag{2.107}$$

The Gibbs energy of the solution per mole of atoms is thus represented by the bond energies and the ideal mixing of bonds plus non-ideal interactions between pairs,

$$G_m = \sum_i \sum_j p_{ij} G_{ij} + \frac{Z}{2} RT \sum_i \sum_j p_{ij} \ln p_{ij} + {}^E G_m \tag{2.108}$$

with G_{ij} the molar bond energy between components i and j, Z the number of bonds per atom, which is divided by two in the equation because two atoms are needed to form one bond, and ${}^E G_m = \sum p_{ij} p_{kl} I_{ijkl}$ the excess Gibbs energy of mixing between bonds. This approach proposed by Guggenheim [2] is called the quasi-chemical method as it is based on the chemical reaction shown in Eq. 2.106.

However, the entropy of mixing in Eq. 2.108 does not reduce to the ideal entropy of mixing for a solution without short-range ordering as defined by Eq. 2.103. An approximated correction may be added for a small degree of short-range ordering as follows:

$$G_m = \sum_i \sum_j p_{ij} G_{ij} + \frac{Z}{2} RT \sum_i \sum_j p_{ij} \ln \frac{p_{ij}}{x_i x_j} + RT \sum x_i \ln x_i + {}^E G_m \tag{2.109}$$

For a random solution defined by Eq. 2.103, Eq. 2.109 becomes

$$G_m = \sum x_i {}^0 G_i + RT \sum x_i \ln x_i + \sum x_i x_j \Delta G_{ij} + {}^E G_m \tag{2.110}$$

with ${}^0 G_i = G_{ii}$, ΔG_{ij} from Eq. 2.107 representing the interaction parameter between components i and j, and ${}^E G_m = \sum x_i x_j x_k x_l I_{ijkl}$ denoting the higher order interactions in comparison with Eq. 2.100.

When short-range ordering exists in a solution, one typically uses the law of mass reaction for the chemical reaction represented by Eq. 2.106 to define the equilibrium among all bonds, i.e.

$$\frac{(p_{ij})^2}{p_{ii}p_{jj}} = e^{-\Delta G_{ij}/kT} \quad 2.111$$

However, this is under the assumption that the chemical activities of all bonds can be represented by their respective probabilities, which is only true for an ideal solution even excluding dilute solutions due to Henry's law, as shown by Eq. 2.78. Preferably, the bond probabilities can be obtained by calculating the driving force for the fluctuation of bond probabilities under constant temperature, pressure, and amount of each component, along with the constraints defined by Eq. 2.104 and Eq. 2.105, and equating the driving force to zero, i.e.

$$\begin{aligned}\frac{1}{N}\left(\frac{\partial G}{\partial \xi}\right)_{T,P,N_n} = \left(\frac{\partial G_m}{\partial p_{ij}}\right)_{T,P,N_n} &= \left(\frac{\partial G_m}{\partial p_{ij}}\right)_{T,P,x_n,p_{kl\neq ij}} - \sum_{kl\neq ij}\left(\frac{\partial G_m}{\partial p_{kl}}\right)_{T,P,x_n,p_{op\neq kl}} \\ &\quad + \left(\frac{\partial G_m}{\partial x_i}\right)_{T,P,x_{q\neq i},p_{kl}} + \left(\frac{\partial G_m}{\partial x_j}\right)_{T,P,x_{q\neq j},p_{kl}} \\ &= 0\end{aligned} \quad 2.112$$

where *op* indicates the indexes different from *ij* and *kl* and $\partial p_{kl}/\partial p_{ij} = -1$ and $\partial x_i/\partial p_{ij} = \partial x_j/\partial p_{ij} = 1$ are used from Eq. 2.104 and Eq. 2.105. Numerical values of p_{ij} can be obtained by minimization of the Gibbs energy under the constraints given by Eq. 2.104 and Eq. 2.105.

The chemical potential of an independent component *i* is defined as in Eq. 2.14 and can be represented by the following equation

$$\mu_i = G_m + \frac{\partial G_m}{\partial x_i} - \sum_{j=1}^{c} x_j\frac{\partial G_m}{\partial x_j} + 2\sum_{j=1}^{c}\frac{\partial G_m}{\partial p_{ij}} - \frac{\partial G_m}{\partial p_{ii}} - 2\sum_{j=1}^{c} x_j\sum_{k=1}^{c}\frac{\partial G_m}{\partial p_{jk}} + \sum_{j=1}^{c} x_j\frac{\partial G_m}{\partial p_{jj}} \quad 2.113$$

The stability of the solution can be derived similarly to Eq. 2.54.

When the bonding between components becomes very strong, distinctive new components may form. They are not independent components and are often called associates. Both the independent and dependent components are collectively called species. The formation of an associate $i_{a_i}j_{b_j}$ consisting of a_i moles of *i* and a_j moles of *j* can be written as

$$a_i i + a_j j = i_{a_i} j_{a_j} \quad 2.114$$

The Gibbs energy of the associate follows the same format as that of a stoichiometric phase, Eq. 2.42,

$$^{0}G_{\,i_{a_i}j_{a_j}} = \sum a_i {}^{0}G_i^{SER} + \Delta_f G_{i_{a_i}j_{a_j}} \quad 2.115$$

The Gibbs energy of the solution is obtained by extending Eq. 2.63 to all species:

$$G_m = \sum y_{i_{a_i}j_{a_j}} {}^{0}G_{i_{a_i}j_{a_j}} + RT\sum y_{i_{a_i}j_{a_j}} \ln y_{i_{a_i}j_{a_j}} + {}^{E}G_m \quad 2.116$$

where $y_{i_{a_i}j_{a_j}}$ is the mole fraction of species $i_{a_i}j_{a_j}$ in the solution, with $a_i = 1$ and $a_j = 0$ for component i and $a_i = 0$ and $a_j = 1$ for component j. The equilibrium amount of each associate $i_{a_i}j_{a_j}$ is obtained by combination of mass balance and the zero driving force for the variation of the amount of the associate, similarly to Eq. 2.112, i.e.

$$\sum_i \sum_j y_{i_{a_i}j_{a_j}} = 1 \qquad 2.117$$

$$x_i = \sum a_i y_{i_{a_i}j_{a_j}} \qquad 2.118$$

$$\left(\frac{\partial G_m}{\partial y_{i_{a_i}j_{a_j}}} \right)_{x_i} = 0 \qquad 2.119$$

Associates are particularly plentiful in the gas phase, and their amounts are significantly affected by pressure. For an ideal gas phase with ${}^E G_m = 0$ and $PV_m = RT$, the effect of pressure is added as follows:

$$G_m = \sum y_{i_{a_i}j_{a_j}} {}^0 G_{i_{a_i}j_{a_j}} + RT \sum y_{i_{a_i}j_{a_j}} \ln y_{i_{a_i}j_{a_j}} + \int_{P_0}^{P} V_m dP$$
$$= \sum y_{i_{a_i}j_{a_j}} {}^0 G_{i_{a_i}j_{a_j}} + RT \sum y_{i_{a_i}j_{a_j}} \ln y_{i_{a_i}j_{a_j}} + RT \ln \frac{P}{P_0} \qquad 2.120$$

where P is the total pressure, and P_0 the reference pressure at which ${}^0 G_{i_{a_i}j_{a_j}}$ is defined, usually chosen to be one atmospheric pressure. Equation 2.120 thus becomes

$$G_m = \sum y_{i_{a_i}j_{a_j}} {}^0 G_{i_{a_i}j_{a_j}} (P = 1 \text{ atm}) + RT \sum y_{i_{a_i}j_{a_j}} \ln y_{i_{a_i}j_{a_j}} + RT \ln P \qquad 2.121$$

where the unit of the total pressure P is atmospheric pressure (atm). The chemical potential of species $i_{a_i}j_{a_j}$ is equal to

$$\mu_{i_{a_i}j_{a_j}} = {}^0 G_{i_{a_i}j_{a_j}} (P = 1 \text{ atm}) + RT \ln y_{i_{a_i}j_{a_j}} P = {}^0 G_{i_{a_i}j_{a_j}} (P = 1 \text{ atm}) + RT \ln P_{i_{a_i}j_{a_j}} \qquad 2.122$$

where $P_{i_{a_i}j_{a_j}}$ is the partial pressure of species $i_{a_i}j_{a_j}$, defined as

$$P_{i_{a_i}j_{a_j}} = y_{i_{a_i}j_{a_j}} P \qquad 2.123$$

Combining Eq. 2.21 and Eq. 2.115 with Eq. 2.122, the relation between the chemical potentials of an associate and its constituents is expressed as

$$\mu_{i_{a_i}j_{a_j}} = a_i \mu_i + a_j \mu_j = a_i {}^0 G_i + a_j {}^0 G_j + RT \ln \left(P_i^{a_i} P_j^{a_j} \right) \qquad 2.124$$

The equilibrium condition for the chemical reaction of an associate forming from its constituents in an ideal gas phase is obtained as

$$\Delta_f G_{i_{a_i}j_{a_j}} + RT \ln \frac{P_i^{a_i} P_j^{a_j}}{P_{i_{a_i}j_{a_j}}} = 0 \qquad 2.125$$

For non-ideal phases, the mole fractions of various associates can be calculated numerically by the minimization of the Gibbs energy under the constraints Eq. 2.117 and Eq. 2.118.

2.3.2 Solutions with long-range ordering

So far, solutions in which a component can occupy any site in a phase have been discussed. In many phases, this is not the case. For example, in the fcc solid solution of Fe and C, Fe atoms take the fcc lattice sites, and C atoms occupy the interstitial sites between the fcc lattice sites. Therefore, Fe atoms do not mix with C atoms on the fcc lattice sites; rather, they develop long-rang ordering by occupying their own distinct sites in the phase. Long-range ordering can also develop when short-range ordering extends to the whole lattice. A new formula for the Gibbs energy of mixing is needed and is obtained by considering the details of how components are distributed and mixed in various sites in a phase.

One way to group various sites in a phase is based on equivalent crystallographic positions in a phase, i.e. Wyckoff positions. Various sets of equivalent positions divide the lattice into subsets of lattices. Each set of equivalent positions forms a sublattice. The distributions of components on each sublattice can be represented by mole fractions of components in the sublattice, commonly referred to as site fractions and defined as

$$y_i^t = \frac{N_i^t}{\sum_j N_j^t} \qquad 2.126$$

$$\sum_i y_i^t = 1 \qquad 2.127$$

where the superscript t denotes the sublattice in which the component resides, and the summation is for all species in sublattice t including the vacancy. Site fractions and mole fractions are related through the mass balance as follows:

$$x_i = \frac{\sum a^t y_i^t}{\sum a^t \left(1 - y_{va}^t\right)} \qquad 2.128$$

where a^t and y_{va}^t are the numbers of sites and the site fraction of vacancies in the sublattice t.

Random solutions form when each component enters all sublattices equally. Mole fractions and site fractions thus become identical. Solutions with both substitutional and interstitial components, like the fcc Fe–C solution mentioned above, can be represented by two sublattices. Stoichiometric compounds have the site fractions equal to unity in each sublattice. When site fractions in a compound deviate from unity, the compound is no longer stoichiometric and develops a composition range of homogeneity. When the composition range is small, the deviations are often referred to as defects. Since many properties of a compound are determined by defects, a distinct field of defect chemistry exists, predominantly for charged species. As will be demonstrated in Section 2.3.4 and the rest of the book, defects can be treated as an integral part of the thermodynamics of a phase with more than one sublattice.

Let us consider a case where there is only one component in each sublattice, which represents one possible stoichiometric composition of the phase and is often called an end-member of the phase. The Gibbs energy of an end-member is the same as that of a phase with a fixed composition as given by Eq. 2.38, Eq. 2.39, or Eq. 2.42. By re-arranging Eq. 2.42, the Gibbs energy of an end-member, ${}^0G_{em}$, is obtained as

$$ {}^0G_{em} = \sum\nolimits_t a^t {}^0G_i^{t,ref} + \Delta_f G_{em} \tag{2.129} $$

where ${}^0G_i^{t,ref}$ represents the Gibbs energy of component i, in a given reference state, which occupies sublattice t in the end-member. For a vacancy, ${}^0G_{Va} = 0$ is defined. The contribution of each end-member to the Gibbs energy of the phase is the product of the site fraction of each component in its respective sublattice and the Gibbs energy of the end-member per mole of formula unit (*mf*), i.e.

$$ {}^0G_{mf} = \sum\nolimits_{em} \left(\prod_t y_i^t {}^0G_{em} \right) \tag{2.130} $$

The ideal mixing in each sublattice is similar to that in a random solution with mole fractions substituted by site fractions. The excess Gibbs energy of mixing consists of two contributions: (i) that due to mixing in one sublattice, with all other sublattices containing only one component, and (ii) that due to simultaneous mixing in more than one sublattice. The Gibbs energy of a solution phase with multi-sublattices can thus be written per mole of formula unit as

$$ G_{mf} = {}^0G_{mf} + RT\sum\nolimits_t a^t \sum\nolimits_i y_i^t \ln y_i^t + {}^EG_{mf} \tag{2.131} $$

with ${}^EG_{mf}$ given by

$$ \begin{aligned} {}^EG_{mf} = {} & \sum_t \prod_{s\neq t} y_l^s \sum_{i>j} \sum_j y_i^t y_j^t L_{i,j:l}^t + \sum_t \prod_{s\neq t} y_l^s \sum_{i>j} \sum_{j>k} \sum_k y_i^t y_j^t y_k^t L_{i,j,k:l}^t \\ & + \sum_t \prod_{s\neq t,u} y_l^s \sum_{i>j} \sum_{j>k} \sum_k y_i^t y_j^t y_m^u y_n^u L_{i,j:m,n:l}^{t,u} \end{aligned} \tag{2.132} $$

The first term in Eq. 2.132 represents the binary interaction between components i and j in sublattice t with sublattice s occupied by component l, with a comma separating interacting components in one sublattice and a colon separating sublattices. The product $\prod_{s\neq t} y_l^s$ runs over all sublattices with one component in each sublattice except sublattice t, in which the interaction is considered. The second term denotes the ternary interaction among i, j, and k in sublattice t with sublattice s occupied by component l. The third term depicts simultaneous interactions in both sublattices t and u, and the product runs over all other sublattices with one component in each sublattice except sublattices t and u, in which the interactions are considered. The third term thus partially reflects the short-range ordering between components in two sublattices. In principle, high-order interaction parameters such as quaternary, quinary, and multiple sublattice interaction parameters could be added, but their contributions to ${}^EG_{mf}$ are small due to the physical insignificance of the co-location of four or five components, indicated by the product of their site fractions in front of the interaction parameters.

In Eq. 2.21, the chemical potential of a stoichiometric compound was defined in terms of a summation of the chemical potentials of individual components in the compound because the relative amounts of components are constrained by the stoichiometry of the compound and the chemical potentials of individual components cannot vary independently. By the same token, the chemical potential of an end-member in a solution can be written as

$$\mu_{em} = G_{em} = \sum\nolimits_t a^t \mu_i^t \tag{2.133}$$

where μ_i^t is the chemical potential of component i that occupies the sublattice t in the end-member, and can be derived using Eq. 2.12:

$$\mu_i^t = a^{t0} G_i^{t,ref} + a^t RT\ln y_i^t + {}^E G_{mf} + \frac{\partial^E G_{mf}}{\partial y_i^t} - \sum_j y_j^t \frac{\partial^E G_{mf}}{\partial y_j^t} \tag{2.134}$$

For constant interaction parameters in Eq. 2.132, Eq. 2.134 for the chemical potential reduces to the following expression from Eq. 2.101:

$$\mu_i^t = a^{t0}\, G_i^t + a^t RT\ln y_i^t + \sum_{j\neq i} \left(y_i^t\right)^2 L_{i,j}^t + \sum_{k>j} \sum_{j\neq i} y_j^t y_k^t \times \left[L_{i,j}^t + L_{i,k}^t - L_{j,k}^t + \left(1 - 2y_i^t\right) L_{i,j,k}^t\right] \tag{2.135}$$

The stability of the solution is defined by $\partial\mu_{em}/\partial N_{em}$, where N_{em} is the number of moles of the end-member in the solution and is given by $N\mu_{em} = N\prod_u y_j^u$ and

$$\frac{\partial N_{em}}{\partial y_j^u} = \frac{N\prod\limits_t y_i^t}{y_j^u\left(1 - \prod\limits_t y_i^t\right)} \tag{2.136}$$

Following Eq. 2.102, one obtains

$$\frac{\partial \mu_{em}}{\partial N_{em}} = \sum_u \frac{\partial\left(\sum\limits_t a^t \mu_i^t\right)}{\partial y_j^u} \frac{\partial y_j^u}{\partial N_{em}} = \frac{\left(1 - \prod\limits_t y_j^t\right)}{N\prod\limits_t y_j^t} \sum_u a^u y_i^u \frac{\partial \mu_i^u}{\partial y_i^u} = \frac{\left(1 - \prod\limits_t y_j^t\right)}{N\prod\limits_t y_j^t} \sum_u a^u y_i^u \times \left\{\frac{RT\left(1 - y_i^u\right)}{y_i^u} - 2\sum_{j\neq i}\left(y_j^u\right)^2 L_{i,j}^u - 2\sum_{k>j}\sum_{j\neq i} y_j^u y_k^u \left[L_{i,j}^u + L_{i,k}^u - L_{j,k}^u + \left(2 - 3y_i^u\right) L_{i,j,k}^u\right]\right\} \tag{2.137}$$

It is self-evident from Eq. 2.126 that a site fraction is only uniquely defined from the mole fraction of the component when the component enters into one sublattice only and does not form any associates. Therefore, in general, the distribution of components on sublattices and different kinds of molecules can only be obtained by equilibrium calculations, and the thermodynamic properties for such a phase thus cannot be represented in a closed form using mole fractions of independent components. This was demonstrated in Section 2.3.1 in the case when short-range ordering exists in solution phases; there the energy minimization procedure was used to obtain the distribution of components over different kinds of bonds and the amounts of individual associates.

2.3.3 Solutions with both short-range and long-range ordering

The short-range ordering in a solution with long-range ordering can take place in each sublattice or between two sublattices. The short-range ordering in one sublattice can be treated as in Section 2.3.1, with mole fractions substituted by site fractions. In the case in which associates form, the relation between mole fractions and site fractions becomes more complicated, as follows:

$$x_i = \frac{\sum a^t \sum_k i_k y_k^t}{\sum a^t \left(1 - y_{va}^t\right)} \qquad 2.138$$

where the summation for k goes over all associates in sublattice t containing component i.

The short-range ordering between two sublattices indicates that a component in one sublattice has different interactions with different components in another sublattice. This results in the local ordering of one component around another component in two neighboring sublattices. Such local ordering involves interactions between two sublattices, shown as the third term in Eq. 2.132.

2.3.4 Solutions with charged species

One special type of solution with both short-range and long-range ordering is solutions with charged species, i.e. ionic solutions, plus electrons and holes. There is an additional constraint on species concentrations to maintain the charge neutrality of such solutions, i.e.

$$0 = \sum_t \sum_i a^t y_i^t v_i^t \qquad 2.139$$

where v_i^t is the valence of species i in sublattice t including its sign, which is positive for cations, negative for anions, and zero for neutral species. Conventional defect chemistry theory is typically based on the ideal mass action laws and applicable to a single set of defects and at very low defect concentrations, i.e. in the limit of ideal solutions. For interacting defects, their concentrations should be replaced by their activities, which can be obtained from thermodynamic principles, as discussed in previous sections. It should be emphasized that in addition to the formation of many more charged species, one component may have different valences. This is particularly the case for the transition metals. Consequently, there can be many more species in an ionic phase than the number of independent components in the system, and their concentrations can be found by equilibrium calculations as discussed in Section 2.3.1.

2.4 Polymer solutions and polymer blends

A polymer solution is a mixture of polymer molecules and solvents, while a polymer blend is a mixture of different polymer molecules. A polymer molecule consists of the same repeating units of one or more monomers; a monomer can be an atom or a small molecule. The number of repeating units is called the degree of polymerization and can be as large as 10^4–10^5. It defines the molecular mass, i.e. the mass of one polymer molecule. There are three typical architectures of polymerization: a linear chain, a branched chain, and a cross-linked polymer. Nearly all polymers are mixtures of molecules with a different degree of polymerization and thus with a molecular mass distribution, complicating the modeling of their thermodynamic properties because of the dependence of the properties on molecular mass.

Gibbs energy functions of polymers with a single molecular mass can be treated as in Section 2.1. For a polymer solution, the ideal entropy of mixing is quite different from that of the atomically random solutions discussed in Section 2.2.1 because the monomers in a polymer molecule are connected to each other and cannot move freely. One common approach to calculating the ideal entropy of mixing is to invoke a lattice model and assume that one monomer occupies a lattice site with a fixed volume. The number of translational states of a single molecule is equal to the number of lattice sites available. In a homogeneous solution, the total number of lattice sites available is

$$n = \sum_i m_i n_i \qquad 2.140$$

where n_i and m_i are the number of molecules i and the number of lattice sites per molecule i, respectively. In its pure state, i.e. before mixing, the number of states of molecule i in terms of the number of lattice sites is

$$w_i = m_i n_i = n\phi_i \qquad 2.141$$

where ϕ_i is the volume fraction of molecule i in the solution. The entropy change per molecule i is thus

$$S_i = k\ln n - k\ln w_i = k\ln\frac{1}{\phi_i} = -k\ln\phi_i \qquad 2.142$$

The total entropy of mixing is the summation for all molecules, normalized to one mole of lattice sites:

$${}^{I}S_m = \frac{N_a}{n}\sum_i n_i S_i = -R\sum_i \frac{\phi_i}{m_i}\ln\phi_i \qquad 2.143$$

where N_a is the Avogadro number. When $m_i = 1$ for all molecules, Eq. 2.143 reduces to Eq. 2.61. Since the m_i values are typically very large numbers for polymers, the entropy of mixing in polymer solutions and blends is thus significantly lower than those in non-polymer solutions, as shown schematically in Figure 2.11 for binary systems with various m_i values.

Similarly to Eq. 2.64, the Gibbs energy of a multi-component random polymer solution or blend can be written as

$$G_m = \sum \frac{\phi_i}{m_i}{}^{0}G_{im} + RT\left(\sum \frac{\phi_i}{m_i}\ln\phi_i + \sum \phi_i\phi_j\chi_{ij}\right) \qquad 2.144$$

where ${}^{0}G_{im}$ is the Gibbs energy of molecule i per mole of lattice sites, and χ_{ij} the unitless interaction parameter between molecules i and j. Other equations shown in Section 2.2.1

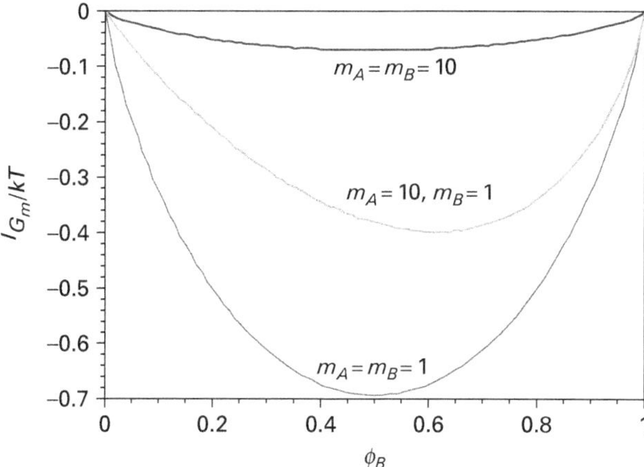

Figure 2.11 *Schematic entropy of mixing in solutions, with the numbers of lattice sites per molecule shown.*

can be derived similarly. It is to be noted that Eq. 2.144 is very similar to the Flory–Huggins solution equation, widely used in the polymer community.

2.5 Elastic, magnetic, and electric contributions to the free energy

Sections 2.1 and 2.2 focused on the thermal and hydrostatic pressure contributions to the Gibbs energy, which are the two prime variables affecting phase stability in typical experimental environments. However, there are other internal and external contributions, which are particularly important for crystalline phases. Two important internal contributions are from the magnetic and electric polarizations of materials, with corresponding external contributions due to magnetic and electric fields. Furthermore, for the non-hydrostatic pressure of solid phases, the PV term in the combined law is to be replaced by the elastic energy calculated from elastic stress and elastic strain. The corresponding work done on a system is as follows [3]:

$$dW_{elastic} = -V\sum_{i,j.k,l}\sigma_{ij}d\varepsilon_{kl} \qquad 2.145$$

$$dW_{magnetic} = -V\sum_i H_i dB_i \qquad 2.146$$

$$dW_{electric} = -V\sum_i E_i dD_i \qquad 2.147$$

where $i, j, k, l = 1, 2, 3$; σ_{ij} and ε_{kl} are the components of stress and strain; H_i and B_i are the components of magnetic field and magnetic induction; E_i and D_i are the components of electric field and electric displacement; and V is the volume of the crystal. The negative signs in front of the equations are due to the fact that the system does work on the surroundings when it expands its volume due to strain, magnetic induction, or electric displacement.

Using the combined law of thermodynamics, Eq. 1.10 can thus be rewritten as follows:

$$dU = TdS - V\left(\sum_{i,j,k,l}\sigma_{ij}d\varepsilon_{kl}+\sum_i H_i dB_i + \sum_i E_i dD_i\right) + \sum \mu_i dN_i - Dd\xi \qquad 2.148$$

A Legendre transformation, similar to that used to obtain the Helmholtz energy and Gibbs energy, Eq. 1.48 and Eq. 1.49, can be made to obtain the following characteristic free energy functions:

$$\begin{aligned} dF &= d(U - TS) \\ &= -SdT - V\left(\sum_{i,j,k,l}\sigma_{ij}d\varepsilon_{kl} + \sum_i H_i dB_i + \sum_i E_i dD_i\right) + \sum \mu_i dN_i - Dd\xi \end{aligned} \qquad 2.149$$

$$\begin{aligned} dF_H &= d\left(U - TS + \sum_i H_i B_i\right) \\ &= -SdT - V\left(\sum_{i,j,k,l}\sigma_{ij}d\varepsilon_{kl} - \sum_i B_i dH_i + \sum_i E_i dD_i\right) + \sum \mu_i dN_i - Dd\xi \end{aligned} \qquad 2.150$$

$$\begin{aligned} dF_E &= d\left(U - TS + \sum_i E_i D_i\right) \\ &= -SdT - V\left(\sum_{i,j,k,l}\sigma_{ij}d\varepsilon_{kl} + \sum_i H_i dB_i - \sum_i D_i dE_i\right) + \sum \mu_i dN_i - Dd\xi \end{aligned} \qquad 2.151$$

$$\begin{aligned} dF_{EH} &= d\left(U - TS + \sum_i H_i B_i + \sum_i E_i D_i\right) \\ &= -SdT - V\left(\sum_{i,j,k,l}\sigma_{ij}d\varepsilon_{kl} - \sum_i B_i dH_i - \sum_i D_i dE_i\right) + \sum_i \mu_i dN_i - Dd\xi \end{aligned} \qquad 2.152$$

$$\begin{aligned} dG &= d\left(U - TS + \sum_{i,j,k,l}\sigma_{ij}\varepsilon_{kl} + \sum_i H_i B_i + \sum_i E_i D_i\right) \\ &= -SdT + V\left(\sum_{i,j,k,l}\varepsilon_{ij}d\sigma_{kl} + \sum_i B_i dH_i + \sum_i D_i dE_i\right) + \sum \mu_i dN_i - Dd\xi \end{aligned} \qquad 2.153$$

From the above equations, it can be seen that the natural variables of the various free energies are $F(T, \varepsilon_{ij}, B_i, D_i, N_i, \xi)$, $F_H(T, \varepsilon_{ij}, H_i, D_i, N_i, \xi)$, $F_E(T, \varepsilon_{ij}, B_i, E_i, N_i, \xi)$, $F_{EH}(T, \varepsilon_{ij}, H_i, E_i, N_i, \xi)$, and $G(T, \sigma_{ij}, H_i, E_i, N_i, \xi)$. Clearly, there can be more combinations when the components of ε_{ij}, D_i, and B_i are partially replaced by their conjugate potentials. The free energies listed above are useful depending on how the system is constrained by the surroundings. For practical applications, the elastic, magnetic, and electric properties are usually considered for phases with fixed compositions, and Eq. 2.153 at equilibrium can then be written as

$$dG = -SdT + V\left(\sum_{i,j,k,l} \varepsilon_{ij} d\sigma_{kl} + \sum_i B_i dH_i + \sum_i D_i dE_i\right) + \mu dN \quad 2.154$$

The corresponding Gibbs–Duhem equation follows from Eq. 2.22

$$0 = -SdT + V\left(\sum_{i,j,k,l} \varepsilon_{ij} d\sigma_{kl} + \sum_i B_i dH_i + \sum_i D_i dE_i\right) - Nd\mu \quad 2.155$$

The general differential form of a molar quantity can be extended from Eq. 2.11 as follows, for a one-component system:

$$dS = \left(\frac{\partial S}{\partial T}\right)_{\sigma,E,H} dT + \left(\frac{\partial S}{\partial \sigma_{kl}}\right)_{E,T,H} d\sigma_{kl} + \left(\frac{\partial S}{\partial E_k}\right)_{\sigma,T,H} dE_k + \left(\frac{\partial S}{\partial H_k}\right)_{\sigma,T,E} dH_k \quad 2.156$$

$$d\varepsilon_{ij} = \left(\frac{\partial \varepsilon_{ij}}{\partial T}\right)_{\sigma,E,H} dT + \left(\frac{\partial \varepsilon_{ij}}{\partial \sigma_{kl}}\right)_{E,T,H} d\sigma_{kl} + \left(\frac{\partial \varepsilon_{ij}}{\partial E_k}\right)_{\sigma,T,H} dE_k + \left(\frac{\partial \varepsilon_{ij}}{\partial H_k}\right)_{\sigma,T,E} dH_k \quad 2.157$$

$$dD_i = \left(\frac{\partial D_i}{\partial T}\right)_{\sigma,E,H} dT + \left(\frac{\partial D_i}{\partial \sigma_{kl}}\right)_{E,T,H} d\sigma_{kl} + \left(\frac{\partial D_i}{\partial E_k}\right)_{\sigma,T,H} dE_k + \left(\frac{\partial D_j}{\partial H_k}\right)_{\sigma,T,E} dH_k \quad 2.158$$

$$dB_i = \left(\frac{\partial B_i}{\partial T}\right)_{\sigma,E,H} dT + \left(\frac{\partial B_i}{\partial \sigma_{kl}}\right)_{E,T,H} d\sigma_{kl} + \left(\frac{\partial B_i}{\partial E_k}\right)_{\sigma,T,H} dE_k + \left(\frac{\partial B_i}{\partial H_k}\right)_{\sigma,T,E} dH_k \quad 2.159$$

The first derivatives in Eq. 2.156 to Eq. 2.159 are the second directives of the Gibbs energy with respect to its natural variables, i.e. potentials, and have their respective nomenclatures as shown in Table 2.1. The limit of stability follows from Eq. 2.18 and can be rewritten as

$$\left(\frac{\partial X_i}{\partial Y_i}\right)_{N,Y_j} = \infty \quad 2.160$$

This means that the derivatives in Eq. 2.156 to Eq. 2.159, i.e. the quantities in Table 2.1, diverge at the limit of stability.

Table 2.1 Physical quantities related to the first derivatives ($\partial S/\partial T$ etc.) in Eq. 2.156 to Eq. 2.159. The table is symmetric because the Maxwell relations are related to the second derivatives of the Gibbs energy with respect to its natural variables.

	T	σ_{kl}	E_k	H_k
S	C/T, heat capacity	α_{kl}, piezocaloric effect	p_k, electrocaloric effect	m_k, magnetocaloric effect
ε_{ij}	α_{ij}, thermal expansion	s_{ijkl}, elastic compliance	d_{ijk}, converse piezoelectricity	q_{ijk}, piezomagnetic moduli
D_i	p_i, pyroelectric coefficients	d_{ikl}, piezoelectric moduli	k_{ik}, permittivities	a_{ik}, magnetoelectric coefficients
B_i	m_i, pyromagnetic coefficients	q_{ikl}, piezomagnetic moduli	a_{ik}, magnetoelectric coefficients	μ_{ik}, permeability

Exercises

1. Derive the relation between the heat capacities at constant pressure and constant volume, i.e.

$$C_V = C_P - \frac{\alpha_V^2 VT}{\kappa_T} = C_P - \alpha_V^2 BVT$$

Note that $C_V = (\partial U/\partial T)_V$ and $C_P = -T(\partial^2 G/\partial T^2)_P$. You may want to represent U as a function of T and V in terms of G and its derivatives with respect to its natural variables, T and P, starting with V and $U = G + TS - PV$ as a function of T and P.

2. Go through the steps in detail needed to derive the following equations and specify which variables are kept constant in the partial derivatives:

$$\mu_j = \left(\frac{\partial G}{\partial N_j}\right)_{T,P,N_{k\neq j}} = \left(\frac{\partial (NG_m)}{\partial N_j}\right)_{T,P,N_{k\neq j}} = G_m + (1 - x_j)\frac{\partial G_m}{\partial x_j} - \sum_{k\neq j} x_k \frac{\partial G_m}{\partial x_k}$$

where $G_m(T, P, N_i)$ is the molar Gibbs energy, $N = \sum_i N_i$ is the total number of moles of components, and $x_i = N_i/N$ is the mole fraction of component i.

3. The Gibbs energy diagram for the Al–Zn system at 800 K is shown below. Obtain necessary data through measurements from the figure and answer the following questions. Note the factor 10^3 at the bottom of the y axis, so the enthalpy values on the y axis need to be multiplied by 1000.

   a. What are the reference states of components Al and Zn shown in the diagram?

   b. Calculate the activities of Al in a metastable, single-phase liquid with $x_{Zn} = 0.2$ when referred to pure liquid Al, pure fcc Al and pure hcp Al, respectively.

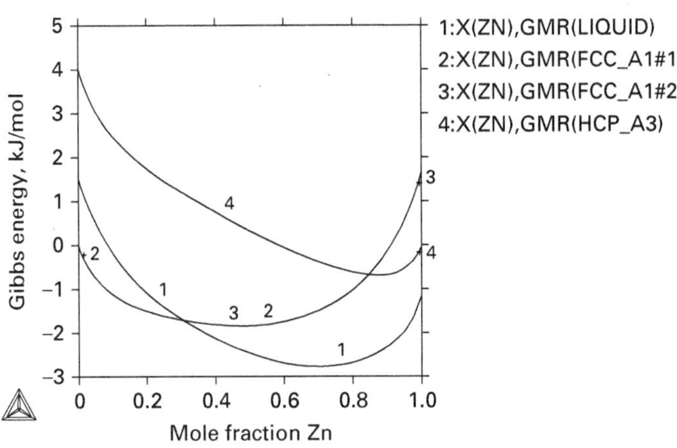

(*cont.*)

4. The Gibbs energy of the fcc Al–Zn solution is given as follows:

$$G_m^{fcc} = x_{Al}\,{}^oG_{Al}^{fcc} + x_{Zn}\,{}^oG_{Zn}^{fcc} + RT(x_{Al}\ln x_{Al} + x_{Zn}\ln x_{Zn}) + 5224x_{Al}x_{Zn}$$

Because the interaction parameter is positive, the fcc phase is unstable under certain conditions, which is why there are two fcc phases in the figure in Exercise 3, and the limit of stability is commonly called the spinodal. Derive the expression for the spinodal and plot it in a $T-x_{Zn}$ diagram.

5. The equation for the Gibbs energy per site, G_m, was given as Eq. 2.144.
   a. Calculate the entropy of mixing per site and the total entropy of mixing for the following three binary systems on a three-dimensional cubic lattice with one ball per site:
      - 300 black 25-ball chains with 300 white 25-ball chains,
      - 300 black 25-ball chains with 7500 white balls,
      - 5000 black balls with 5000 white balls.
   b. Under what values of χ_{ij} can a solution become unstable? Explain.
   c. Assuming that the three solutions discussed in part a above have the same values for χ_{ij}, which solution has the strongest tendency to become unstable, which has the weakest tendency to become unstable, and which is in the middle? Explain.

6. Given $S(T,P,N_i) = NS_m(T,P,x_i)$, where S is the entropy of a system, N the number of moles in the system, and S_m the molar entropy, and $S_i = (\partial S/\partial N_i)_{T,P,N_{j\neq i}}$, where S_i is the partial entropy of component i, show that

$$S_i = S_m + \left(\frac{\partial S_m}{\partial x_i}\right)_{T,P,x_{k\neq i}} - \sum_{j=1}^{c} x_j\left(\frac{\partial S_m}{\partial x_j}\right)_{T,P,x_{k\neq j}}$$

7. An α solution in the *A*–*B* system has $a_B = 0.9$ at 1000 K when pure α-*B* is used as the reference state. Calculate a_B referred to another state of *B*, called β-*B*, which is more stable than α-*B* by 1200 J/mol. Illustrate with a G_m diagram.

8. Assuming that the Gibbs energy of an *A*–*B* binary solution is represented by the equation
$G_m = x_A{}^0G_A + x_B{}^0G_B + RT\ (x_A\ln x_A + x_B\ln x_B) - 1000x_Ax_B$ (J/mol)
where ${}^0G_A$ and ${}^0G_B$ are the Gibbs energies of pure *A* and pure *B*, respectively. Do the following.
   a. Plot G_m as a function of x_B at 1000 K with ${}^0G_A = -2000$ and ${}^0G_B = -1000$ (J/mol).
   b. Derive expressions for partial quantities, μ_A and μ_B, and show $G_m = \mu_Ax_A + \mu_Bx_B$.
   c. Calculate the values of μ_A and μ_B at $x_B = 0.6$ and $T = 1000$ K.

(*cont.*)

d. Draw a tangent at $x_B = 0.6$ in your G_m versus x_B diagram and show that μ_A and μ_B are the intercepts of the tangent at $x_B = 0$ and $x_B = 1$, respectively.

9. What is the free energy of mixing of 2 moles of polystyrene (molar mass 2×10^5 g/mol) with 1×10^4 liter of toluene (molar mass 92.14 g/mol) at 25 °C, with Flory interaction parameter $\chi = 0.37$? The densities of polystyrene and toluene are 1.06 and 0.87 g/cm^3, respectively. Assume no volume change upon mixing.

10. At $T = 0$ K, the entropic contributions to the free energy of mixing disappear. With $\chi = A + B/T$ where A and B are both constants, sketch the composition dependence of the free energy for cases where $B < 0,\ B = 0,$ and $B > 0$, and discuss whether any of those situations leads to a stable mixture at $T = 0$ K. Does your answer depend on whether polymer solutions, or polymer blends, or metallic solutions are considered?

11. Plot on a single graph the composition dependence of the free energy of mixing per mole of atoms at 1000 K in a binary metallic solution with interaction parameter $L_{AB} = -20000, 0, 20000$ (J/mol). Which choices of the interaction parameter make the solution unstable at what composition range at 1000 K? This limit of instability is often called the spinodal. Calculate and plot the spinodal as a function of temperature for the choices you find.

12. The Gibbs energy of the Al–Cu binary fcc solution is represented by the following equation,

$$G_m = x_{Al}{}^0G_{Al} + x_{Cu}{}^0G_{Cu} + \ RT\ (x\ _{Al}\ln x\ _{Al} + x_{Cu}\ln x_{Cu}) + x_{Al}x_{Cu}(-50000 + 2T)$$

where ${}^0G_{Al}$ and ${}^0G_{Cu}$ are the Gibbs energies of pure fcc Al and fcc Cu, respectively. Do the following.

b. Derive expressions for the partial quantities G_{Al} and G_{Cu} and show $G_m = G_{Al}x_{Al} + G_{Cu}x_{Cu}$.

c. Calculate the partial quantities G_{Al} and G_{Cu} at 500 K and $x_{Al} = 0.4$.

d. Plot G_m, G_{Al}, and G_{Cu}as a function of x_{Al} at 500 K.

e. Validate the Gibbs–Duhem equation using the above information, at constant temperatures.

f. Discuss the applicability of Henry's law and Rault's law based on your results.

g. Will the solution become unstable at any temperatures? Why or why not?

13. A complex oxide can be represented by the formula for two sublattices, i.e. (A,B)(O,Va)$_b$ with a small oxygen deficiency in the second sublattice. At one temperature, the composition of the oxide is measured in terms of mole fraction as $x_A = 0.21$ and $x_B = 0.13$. Calculate the value of b, i.e. the number of sites of the second sublattice, and the site fractions of A and B in the first sublattice and of O and Va in the second sublattice.

(*cont.*)

14. The sublattice model $(Ca^{2+},Zr^{4+})_1(O^{2-})_{1.5}(O^{2-},Va)_{0.5}$ is used to describe the doping of Ca into ZrO_2. Show in a composition square where you would expect to find such a solid solution. Calculate the entropy of mixing, assuming random mixing in the first and third sublattices for equal amounts of O and Va in the third sublattice. Compare with the ideal entropy of mixing when the second and third sublattices are combined into one sublattice.
15. Consider the contents of an expandable vessel as a system. We have enclosed a certain amount of water in the vessel. Then we vary T and P by actions from the outside and studies what happens to V in an attempt to decide whether the system behaves as a unary system. Due to its larger volume, it is easy to see when a gas phase forms. Discuss what we would expect to happen. Assume that the wall of the vessel acts as a catalyst for the dissociation of H_2O into H_2 and $\frac{1}{2}O_2$.
16. The stable element reference (SER) state is widely used in thermodynamic databases. Describe in detail its definition. From a database using such a reference state, one obtains the following Gibbs energy values in joules per mole of atoms at 1273 K: fcc-Fe, 62287; C in graphite, 20089; Fe_3C, −52379. Calculate the standard Gibbs energy of formation of Fe_3C at 1273 K. Based on the value you obtain, should Fe_3C be stable at this temperature? Why or why not? Explain your results in detail in the context of the stable and metastable phase equilibria of Fe–C at 1273 K using Thermo-Calc.
17. The sublattice model $(Y^{3+},Zr^{4+})_1(O^{2-})_{1.5}(O^{2-},Va)_{0.5}$ is used to describe the doping of Y into ZrO_2. Show in a composition square where you would expect to find such a solid solution. Calculate the entropy of mixing, assuming random mixing in the first and third sublattices for equal amounts of O and Va in the third sublattice. Compare with the ideal entropy of mixing when the second and third sublattices are combined into one sublattice.
18. Show that $\partial\, {}^EG_i/\partial x_j = \partial\, {}^EG_j/\partial x_i$ with ${}^EG$ the excess Gibbs energy.

3 Phase equilibria in heterogeneous systems

3.1 General condition for equilibrium

A system is heterogeneous if some properties have different values in different portions of the system when the system is at equilibrium. Two scenarios may exist, where variations of the properties can be either continuous or discontinuous. In the scenario of continuous variations, the gradients of the variations must be coupled so that the system remains at equilibrium. The number of independent variables is thus reduced. These gradients must also be constrained along the boundaries between the system and the surroundings. This type of constrained equilibrium is not discussed in the book as it involves heterogeneous boundary conditions between the system and the surroundings and depends on the morphology of the system.

In the second scenario, with discontinuous variations, the properties have different values in different portions of the system, but remain homogenous within each portion. The system is in equilibrium as each portion is in equilibrium with all other portions of the system. The homogeneous portions represent different phases in the system, with the properties in each phase being homogeneous at equilibrium. In the previous chapter, it was been shown that all potentials are homogeneous in a homogeneous system.

For a heterogeneous system, the same conclusion can be obtained. If the temperature is inhomogeneous, heat can be conducted from high temperature locations to low temperature locations, and this process is irreversible based on the second law of thermodynamics because it increases the internal entropy of the system. If the pressure is inhomogeneous, the amounts of lower molar volume phases will increase to reduce the internal energy of the system. If the chemical potential of a component is inhomogeneous, the chemical potential difference of the component will drive that component to locations with a lower chemical potential in order to decrease the internal energy of the system. Therefore, it can be concluded that all potentials are homogeneous in a heterogeneous system at equilibrium, and the variables that are not homogeneous are thus their conjugate molar quantities. Under certain special circumstances, to be discussed later in this book, some molar quantities may also have the same values in difference phases.

In a system at equilibrium, with c independent components, there are $c + 2$ pairs of conjugate variables, based on Eq. 1.10 or Eq. 1.49, though more can be added, as shown by Eq. 2.148, depending on experimental conditions. For simplicity, most discussions in this book are limited to systems with $c + 2$ pairs of conjugate variables unless otherwise specified, where the number "2" represents the variables T, $(-P)$ or their conjugate variables S, V.

For a system under constant temperature, pressure, and number of moles of each independent component, the equilibrium condition is derived from Eq. 1.49 as

$$dG = -SdT - Vd(-P) + \sum \mu_i dN_i - Dd\xi = -Dd\xi = 0 \qquad 3.1$$

Consequently, the equilibrium state is defined by minimization of the Gibbs energy of the system at constant T, P, and N_i because the second derivatives need to be positive for the equilibrium system to be stable, as stipulated by Eq. 2.53. For heterogeneous systems with two or more phases, the Gibbs energy of the system is the weighted sum of the Gibbs energies of the individual phases, i.e.

$$\frac{G}{N} = G_m = \sum_{\beta} f^{\beta} G_m^{\beta} \qquad 3.2$$

where f^{β} and G_m^{β} are the mole fraction and molar Gibbs energy of the phase β, respectively, and the summation goes over all phases in the system; f^{β} is equal to zero for phases not present in the equilibrium state.

The minimization of the Gibbs energy of the system is carried out under the following mass balance conditions:

$$x_i = \sum_{\beta} f^{\beta} x_i^{\beta} = \sum_{\beta} f^{\beta} \frac{\sum_{\beta-t} a^{\beta-t} \sum_k i_k^{\beta-t} y_k^{\beta-t}}{\sum_{\beta-t} a^{\beta-t}\left(1 - y_{va}^{\beta-t}\right)} \qquad 3.3$$

$$\sum_i x_i = 1 \qquad 3.4$$

$$\sum_k y_k^{\beta-t} = 1 \qquad 3.5$$

where $a^{\beta-t}$ and $y_k^{\beta-t}$ are the number of sites in sublattice t in the β phase and the corresponding site fraction of species k in the sublattice, respectively, and $i_k^{\beta-t}$ is the stoichiometry of the component i in the species k, as used in Eq. 2.138. The summation in Eq. 3.5 runs over species for each sublattice. For phases containing ionic species, electroneutrality also needs to be maintained, i.e. Eq. 2.139 is applied to each phase. This minimization problem of the Gibbs energy under the constraint of mass conservation can be solved by means of a range of algorithms. It should be noted that the mole fractions of phases and site fractions of species are bounded between 0 and 1.

This minimization procedure must have the result that potentials are homogeneous in the system as discussed above. Since the present book deals with the thermodynamics of materials, the chemical potential of each component is of particular interest and must be homogenous in all phases of the system at equilibrium, i.e.

$$\mu_i^{\alpha} = \mu_i^{\beta} = \mu_i^{\gamma} \ldots \qquad 3.6$$

For phases in which the chemical potentials of individual components cannot be evaluated by stoichiometry, the combined chemical potentials can be used to relate individual potentials as shown by Eq. 2.21 and Eq. 2.133. As proved in Section 2.2.1, see Eq. 2.68, the chemical potential of a component in a solution is represented by the intercept on the Gibbs energy axis of the multi-dimensional tangent surface of the Gibbs energy of the solution, plotted with respect to the mole fractions of the independent components. The Gibbs energy functions of all phases in equilibrium must thus share the same tangent surface. This is usually referred to as the common tangent construction for phases at equilibrium. Any phase with its Gibbs energy curve above the tangent surface is not stable under the given composition of the system.

3.2 Gibbs phase rule

The Gibbs–Duhem equation, i.e. Eq. 1.55, states that only $c + 1$ potentials are independent in a homogeneous system with c independent components and the additional two variables of temperature and pressure. In a heterogeneous system at equilibrium, this equation can be applied to individual phases as each phase is homogeneous. Noting that each potential has the same value in all phases at equilibrium, Eq. 1.55 can be written as follows for each individual phase, β, in the system at equilibrium:

$$0 = -S^{\beta}dT - V^{\beta}d(-P) - \sum N_i^{\beta}d\mu_i \qquad 3.7$$

For a system with p phases at equilibrium, there are p such equations relating the potentials in the system. The number of independent potentials thus becomes

$$\upsilon = c + 2 - p \qquad 3.8$$

Equation 3.8 is called the Gibbs phase rule. It dictates the number of potentials that can change independently, for a given number of phases co-existing at equilibrium, commonly called the degree of freedom of the system at equilibrium. It stipulates that the maximum number of phases which can co-exist in a system at equilibrium is obtained by setting $\upsilon = 0$. This is called an invariant equilibrium due to the zero degrees of freedom,

$$p_{max} = c + 2 \qquad 3.9$$

There are thus a maximum of three phases in a one-component system, four phases in a binary system, five phases in a ternary system, and so on, that can co-exist simultaneously at equilibrium with all potentials in the system at fixed values. This should not be confused with the total number of phases that could exist, but not co-exist, in a system, which of course is not limited by the Gibbs phase rule.

It should be emphasized that the degree of freedom, υ, refers to the number of potentials only, not to molar quantities of the system, because molar quantities are generally not homogeneous in a heterogeneous system. For example, in a system at equilibrium with $\upsilon = 0$, the amount of each component can be varied, while keeping the number of phases at $p_{max} = c + 2$. This can be done by changing the amount of each phase in the system through the mass balance equation, Eq. 3.3, without altering the composition of each phase and thus the chemical potentials in the system. As mentioned at the beginning of Section 3.1, the number of independent variables in a system at equilibrium, i.e. the sum of numbers of independent potentials and independent molar quantities, is $c + 2$, with the maximum number of independent potentials determined by the Gibbs phase rule, Eq. 3.8.

3.3 Potential phase diagrams

The Gibbs phase rule can be further understood through the $(c + 2)$-dimensional space of potentials consisting of T, $-P$, and μ_i with i ranging from 1 to c. Each phase is a $(c + 1)$-dimensional feature in this $(c + 2)$-dimensional space, characterized by Eq. 3.7. The directions of this $(c + 1)$-dimensional feature are represented by their molar quantities as shown by the following equations, obtained from Eq. 3.7,

$$\left(\frac{\partial \mu_i}{\partial T}\right)_{P,\mu_{j\neq i}} = -\frac{S^\beta}{N_i^\beta} = -\frac{S_m^\beta}{x_i^\beta} \quad 3.10$$

$$\left(\frac{\partial \mu_i}{\partial(-P)}\right)_{T,\mu_{j\neq i}} = -\frac{V^\beta}{N_i^\beta} = -\frac{V_m^\beta}{x_i^\beta} \quad 3.11$$

$$\left(\frac{\partial(-P)}{\partial T}\right)_{\mu i} = -\frac{S^\beta}{V^\beta} = -\frac{S_m^\beta}{V_m^\beta} \quad 3.12$$

As can be seen, all the direction derivatives are negative, indicating that the $(c + 1)$-dimensional feature is convex. The intercept of any two $(c + 1)$-dimensional features is thus a c-dimensional feature. On this c-dimensional feature, these two phases are in equilibrium with each other because each potential has the same value in both phases. This feature thus represents a two-phase equilibrium. By the same token, the intercept of any three $(c + 1)$-dimensional features is a $(c - 1)$-dimensional feature in the $(c + 2)$-dimensional space of potentials and represents a three-phase equilibrium. This continues until the number of phases reaches $c + 2$, with all $c + 2$ potentials completely determined, and the dimension of their intercepts becomes zero.

Those $(c + 1)$- to zero-dimensional geometrical features in the $(c + 2)$-dimensional space of potentials thus denote one-phase, two-phase, three-phase to $(c + 2)$-phase equilibria of the system with the dimensionality of the feature and the number of phases in equilibrium related by Eq. 3.8, i.e. the Gibbs phase rule. Their arrangements in the $(c + 2)$-dimensional space of potentials thus depict the phase relations in the system and are commonly called phase diagrams. Since all the diagram axes in the phase diagram discussed above are potentials, the diagram is called a potential phase diagram in order to differentiate it from phase diagrams, in which some or all diagram axes are the conjugate molar quantities. Both potential and molar phase diagrams are discussed in this chapter.

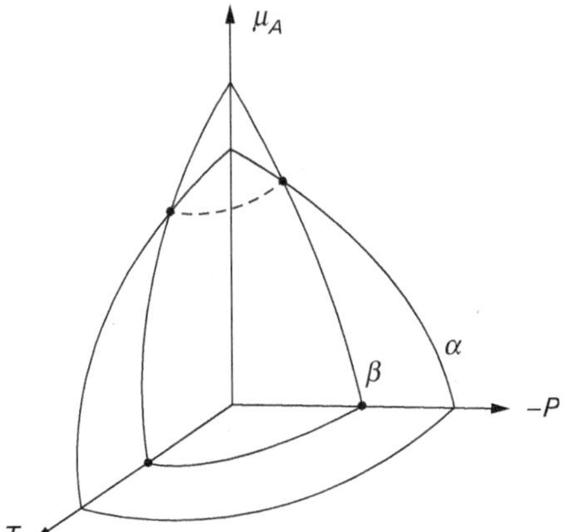

Figure 3.1 *Gibbs energy surfaces of two phases and their intersection (dashed line), representing two-phase equilibrium.*

3.3.1 Potential phase diagrams of one-component systems

As the physical vision of human beings is limited to three dimensions, only one-component systems can be completely visualized as shown in Figure 2.1 for one phase where any two of the three potentials can change independently. From the Gibbs phase rule, when two phases are in equilibrium, only one potential can vary freely if the two-phase equilibrium is to be maintained. When three phases are in equilibrium, the degree of freedom is zero, and all three potentials are fixed.

For a two-phase equilibrium, two surfaces intersect each other as depicted by the dashed line in Figure 3.1. This two-phase equilibrium line is obtained by applying Eq. 3.7 to both phases in the one-component system. Since one of the potentials is dependent on the other two, one can eliminate it by dividing the equation by its conjugate molar quantity and subtracting the two equations, resulting in the following three equations:

$$0 = \left(\frac{S^\alpha}{N_A^\alpha} - \frac{S^\beta}{N_A^\beta}\right) dT + \left(\frac{V^\alpha}{N_A^\alpha} - \frac{V^\beta}{N_A^\beta}\right) d(-P) = \Delta S_m^{\alpha\beta} dT + \Delta V_m^{\alpha\beta} d(-P) \tag{3.13}$$

$$0 = \left(\frac{S^\alpha}{V^\alpha} - \frac{S^\beta}{V^\beta}\right) dT + \left(\frac{N_A^\alpha}{V^\alpha} - \frac{N_A^\beta}{V^\beta}\right) d\mu_A = \Delta\left(\frac{S_m}{V_m}\right)^{\alpha\beta} dT + \Delta\left(\frac{1}{V_m}\right)^{\alpha\beta} d\mu_A \tag{3.14}$$

$$0 = \left(\frac{V^\alpha}{S^\alpha} - \frac{V^\beta}{S^\beta}\right) d(-P) + \left(\frac{N_A^\alpha}{S^\alpha} - \frac{N_A^\beta}{S^\beta}\right) d\mu_A = \Delta\left(\frac{V_m}{S_m}\right)^{\alpha\beta} d(-P) + \Delta\left(\frac{1}{S_m}\right)^{\alpha\beta} d\mu_A \tag{3.15}$$

The directions of the two-phase equilibrium line can thus be obtained as

$$\frac{d(-P)}{dT} = -\frac{\Delta S_m^{\alpha\beta}}{\Delta V_m^{\alpha\beta}} \tag{3.16}$$

$$\frac{dT}{d\mu_A} = -\Delta\left(\frac{1}{V_m}\right)^{\alpha\beta} \Big/ \Delta\left(\frac{S_m}{V_m}\right)^{\alpha\beta} \quad 3.17$$

$$\frac{d(-P)}{d\mu_A} = -\Delta\left(\frac{1}{S_m}\right)^{\alpha\beta} \Big/ \Delta\left(\frac{V_m}{S_m}\right)^{\alpha\beta} \quad 3.18$$

These three equations define the mathematical forms of the two-phase equilibrium line in the two-dimensional $T-(-P)$, μ_A–T and μ_A–$(-P)$ spaces, respectively, and can thus be plotted as two-dimensional diagrams. Equation 3.16 is commonly called the Clausius–Clapeyron equation in the literature. One may thus call all three equations above generalized Clausius–Clapeyron equations. At equilibrium, the chemical potentials of the components in both phases are equal to each other and so are their Gibbs energies. One thus has

$$G_m^\alpha - G_m^\beta = 0 = \Delta G_m = \Delta H_m^{\alpha\beta} - T\Delta S_m^{\alpha\beta} \quad 3.19$$

The Clausius–Clapeyron equation, Eq. 3.16, can be rewritten as

$$\frac{d(-P)}{dT} = -\frac{\Delta H_m^{\alpha\beta}}{T\Delta V_m^{\alpha\beta}} \quad 3.20$$

As an example, three potential phase diagrams of pure Fe are shown in Figure 3.2. There are four phases in the system, bcc, fcc, hcp, and liquid. In the literature, the high temperature and low temperature bcc phases are usually denoted by δ (high temperature) and α (low temperature), the fcc and hcp phases by γ and ε, and the liquid phase by L, respectively. In these figures, the two-dimensional areas are single-phase regions where two potentials can change independently with the system remaining as single-phase. The lines denote two-phase equilibrium regions where only one potential can vary independently if the two-phase equilibrium is to be maintained. The points where three two-phase equilibrium lines meet represent the invariant three-phase equilibria with three potentials fixed.

Based on the discussions in Section 2.1, the enthalpy and entropy of a phase increase monotonically with temperature, and phases that are stable at higher temperatures have higher enthalpy and entropy than phases that are stable at lower temperatures. Consequently, the two-phase equilibrium lines in a T–$(-P)$ potential phase diagram have negative slopes if the phase stable at higher temperatures also has larger molar volume than the phase stable at lower temperatures (note that if P is plotted instead of $-P$, the slope is positive). This is the case for the two-phase equilibrium lines of δ/L, γ/L, and γ/δ at high temperatures, and ε/γ shown in Figure 3.2a. On the other hand, the two-phase equilibrium lines of α/ε and α/γ at low temperatures have positive slopes, indicating that ε and γ have smaller molar volume than α, as ε and γ are more stable at higher pressures than α at constant temperatures. It is thus evident that the phase that is stable at higher pressure can have either higher or lower entropy than the phase that is stable at lower pressure, and the phase that is stable at higher temperature can have either higher or lower volume than the phase that is stable at lower temperature. This is the property anomaly discussed in Section 1.3.

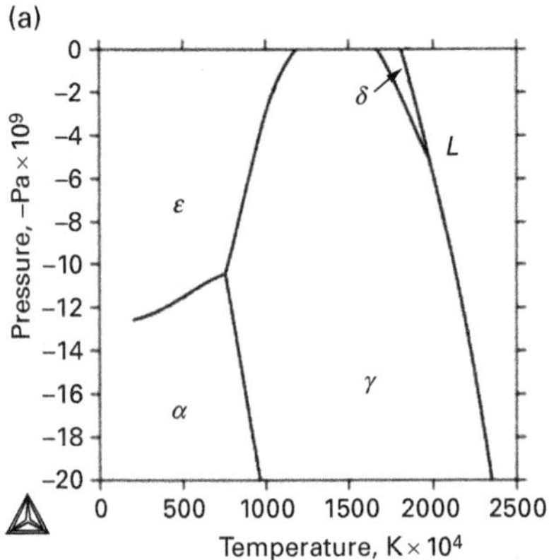

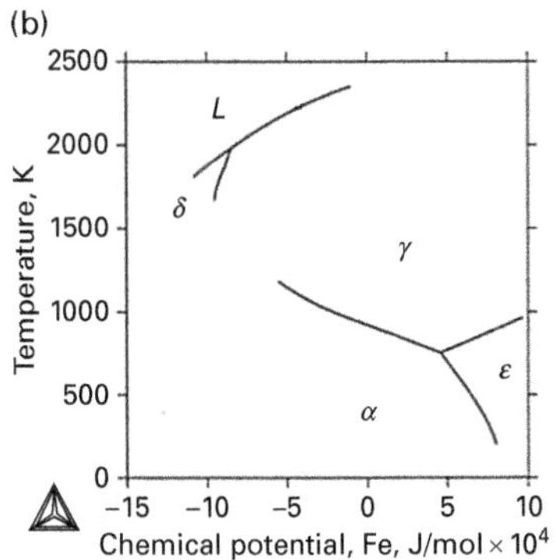

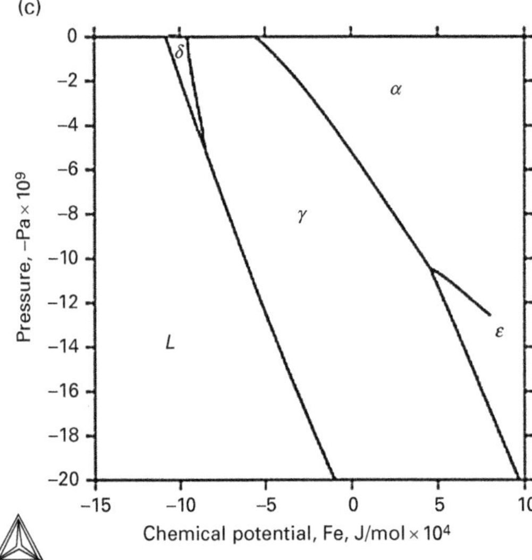

Figure 3.2 *T–$(-P)$, μ_A–T, and μ_A–$(-P)$ potential phase diagrams of pure Fe.*

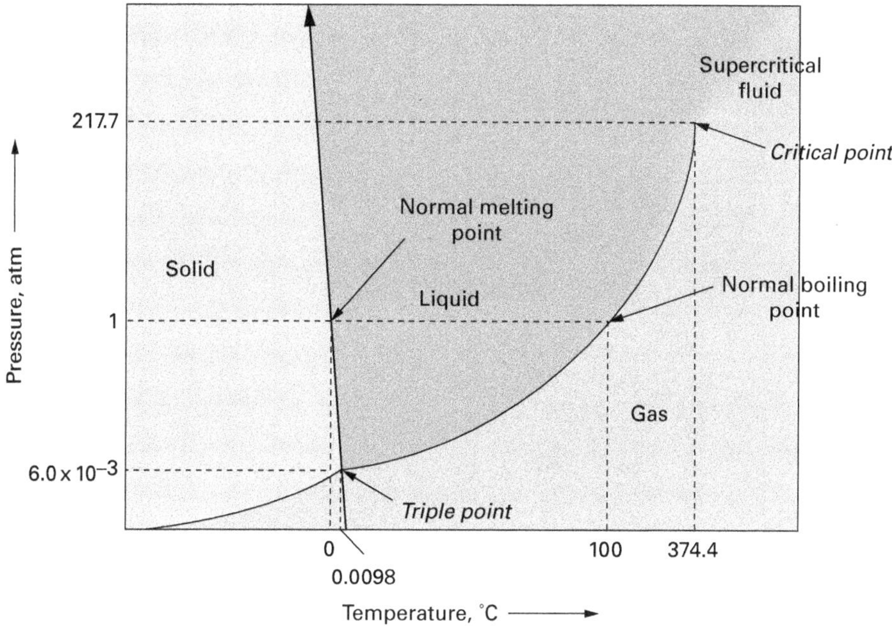

Figure 3.3 *P–T phase diagram of* H_2O.

Another useful example of a potential phase diagram is the pressure–temperature phase diagram of H_2O, shown in Figure 3.3, with three phases: ice, water, and vapor. It is known that solid ice has many polymorphic structures at high pressures, which are not included in this diagram. As in the pure Fe potential phase diagram discussed above, the single-phase regions of ice, water, and vapor are represented by the two-dimensional areas with two degrees of freedom based on the Gibbs phase rule. The lines are for the two-phase regions of ice–water, ice–vapor, and water–vapor, and the three-phase equilibrium has zero degrees of freedom, represented by a point at 273.16 K and 611.73 Pa.

There are two features in Figure 3.3 which are different from those of Fe shown in Figure 3.2a. The first feature is that the slope of the liquid–solid two-phase equilibrium line in Figure 3.3 has the opposite sign to that in Figure 3.2a, because solid ice has a larger molar volume than liquid water, while the molar volume of liquid Fe is larger than the molar volumes of fcc-Fe and bcc-Fe. The second feature is that the two-phase equilibrium line of water–vapor ends at 647 K and 22.064×10^6 Pa. Beyond this point, the difference between vapor and water disappears when the pressure and temperature are changed, i.e. it behaves as one phase. This point is a critical point, as discussed in Section 1.3. However, it should be noted that it does not represent an invariant reaction as the degree of freedom based on the Gibbs phase rule is equal to one and not zero. On the other hand, both the temperature and pressure of the critical point are invariant due to the two constraints introduced by the limit of stability of a single phase, i.e. the second and third derivatives of temperature with respect to entropy or pressure with respect to volume are zero.

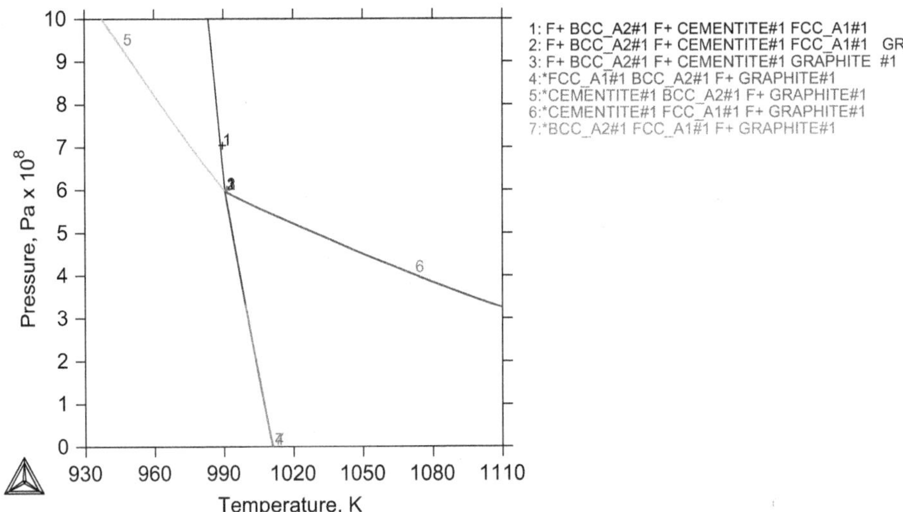

Figure 3.4 *Projected potential phase diagram of the Fe–C system with fcc, bcc, Fe_3C, and graphite phases.*

3.3.2 Potential phase diagrams of two-component systems

From the Gibbs–Duhem equation (see Eq. 1.55), a single phase in a two-component system has three independent potentials, out of the four potentials T, $-P$, μ_A, and μ_B, and is a three-dimensional geometric feature in a four-dimensional space. Alternatively, it can be represented by a three-dimensional feature bassed on three independent potentials. A two-phase equilibrium is thus a two-dimensional surface, in this three-dimensional space, created by the intercept of two three-dimensional features, a three-phase equilibrium is a one-dimensional line, and a four-phase equilibrium is a zero-dimensional point. This is shown in Figure 3.4 for the Fe–C binary system involving four phases: fcc, bcc, Fe_3C (cementite) and graphite.

The two-phase equilibrium surfaces are obtained by choosing any of the four potentials as the dependent one and solving the Gibbs–Duhem equations for both phases, resulting in the following four equations:

$$\begin{aligned}0 &= \left(\frac{S^\alpha}{N_A^\alpha} - \frac{S^\beta}{N_A^\beta}\right)dT + \left(\frac{V^\alpha}{N_A^\alpha} - \frac{V^\beta}{N_A^\beta}\right)d(-P) + \left(\frac{N_B^\alpha}{N_A^\alpha} - \frac{N_B^\beta}{N_A^\beta}\right)d\mu_B \\ &= \Delta S_{mA}^{\alpha\beta}dT + \Delta V_{mA}^{\alpha\beta}d(-P) + \Delta z_B^{\alpha\beta}d\mu_B\end{aligned} \qquad 3.21$$

$$\begin{aligned}0 &= \left(\frac{S^\alpha}{N_B^\alpha} - \frac{S^\beta}{N_B^\beta}\right)dT + \left(\frac{V^\alpha}{N_B^\alpha} - \frac{V^\beta}{N_B^\beta}\right)d(-P) + \left(\frac{N_A^\alpha}{N_B^\alpha} - \frac{N_A^\beta}{N_B^\beta}\right)d\mu_A \\ &= \Delta S_{mB}^{\alpha\beta}dT + \Delta V_{mB}^{\alpha\beta}d(-P) + \Delta z_A^{\alpha\beta}d\mu_A\end{aligned} \qquad 3.22$$

$$0 = \left(\frac{S^\alpha}{V^\alpha} - \frac{S^\beta}{V^\beta}\right)dT + \left(\frac{N_A^\alpha}{V^\alpha} - \frac{N_A^\beta}{V^\beta}\right)d\mu_A + \left(\frac{N_B^\alpha}{V^\alpha} - \frac{N_B^\beta}{V^\beta}\right)d\mu_B$$
$$= \Delta\left(\frac{S_m}{V_m}\right)^{\alpha\beta} dT + \Delta\left(\frac{1}{V_{mA}}\right)^{\alpha\beta} d\mu_A + \Delta\left(\frac{1}{V_{mB}}\right)^{\alpha\beta} d\mu_B \qquad 3.23$$

$$0 = \left(\frac{V^\alpha}{S^\alpha} - \frac{V^\beta}{S^\beta}\right)d(-P) + \left(\frac{N_A^\alpha}{S^\alpha} - \frac{N_A^\beta}{S^\beta}\right)d\mu_A + \left(\frac{N_B^\alpha}{S^\alpha} - \frac{N_B^\beta}{S^\beta}\right)d\mu_B$$
$$= \Delta\left(\frac{V_m}{S_m}\right)^{\alpha\beta} d(-P) + \Delta\left(\frac{1}{S_{mA}}\right)^{\alpha\beta} d\mu_A + \Delta\left(\frac{1}{S_{mB}}\right)^{\alpha\beta} d\mu_B \qquad 3.24$$

A three-phase equilibrium line is represented by the intercept of two two-phase surfaces by applying any of the above four equations to two two-phase equilibria. Let us use Eq. 3.21 as an example:

$$0 = \Delta S_{mA}^{\alpha\beta} dT + \Delta V_{mA}^{\alpha\beta} d(-P) + \Delta z_B^{\alpha\beta} d\mu_B \qquad 3.25$$

$$0 = \Delta S_{mA}^{\alpha\gamma} dT + \Delta V_{mA}^{\alpha\gamma} d(-P) + \Delta z_B^{\alpha\gamma} d\mu_B \qquad 3.26$$

It is self-evident that the two-phase equilibrium surface between β and γ is not independent and can be obtained by the subtraction of Eq. 3.25 and Eq. 3.26:

$$0 = \left(\Delta S_{mA}^{\alpha\beta} - \Delta S_{mA}^{\alpha\gamma}\right)dT + \left(\Delta V_{mA}^{\alpha\beta} - \Delta V_{mA}^{\alpha\gamma}\right)d(-P) + \left(\Delta z_B^{\alpha\beta} - \Delta z_B^{\alpha\gamma}\right)d\mu_B$$
$$= \Delta S_{mA}^{\gamma\beta} dT + \Delta V_{mA}^{\gamma\beta} d(-P) + \Delta z_B^{\gamma\beta} d\mu_B \qquad 3.27$$

Eliminating one of three potentials in Eq. 3.25 and Eq. 3.26, one can obtain three equations for the three-phase equilibrium line:

$$\frac{d(-P)}{dT} = -\frac{\Delta S_{mA}^{\alpha\beta}/\Delta z_B^{\alpha\beta} - \Delta S_{mA}^{\alpha\gamma}/\Delta z_B^{\alpha\gamma}}{\Delta V_{mA}^{\alpha\beta}/\Delta z_B^{\alpha\beta} - \Delta V_{mA}^{\alpha\gamma}/\Delta z_B^{\alpha\gamma}} \qquad 3.28$$

$$\frac{dT}{d\mu_B} = -\frac{\Delta z_B^{\alpha\beta}/\Delta V_{mA}^{\alpha\beta} - \Delta z_B^{\alpha\gamma}/\Delta V_{mA}^{\alpha\gamma}}{\Delta S_{mA}^{\alpha\beta}/\Delta V_{mA}^{\alpha\beta} - \Delta S_{mA}^{\alpha\gamma}/\Delta V_{mA}^{\alpha\gamma}} \qquad 3.29$$

$$\frac{d(-P)}{d\mu_B} = -\frac{\Delta z_B^{\alpha\beta}/\Delta S_{mA}^{\alpha\beta} - \Delta z_B^{\alpha\gamma}/\Delta S_{mA}^{\alpha\gamma}}{\Delta V_{mA}^{\alpha\beta}/\Delta S_{mA}^{\alpha\beta} - \Delta V_{mA}^{\alpha\gamma}/\Delta S_{mA}^{\alpha\gamma}} \qquad 3.30$$

Equations 3.28 to 3.30 can be referred to as generalized Clausius–Clapeyron equations for binary systems. Similar equations can be derived for T–$(-P)$–μ_A, T–μ_A–μ_B, and $(-P)$–μ_A–μ_B potential phase diagrams from Eq. 3.22 to Eq. 3.24, and are listed below.

- Generalized Clausius–Clapeyron equations for a three-phase equilibrium in T–$(-P)$–μ_A potential phase diagrams:

$$\frac{d(-P)}{dT} = -\frac{\Delta S_{mB}^{\alpha\beta}/\Delta z_A^{\alpha\beta} - \Delta S_{mB}^{\alpha\gamma}/\Delta z_A^{\alpha\gamma}}{\Delta V_{mB}^{\alpha\beta}/\Delta z_A^{\alpha\beta} - \Delta V_{mB}^{\alpha\gamma}/\Delta z_A^{\alpha\gamma}} \qquad 3.31$$

$$\frac{dT}{d\mu_A} = -\frac{\Delta z_A^{\alpha\beta}/\Delta V_{mB}^{\alpha\beta} - \Delta z_A^{\alpha\gamma}/\Delta V_{mB}^{\alpha\gamma}}{\Delta S_{mB}^{\alpha\beta}/\Delta V_{mB}^{\alpha\beta} - \Delta S_{mB}^{\alpha\gamma}/\Delta V_{mB}^{\alpha\gamma}} \quad 3.32$$

$$\frac{d(-P)}{d\mu_A} = -\frac{\Delta z_A^{\alpha\beta}/\Delta S_{mB}^{\alpha\beta} - \Delta z_A^{\alpha\gamma}/\Delta S_{mB}^{\alpha\gamma}}{\Delta V_{mB}^{\alpha\beta}/\Delta S_{mB}^{\alpha\beta} - \Delta V_{mB}^{\alpha\gamma}/\Delta S_{mB}^{\alpha\gamma}} \quad 3.33$$

- Generalized Clausius–Clapeyron equations for a three-phase equilibrium in T–μ_A–μ_B potential phase diagrams:

$$\frac{dT}{d\mu_A} = -\frac{\Delta(1/V_{mA})^{\alpha\beta}/\Delta(1/V_{mB})^{\alpha\beta} - \Delta(1/V_{mA})^{\alpha\gamma}/\Delta(1/V_{mB})^{\alpha\gamma}}{\Delta(S_m/V_m)^{\alpha\beta}/\Delta(1/V_{mB})^{\alpha\beta} - \Delta(S_m/V_m)^{\alpha\gamma}/\Delta(1/V_{mB})^{\alpha\gamma}} \quad 3.34$$

$$\frac{dT}{d\mu_B} = -\frac{\Delta(1/V_{mB})^{\alpha\beta}/\Delta(1/V_{mA})^{\alpha\beta} - \Delta(1/V_{mB})^{\alpha\gamma}/\Delta(1/V_{mA})^{\alpha\gamma}}{\Delta(S_m/V_m)^{\alpha\beta}/\Delta(1/V_{mA})^{\alpha\beta} - \Delta(S_m/V_m)^{\alpha\gamma}/\Delta(1/V_{mA})^{\alpha\gamma}} \quad 3.35$$

$$\frac{d\mu_A}{d\mu_B} = -\frac{\Delta(1/V_{mB})^{\alpha\beta}/\Delta(S_m/V_m)^{\alpha\beta} - \Delta(1/V_{mB})^{\alpha\gamma}/\Delta(S_m/V_m)^{\alpha\gamma}}{\Delta(1/V_{mA})^{\alpha\beta}/\Delta(S_m/V_m)^{\alpha\beta} - \Delta(1/V_{mA})^{\alpha\gamma}/\Delta(S_m/V_m)^{\alpha\gamma}} \quad 3.36$$

- Generalized Clausius–Clapeyron equations for a three-phase equilibrium in $(-P)$–μ_A–μ_B potential phase diagrams:

$$\frac{d(-P)}{d\mu_A} = -\frac{\Delta(1/S_{mA})^{\alpha\beta}/\Delta(1/S_{mB})^{\alpha\beta} - \Delta(1/S_{mA})^{\alpha\gamma}/\Delta(1/S_{mB})^{\alpha\gamma}}{\Delta(V_m/S_m)^{\alpha\beta}/\Delta(1/S_{mB})^{\alpha\beta} - \Delta(V_m/S_m)^{\alpha\gamma}/\Delta(1/S_{mB})^{\alpha\gamma}} \quad 3.37$$

$$\frac{d(-P)}{d\mu_B} = -\frac{\Delta(1/S_{mB})^{\alpha\beta}/\Delta(1/S_{mA})^{\alpha\beta} - \Delta(1/S_{mB})^{\alpha\gamma}/\Delta(1/S_{mA})^{\alpha\gamma}}{\Delta(V_m/S_m)^{\alpha\beta}/\Delta(1/S_{mA})^{\alpha\beta} - \Delta(V_m/S_m)^{\alpha\gamma}/\Delta(1/S_{mA})^{\alpha\gamma}} \quad 3.38$$

$$\frac{d\mu_A}{d\mu_B} = -\frac{\Delta(1/S_{mB})^{\alpha\beta}/\Delta(V_m/S_m)^{\alpha\beta} - \Delta(1/S_{mB})^{\alpha\gamma}/\Delta(V_m/S_m)^{\alpha\gamma}}{\Delta(1/S_{mA})^{\alpha\beta}/\Delta(V_m/S_m)^{\alpha\beta} - \Delta(1/S_{mA})^{\alpha\gamma}/\Delta(V_m/S_m)^{\alpha\gamma}} \quad 3.39$$

3.3.3 Sectioning of potential phase diagrams

Based on the Gibbs–Duhem equation (see Eq. 1.55), a single-phase equilibrium in a system with more than two independent components has more than three independent potentials. There is no problem in representing them using the mathematical formulas discussed so far, but it is not possible for us to visualize graphically the full potential phase diagrams for multi-component systems with more than two independent components. In principle, there are two options. One option is to project the multi-dimensional potential phase diagram onto a two- or three-dimensional diagram, and another option is to section the multi-dimensional potential phase diagram by fixing the values of some potentials.

The projection approach was used for one-component systems in Section 3.3.1. Since a two-phase equilibrium in a one-component potential phase diagram is one-dimensional, the projection does not lose any information, and the same is true for a three-phase equilibrium in a one-component potential phase diagram. In a binary system, the projections of three-phase and four-phase equilibria do not lose any information, while the projections of two-phase equilibria become two-dimensional and cannot retain all the information as the original dimensionality of these two-phase equilibria is three. Consequently, sectioning at fixed values of some potentials is necessary in order to visualize the phase relations in systems with two or more components. The Gibbs phase rules shown in Eq. 3.8 and Eq. 3.9 are thus modified to

$$\upsilon = c + 2 - p - n_s = (c - n_s) + 2 - p \tag{3.40}$$

$$p_{max} = (c - n_s) + 2 \tag{3.41}$$

where n_s is the number of potentials fixed in sectioning. As can be seen in the last part of Eq. 3.40, the number of sections is equivalent to the reduction of the effective number of independent components. Therefore, any multi-component system with $n_s = c - i$ behaves like an i-component system. The equations presented in Sections 3.3.1 and 3.3.2 are thus directly applicable to multi-component systems with $n_s = c - 1$ and $n_s = c - 2$, respectively.

A common practice in experiments is to fix the pressure, temperature, or chemical potentials of volatile components as they are usually the variables controlled experimentally. In a binary system, the potential phase diagram at constant pressure can be represented by any two of the three potentials, i.e. two chemical potentials and temperature, with the remaining potential being dependent, and has morphology identical to that of a one-component system. The Gibbs–Duhem equation under such conditions becomes

$$0 = -SdT - N_A d\mu_A - N_B d\mu_B \tag{3.42}$$

The corresponding two-phase Clausius–Clapeyron equations are written as

$$\frac{dT}{d\mu_A} = -\frac{\Delta z_A^{\alpha\beta}}{\Delta S_{mB}^{\alpha\beta}} \tag{3.43}$$

$$\frac{dT}{d\mu_B} = -\frac{\Delta z_B^{\alpha\beta}}{\Delta S_{mA}^{\alpha\beta}} \tag{3.44}$$

$$\frac{d\mu_A}{d\mu_B} = \frac{\Delta(1/S_{mB})^{\alpha\beta}}{\Delta(1/S_{mA})^{\alpha\beta}} \tag{3.45}$$

As an example, the T–μ_C potential phase diagram for the Fe–C binary system at one atmospheric pressure is shown in Figure 3.5.

In a ternary system, two potentials need to be fixed in order to obtain two-dimensional potential phase diagrams. When the pressure and the chemical potential of one species are fixed, the system behaves like the binary system discussed above. When the system temperature and pressure are fixed, the Gibbs–Duhem equation is written as

$$0 = -N_A d\mu_A - N_B d\mu_B - N_C d\mu_C \tag{3.46}$$

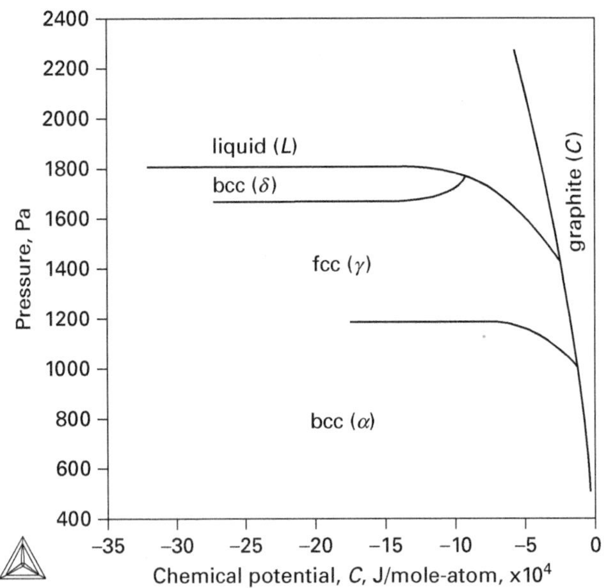

Figure 3.5 *T–μ_C potential phase diagram for the Fe–C binary system at* $P = 1$ atm.

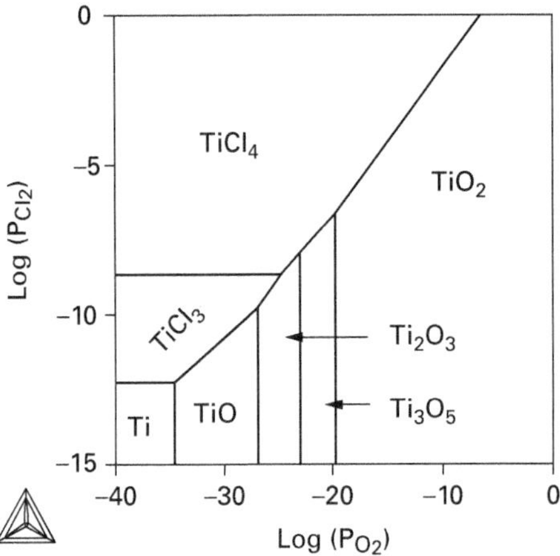

Figure 3.6 *Ti–O–Cl potential phase diagram at* 600 °C *and one atmospheric pressure.*

Taking the component A as the dependent element, the two-phase Clausius–Clapeyron equation is simplified as

$$\frac{d\mu_B}{d\mu_C} = \frac{d(\ln a_B)}{d(\ln a_C)} = -\frac{\Delta z_C^{\alpha\beta}}{\Delta z_B^{\alpha\beta}} \qquad 3.47$$

When the two phases in equilibrium are stoichiometric phases, the two-phase equilibrium is thus a straight line. For example, the Ti–O–Cl potential phase diagram at 600 °C and one atmospheric pressure is shown in Figure 3.6. Since both O and Cl are volatile components, their activities are usually represented by their partial pressures, with the pure O_2 and Cl_2 gas as their respective reference states at the given temperature and pressure.

For systems with more than three components, the chemical potentials of one or more components must be fixed in order to obtain a two-dimensional potential phase diagram similar to the potential phase diagrams discussed above.

3.4 Molar phase diagrams

The potential phase diagrams discussed in Section 3.3 present information on which phases are in equilibrium under given values of potentials, but do not have any information on the properties of the phases in equilibrium. On the other hand, there are direct relations between potentials and their conjugate molar quantities for each phase at equilibrium, given by Eq. 2.2 to Eq. 2.5. One can thus substitute the potentials by their conjugate molar quantities in the potential phase diagrams, as molar quantities provide more information on the properties of the phases in the system. This is particularly true when chemical potentials are replaced by compositions, as the compositions of the system are often the variables controlled and measured in experiments instead of chemical potentials.

3.4.1 Tie-lines and lever rule

It is self-evident from Eq. 2.2 to Eq. 2.4 that while potentials are homogeneous in all phases in a heterogeneous system at equilibrium, the molar quantities usually have different values in the individual phases. This is also stipulated in various Clausius–Clapeyron equations such as Eq. 3.16 to Eq. 3.18 and Eq. 3.28 to Eq. 3.39. The difference in molar quantities thus increases the dimensionality of the phase region by the number of potentials that have been replaced by their conjugate molar quantities. The maximum dimensionality of a phase region is the dimensionality of the phase diagram under consideration. This thus creates a finite space between phases in equilibrium in the phase diagram when some axes are molar quantities.

For an equilibrium system under constant T, P, and N_i, the potentials in the system and their conjugate molar quantities in each phase are all uniquely defined. In a phase diagram with one or more potentials replaced by their conjugate molar quantities, two phases in equilibrium in a system with c independent components are connected by a c-dimensional line in a $(c+2)$-dimensional space or its $(c+1)$-dimensional projection, as discussed in Section 3.3. These lines are called tie-lines and collectively represent a two-phase equilibrium region. For a k-phase equilibrium, there are in total $C_k^2 = k(k-1)/2$ tie-lines connecting every two phases, with $k-1$ of them independent because the number of independent tie-lines increases by one with each new phase added. For the invariant equilibrium, with zero degrees of freedom, the number of phases in equilibrium is $c+2$, as shown by Eq. 3.9, corresponding to $C_{c+2}^2 = (c+2)(c+1)/2$ tie-lines with $c+1$ of them independent.

Inside the space encapsulated by the tie-lines, the axis variables of the phase diagram (a mixture of potentials and molar quantities) can be changed independently without changing the phases in equilibrium and their properties. Accordingly, only the relative amounts of individual phases are adjusted, to maintain the conservation of the molar quantities in the system specified by the molar quantity axes of the phase diagram. The geometric feature circumscribing the space encapsulated by the tie-lines no longer represents any phase regions, but a boundary between the neighboring phase regions. Its characteristics will be discussed in more detail in the next few sections. As the properties in each phase are homogeneous, the values of the molar quantities of a system are simply the sum of individual phases and can be represented by the following equation:

$$A_m = \sum_{\alpha} f^{\alpha} A_m^{\alpha} \qquad 3.48$$

where A_m and A_m^{α} represent the values of a molar quantity of the system and of the α phase, respectively, f^{α} is the mole fraction of the α phase, and the summation goes over all phases in equilibrium with each other. With $\sum_{\alpha} f^{\alpha} = 1$, Eq. 3.48 can be re-arranged into the following equation:

$$\sum_{\alpha} f^{\alpha} \left(A_m - A_m^{\alpha} \right) = 0 \qquad 3.49$$

Equation 3.49 is commonly referred to as the lever rule. For a two-phase equilibrium of α and β, it becomes

$$f^{\alpha} = \frac{A_m^{\beta} - A_m}{A_m^{\beta} - A_m^{\alpha}} \qquad 3.50$$

$$f^{\beta} = \frac{A_m^{\alpha} - A_m}{A_m^{\alpha} - A_m^{\beta}} \qquad 3.51$$

For a phase diagram with number of axes $n = (c - n_s) + 1$, the number of possible axes that are molar quantities is thus $k \leq n$. There are thus k equations similar to Eq. 3.49 with one for each molar quantity, A_{mi}, resulting in the following $k + 1$ equations:

$$\sum_{\alpha} f^{\alpha} \left(A_{mi} - A_{mi}^{\alpha} \right) = 0 \qquad 3.52$$

$$1 - \sum_{\alpha} f^{\alpha} = 0 \qquad 3.53$$

The summations in Eq. 3.52 and Eq. 3.53 go over the phases in equilibrium, and the amount of each phase is obtained by solving these $k + 1$ equations simultaneously along with the equilibrium conditions.

3.4.2 Phase diagrams with both potential and molar quantities

Based on the Gibbs phase rule discussed in Sections 3.2 and 3.3.3, the dimensionalities of phase regions in a potential phase diagram are given by Eq. 3.8 or Eq. 3.40.

As each potential is substituted by its conjugate molar quantity, the dimensionalities of phase regions increase by one until the phase region reaches the dimensionality of the phase diagram. The axes of this phase diagram now consist of both potentials and molar quantities. The dimensionality of a phase region can thus be represented by the following equation based on Eq. 3.40:

$$\upsilon_m = (c - n_s) + 2 - p + n_m \leq (c - n_s) + 1 \quad 3.54$$

where n_m is the number of molar axes. This equation is applicable to phase regions with more than $n_m + 1$ phases. For phase regions with $n_m + 1$ phases or fewer, i.e. $p \leq n_m + 1$, the dimensionalities are the same as the phase diagram, i.e. $\upsilon_m = (c - n_s) + 1$, and no longer vary with the number of molar quantity axes. When all $(c - n_s) + 1$ potentials are substituted by their conjugate molar quantities, one obtains a complete molar phase diagram, to be discussed in Section 3.4.3, and all phase regions have the same dimensionality, $(c - n_s) + 1$.

For the sake of graphic visualization, let us examine a two-dimensional phase diagram of a one-component system. Topologically, it is equivalent to a multi-component system with $n_s = c - 1$. In Figure 3.2, three two-dimensional phase diagrams were shown for pure Fe. In principle, one can use any one of them to illustrate mixed potential and molar phase diagrams. For practical purposes, one selects the T–$(-P)$ potential diagram because the temperature and pressure are the two typical variables controlled in experiments on one-component systems. The conjugate molar quantities of $-P$ and T are molar volume and molar entropy, respectively. For stable phases, any two conjugate variables change in the same direction, as illustrated by Eq. 1.35 and Eq. 2.16, i.e. the phase stable at higher T has higher molar entropy, and the phase stable at higher $-P$, i.e. lower pressure, has higher molar volume. Let us first substitute $-P$ by V_m as shown in Figure 3.7a. The dimensionality of a single-phase region remains unchanged because $p = 1 < 2 = n_m + 1$. The dimensionality of two-phase regions is changed from 1 to 2 due to $\upsilon_m = 3 - 2 + 1 = 2$ from Eq. 3.54.

As both phases in a two-phase equilibrium have the same temperature, all tie-lines, depicted as parallel vertical lines in Figure 3.7a, are perpendicular to the temperature axis. When the molar volume of the system changes from one end of a tie-line to the other end at a constant temperature, the mole fraction of one phase increases from 0 to 1, and the mole fraction of the other phase decreases from 1 to 0. The tie-lines at various temperatures combine together to form a two-dimensional two-phase region. The two curves at the two ends of the tie-lines represent the boundaries between the single-phase and two-phase regions and are no longer phase regions themselves. They are thus called phase boundaries.

By the same token, by changing the temperature at a constant molar volume of the system, the system will locate on different tie-lines, with the amounts of the two phases determined by the lever rule. It is thus clear that the system maintains the two-phase equilibrium state with both T and V_m changing independently inside the two-dimensional two-phase region. This seems in contradiction to the Gibbs phase rule $\upsilon = 3 - p = 1$ from Eq. 3.8, but it is not because the Gibbs phase rule applies strictly to potential phase diagrams only, while the T–V_m phase diagram has for one of its axes a

(a)

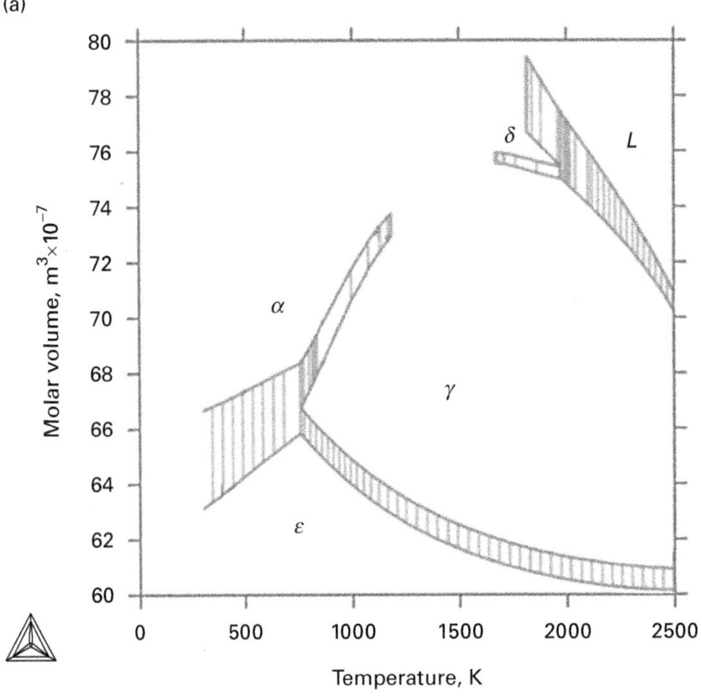

(b)

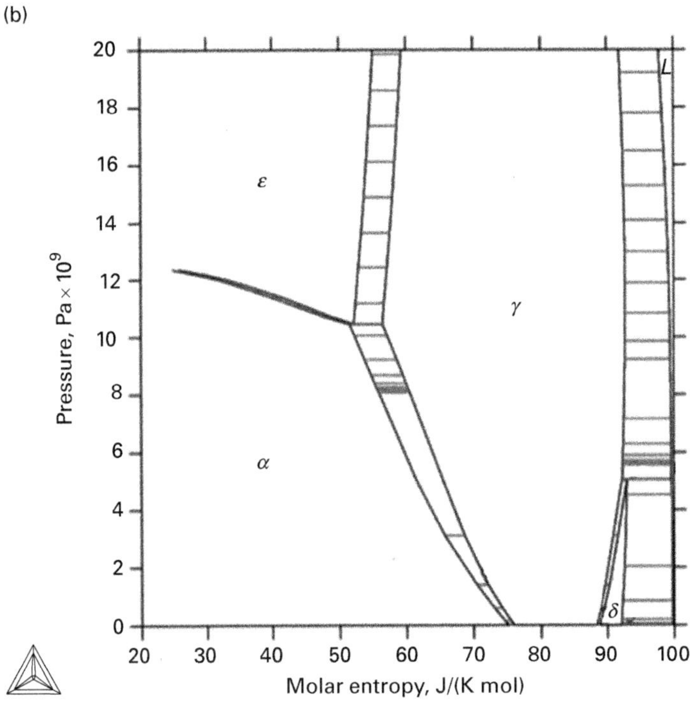

Figure 3.7 *T–V_m and S_m–($-P$) phase diagrams of Fe.*

molar quantity. As an alternative, one can consider Eq. 3.54 as a modified Gibbs phase rule in describing the dimensionality of a phase region in a mixed potential and molar phase diagram with $p \geq n_m + 1$.

The two three-phase equilibria in pure Fe are also represented by tie-lines connecting all three phases. The dimensionality of three-phase regions is $v_m = 3 - 3 + 1 = 1$ from Eq. 3.54, and the three two-phase tie-lines for a three-phase equilibrium thus overlap each other with their three molar volumes on the same tie-line.

Let us examine the two three-phase equilibria in more detail. In the $\gamma/\alpha/\varepsilon$ three-phase equilibrium at $T_E = 756.6$ K and $(-P_E) = -1.046 \times 10^{10}$ Pa, the molar volumes of α, γ, and ε are 6.837, 6.677, and 6.582 $\times 10^{-6}\text{m}^3/\text{mol}$, respectively. There are two two-phase regions at higher temperatures and one two-phase region at lower temperatures. This is also shown in the potential phase diagram of Figure 3.2a with two two-phase curves entering into and one two-phase curve leaving the three-phase equilibrium point with decreasing temperature. Consider a system with fixed molar volume equal to that of the γ phase, i.e. $V_m(T_E) = 6.677 \times 10^{-6}$ m^3/mol. At $T > T_E$, the system is in the single γ phase region. With a decrease in temperature across T_E, it enters into the $\alpha + \varepsilon$ two-phase region. This transformation can be written as follows and is called a eutectoid reaction,

$$\gamma \rightarrow \alpha + \varepsilon \tag{3.55}$$

if the high temperature phase is a liquid phase. If the system molar volume is between V_m^γ and V_m^α at 756.6 K, with a decrease of temperature the system first moves from the single γ phase region to the $\alpha + \gamma$ two-phase region when the $\gamma/(\alpha + \gamma)$ phase boundary is crossed. When the temperature reaches T_E, the eutectoid transformation takes place in the remaining γ phase. The α formed prior to the eutectoid transformation is called proeutectoid α. By the same token, when the system molar volume is between V_m^γ and V_m^ε at 756.6 K, with a decrease in temperature proeutectoid ε would form followed by the eutectoid transformation.

On the other hand, the $L/\delta/\gamma$ three-phase equilibrium at $T_P = 1977.9$ K and $(-P_P) = -5.111 \times 10^9$ Pa has different characteristics; the subscript P will be defined shortly. There is one two-phase equilibrium above, and there are two two-phase equilibria below the invariant temperature, shown in the potential phase diagram of Figure 3.2a with one two-phase curve entering into and two two-phase curves leaving the three-phase equilibrium point with decreasing temperature. The molar volumes of L, δ, and γ at T_P are 7.735, 7.542, and 7.498 $\times 10^{-6}$ m^3/mol, respectively. For $T > T_P$, the two-phase region is L+γ. If the system molar volume is between 7.735 and 7.498 $\times 10^{-6}$ m^3/mol, i.e. $V_m^L(T_P)$ and $V_m^\gamma(T_P)$, when the temperature reaches T_P L and γ are combined to form δ, with the transformation written as

$$L + \gamma \rightarrow \delta \tag{3.56}$$

This type of reaction is called a peritectic reaction or peritectoid reaction when the high temperature phase by a solid phase, denoted by the subscript P. At $T < T_P$, one or both high temperature phases may no longer be present in equilibrium, depending on the value of the system molar volume. For $V_m = V_m^\delta(T_P) = 7.542 \times 10^{-6}$ m^3/mol, the peritectic reaction, Eq. 3.56, can come to completion with no L and γ left, as the temperature decreases. For $V_m^{\gamma}(T_P) = 7.498 < V_m < V_m^{\delta}(T_P) = 7.542\ (10^{-6}\ \text{m}^3/\text{mol})$, the liquid phase is consumed, and the system enters the $\gamma + \delta$ two-phase region. On the other hand, for $V_m^{\delta}(T_P) = 7.542 < V_m < V_m^{L}(T_P) = 7.735\ (10^{-6}\ \text{m}^3/\text{mol})$, the γ phase is consumed instead, and the system enters into the $L + \delta$ two-phase equilibrium region upon cooling.

Let us now replace T by S_m to obtain the $(-P)$–S_m phase diagram shown in Figure 3.7b. The morphology of this phase diagram is identical to the T–V_m phase diagram just discussed, but with all tie-lines perpendicular to the pressure axis. The transformations at the two three-phase equilibria with $(-P)$ decreasing or P increasing are as follows:

$$\gamma + \alpha \rightarrow \varepsilon \tag{3.57}$$

$$\delta \rightarrow L + \gamma \tag{3.58}$$

To visualize two-dimensional phase diagrams of binary systems, one usually keeps the pressure constant. One type of commonly used binary phase diagram is the temperature–composition (T–x) phase diagram. As an example, let us re-plot the T–μ_C potential diagram shown in Figure 3.5 as a T–x_C mixed potential and molar phase diagram by replacing the chemical potential of C by its mole fraction. The T–x_C phase diagram thus obtained is shown in Figure 3.8. In this phase diagram there are one peritectic, one eutectic, and one eutectoid reactions as follows:

$$L + \delta \rightarrow \gamma \tag{3.59}$$

$$L \rightarrow \gamma + C \tag{3.60}$$

$$\gamma \rightarrow \alpha + C \tag{3.61}$$

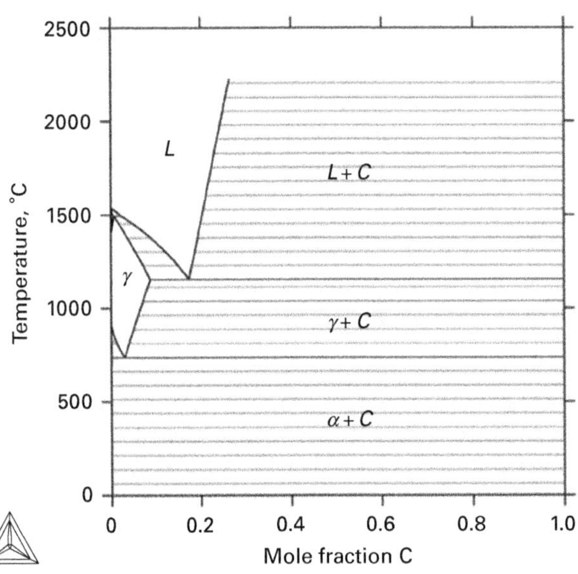

Figure 3.8 *T–x_C phase diagram of the Fe–C binary system.*

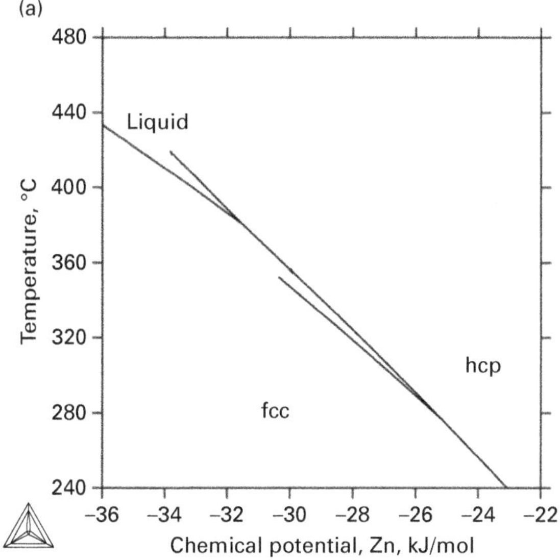

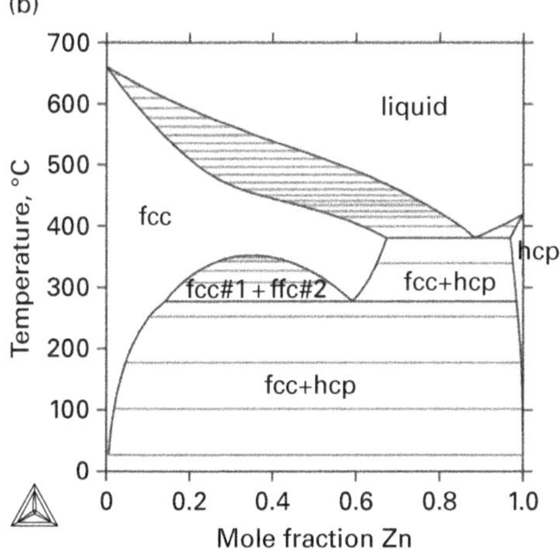

Figure 3.9 *T–μ_{Zn} potential phase diagram (a) and T–x_{Zn} mixed potential and molar phase diagram (b) of the Al–Zn binary system.*

In Figure 2.9 in Section 2.2.2, the formation of miscibility gaps due to repulsive interactions between components was illustrated. One example is shown in Figure 3.9 for the Al–Zn binary system in terms of both a T–μ_{Zn} potential phase diagram and a T–x_{Zn} mixed potential and molar phase diagram.

In Figure 3.9, there are one eutectic reaction and one eutectoid reaction as follows:

$$L \rightarrow fcc + hcp \qquad 3.62$$

$$fcc\#1 \rightarrow fcc\#2 + hcp \qquad 3.63$$

The eutectoid reaction, Eq. 3.63, is also termed a monotectoid reaction because the fcc phase appears on both sides of the reaction with different compositions, *fcc*#1 and *fcc*#2, due to the miscibility gap. The highest temperature of the miscibility gap is called the consolute point, as discussed in Section 2.2.2, and can be clearly seen in the T–μ_{Zn} potential phase diagram shown in Figure 3.9. This is a critical point, marking the limit of instability as shown in Figure 2.8.

When there is only one phase on either side of the reaction, i.e. both phases have the same composition, the reaction is called a congruent reaction. One example is shown in Figure 3.10 for the T–x_{SiO_2} mixed potential and molar phase diagram of the CaO–SiO_2 pseudo-binary system with two congruent reactions as follows:

$$L \rightarrow CaSiO_3 \tag{3.64}$$

$$L \rightarrow Ca_2SiO_4 \tag{3.65}$$

They are not invariant reactions, based on the Gibbs phase rule. In Figure 3.10, it is noted that there is a miscibility gap in the liquid phase close to the SiO_2 side and there are four eutectic reactions, one being monotectic involving two liquid phases due to the miscibility gap, and three peritectic reactions.

Let us generalize the above discussion to phase diagrams with $(c - n_s) + 1$ axes. In such a phase diagram, the maximum number of phases is given by Eq. 3.41 as $p_{max} = (c - n_s) + 2$. The number of phases on either side of an invariant reaction can vary from one phase to $p_{max} - 1 = (c - n_s) + 1$ phases, with the remaining phases on the other side of the reaction, typically with the potential decreasing from left to right. The invariant reaction with one phase on the left of the reaction is named a eutectic or

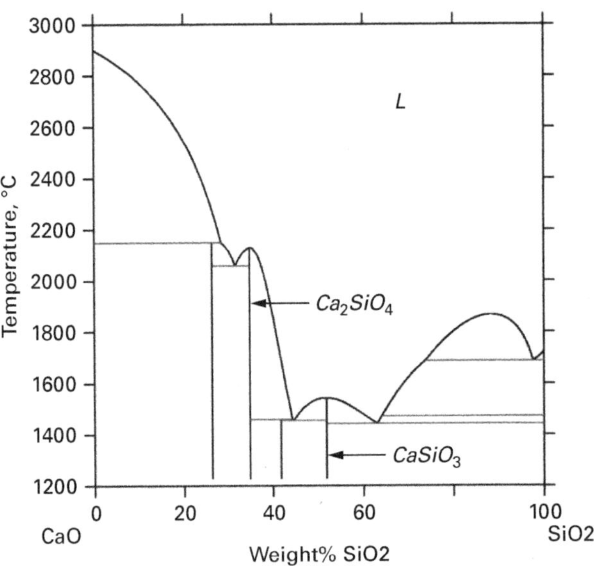

Figure 3.10 *T–x_{SiO_2} mixed potential and molar phase diagram of the CaO–SiO_2 pseudo-binary system.*

eutectoid reaction, depending on whether the phase on the left of the reaction is liquid or solid. The remaining invariant reactions are named peritectic or peritectoid reactions, according to whether they occur with or without a liquid phase.

3.4.3 Phase diagrams with only molar quantities

When all $(c - n_s) + 1$ potentials in a potential phase diagram are replaced by their conjugate molar quantities, one obtains a molar phase diagram with molar quantities on all axes of the phase diagram. For regions with the number of phases $p \leq n_m + 1 = (c - n_s) + 2 = p_{max}$ (see Eq. 3.41), the phase regions have the same dimensionality as that of the phase diagram, i.e. all phase regions have the same dimensionality, $(c - n_s) + 1$, and any geometric features with lower dimensionalities, i.e. from 0 to $(c - n_s)$, are not phase regions but phase boundaries between neighboring phase regions. For the sake of graphic visualization, the molar phase diagram of pure Fe is shown in Figure 3.11; it was obtained by combining the two mixed phase diagrams in Figure 3.7.

In this molar phase diagram, all one-, two-, and three-phase regions are two-dimensional, the same as the dimensionality of the phase diagram. A two-phase region is made up of tie-lines connecting the two phases in equilibrium, while a three phase-region is surrounded by three two-phase tie-lines, i.e. a tie-triangle. The amount of each phase in the tie-triangle can be obtained using the lever rule represented by Eq. 3.52 and Eq. 3.53. As can be seen, phase boundaries between a one-phase region

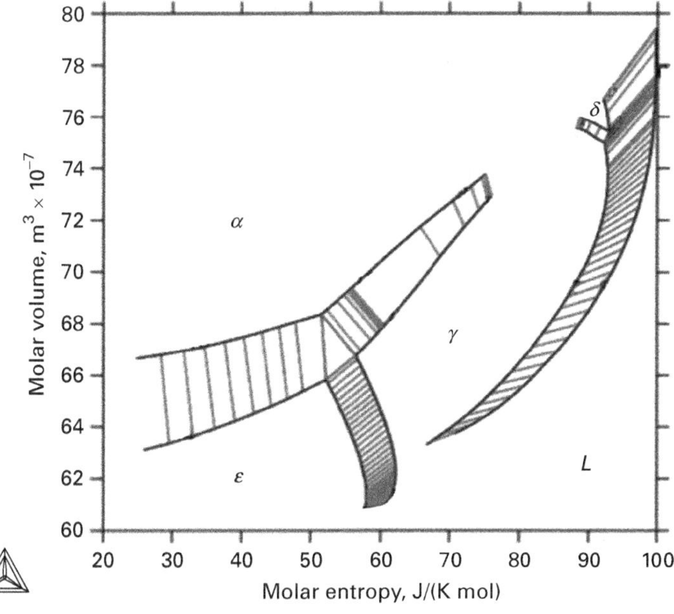

Figure 3.11 *Molar phase diagram of Fe.*

and a two-phase region are one-dimensional. When the system crosses such a phase boundary the number of phases changes by one, from two to one or vice versa. Phase boundaries between two- and three-phase regions are represented by two-phase tie-lines. When the system crosses such a phase boundary, the number of phases also changes by one. The lowest-dimensional phase boundaries are points between one-phase and three-phase regions that are zero-dimensional and the intercept of four one-dimensional phase boundaries. When the system crosses such a phase boundary, the number of phases changes by two.

For multi-component systems, the phase relations cannot be directly visualized. By representing the system of equations in terms of equilibrium conditions and lever rules on a phase boundary, using phases separately from the two adjacent phase regions, Palatnik and Landau [4] postulated that the difference between the number of unknowns and equations gives the dimensionality of the phase boundary and derived the following relationship:

$$D^+ + D^- = r - b = (c - n_s) + 1 - b \qquad 3.66$$

where D^+ and D^- are the numbers of phases added and removed when the phase boundary is crossed, and r and b are the dimensionalities of the phase diagram and the phase boundary, respectively. They termed Eq. 3.66 the contact rule, called the MPL boundary rule by Hillert [1].

By the same token, Eq. 3.66 is applicable to any phase boundary where the two adjacent phase regions have the same dimensionality as the phase diagram, even in phase diagrams with a mixture of potentials and molar quantities as the diagram axes. This can be understood because the potentials are homogeneous in all phases in equilibrium on the phase boundary. The phase boundary is thus equivalent to those in a complete molar phase diagram with the number of components equal to the number of molar axes in a mixed potential and molar phase diagram minus one, i.e.

$$c' = n_m \leq c - n_s \qquad 3.67$$

The last part of Eq. 3.67 stems from the discussion related to Eq. 3.54 when all c'+1 potentials are replaced by their conjugate molar quantities, which is analogous to a molar phase diagram with c independent components and n_s potentials fixed.

For a two-dimensional phase diagram with $r = 2$, the phase boundary can be either zero- or one-dimensional. As shown in Figure 3.11, the basic element of a molar phase diagram is a joint of four one-dimensional phase boundary lines. When a phase boundary line is crossed, the number of phases is either increased or decreased by one. The joint of four one-dimensional phase boundary lines is zero-dimensional. The number of phases differs by two between the phase regions across the zero-dimensional join. Two scenarios are possible.

- Two phases are added or removed, i.e. $D^+ = 2$ and $D^- = 0$ or $D^+ = 0$ and $D^- = 2$, and the number of phases in the two-phase regions differs by two.

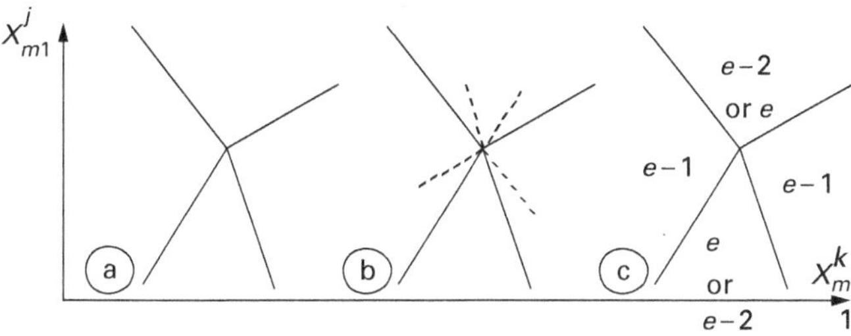

Figure 3.12 *Schematic molar phase diagram, demonstrating Schreinemakers' rule, from* [1] *with permission from Cambridge University Press.*

- One phase is added, one phase is removed, i.e. $D^+ = D^- = 1$, and the phase regions have the same number of phases.

By combining the contact rules for both zero- and one-dimensional phase boundaries, it is evident that the above two scenarios co-exist in a joint of four-phase regions, with some two-phase regions following the first scenario and other two-phase regions following the second scenario. Based on Schreinemakers' rule generalized by Hillert [5], each of the two-phase regions with the same number of phases contains one extrapolation of the phase boundaries, while the other two-phase regions contain either zero or two extrapolations of the phase boundaries. This can be observed for all the zero-dimensional phase boundaries in Figure 3.11 and is further schematically illustrated in Figure 3.12 for general cases.

3.4.4 Projection and sectioning of phase diagrams with potential and molar quantities

As discussed in Section 3.3.3, projections of high-dimensional phase diagrams usually do not keep all the information. However, there is one type of projection widely used in the literature, i.e. the liquidus surface in ternary systems under constant pressure with temperature and mole fractions of two components as its axes. The projection along the temperature axis reveals the composition regions for primary phases that solidify from liquid upon cooling. These regions are separated by univariant lines of three-phase equilibria. The projections along one of the two mole fractions show the temperature as a function of composition on the univariant three-phase equilibrium lines and also depict whether a four-phase equilibrium is peritectic or eutectic. There are four scenarios for the three univariant three-phase equilibrium lines to meet at the four-phase equilibrium, as depicted in Figure 3.13 and discussed individually below.

The first scenario is that with decreasing temperature all three univariant lines merge into the four-phase equilibrium. This indicates that the liquid phase does not exist at temperatures below the four-phase invariant reaction. This invariant reaction is thus a ternary eutectic reaction with the liquid completely transformed to three solid phases upon cooling, i.e.

$$L \rightarrow \alpha + \beta + \gamma \qquad 3.68$$

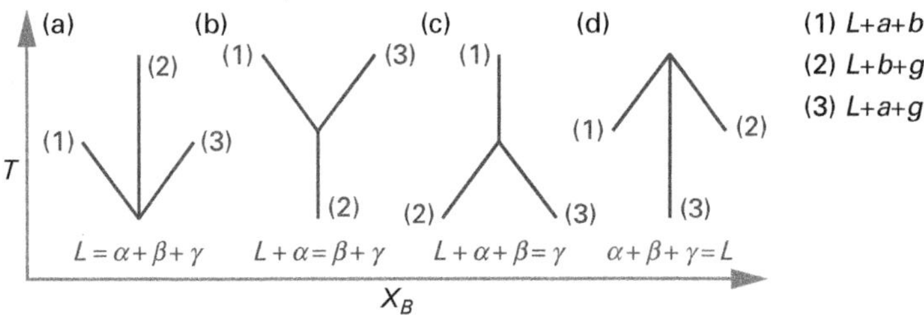

Figure 3.13 *Schematic diagram showing four options for three univariant three-phase equilibrium lines to meet at the invariant four-phase equilibrium.*

In the second scenario, two univariant lines merge into and one leaves the four-phase equilibrium with decreasing temperature. This means that one solid phase at higher temperatures is no longer stable at lower temperatures, and it must react with the liquid phase to form the remaining two solid phases. The four-phase invariant reaction is thus peritectic. The solid phase common to both univariant lines at high temperatures reacts with the liquid phase. Assuming that this phase is α, the four-phase invariant reaction becomes

$$L + \alpha \rightarrow \beta + \gamma \tag{3.69}$$

In the third scenario, one univariant line points to and two leave the four-phase equilibrium with decreasing temperature. A new phase forms at low temperatures from the three high temperature phases, for example liquid, α, and β, with the four-phase invariant reaction as

$$L + \alpha + \beta \rightarrow \gamma \tag{3.70}$$

The fourth scenario is the inverse of the first scenario, indicating the formation of liquid from solid phases upon cooling, i.e.

$$\alpha + \beta + \gamma \rightarrow L \tag{3.71}$$

This case has not been observed in reality.

As an example, the liquidus projections of the Al–Fe–Si ternary system are shown in Figure 3.14 in two formats [6], i.e. (a) three-dimensional liquidus surface with the isotherms showing the liquidus contours; (b) conventional projection onto the composition axes with the temperature decrease shown by contours; (c) projection onto the temperature and weight fraction of Si. The first to third scenarios of invariant reactions discussed above can clearly be identified and are listed in Table 3.1. It is evident that Figure 3.14c provides the easiest route to visualize the types of invariant reaction shown by Figure 3.13.

In contrast to projections, sectioning is used more often to understand phase relations in multi-component systems. Sectioning of a potential phase diagram is relatively

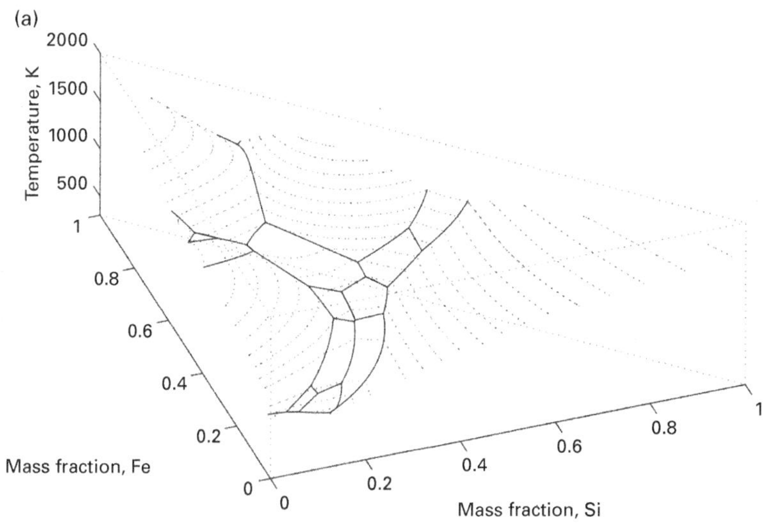

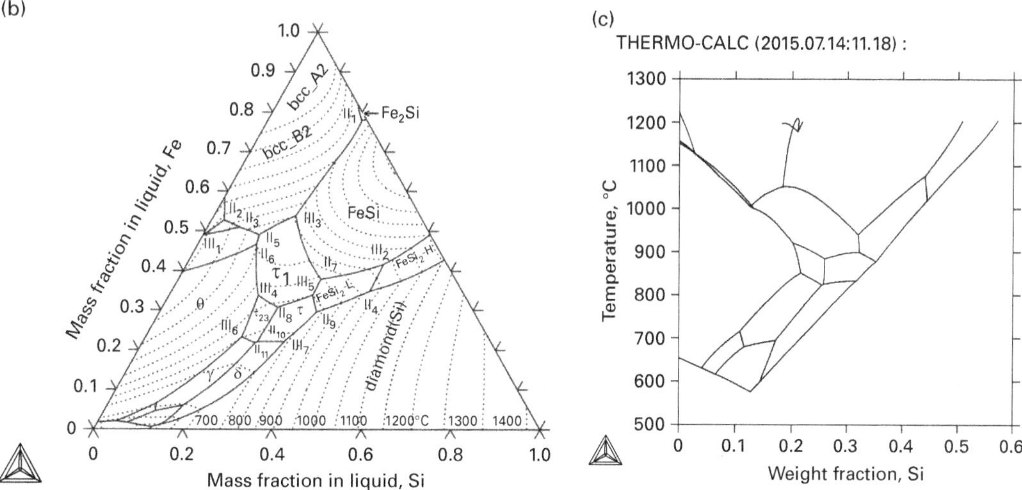

Figure 3.14 *Liquidus of the Al–Fe–Si ternary system [6]: (a) three-dimensional presentation of the liquidus; (b) projection onto the composition triangle with isotherms (dotted lines) superimposed and their temperatures indicated close to the horizontal axis; (c) projection onto the temperature and weight fraction of Si, from* [6] *with permission from Springer.*

simple as the resulting phase diagram behaves like a system with one component less. The same is true if potentials are sectioned in phase diagrams with both potential and molar quantities, as the section is along the tie-lines of the fixed potentials. As an example, Figure 3.15 shows the ternary Al–Fe–Si potential and molar phase diagrams sectioned at $T = 1273$ K and $P = 1$ atm, commonly referred to as an isothermal section. It is evident that the geometric features of both phase diagrams are identical to those

Table 3.1 *Invariant liquidus reactions of the Al–Fe–Si ternary system with the composition of the liquid phase [6]*

Reaction	T, °C	w_{Fe}, %	w_{Si}, %
$L + Fe_2Si \Leftrightarrow bcc + FeSi$	1178	77.9	21.0
$L + Al_5Fe_2 + Al_5Fe_4 \Leftrightarrow Al_2Fe$	1155	49.0	0.16
$L + Al_5Fe_4 \Leftrightarrow Al_2Fe + bcc$	1127	52.7	2.81
$L + FeSi + FeSi_2\text{-H} \Leftrightarrow FeSi_2\text{-L}$	1076	41.6	44.0
$L + Al_2Fe \Leftrightarrow Al_5Fe_2 + bcc$	1073	51.0	7.19
$L + bcc \Leftrightarrow Al_5Fe_2 + \tau_1$	1050	53.8	18.4
$L + FeSi_2\text{-H} \Leftrightarrow Si + FeSi_2\text{-L}$	1019	34.8	44.4
$L + bcc \Leftrightarrow Al_5Fe_2 + \tau_1$	1004	49.0	12.6
$L + Al_5Fe_2 \Leftrightarrow \tau_1 + \theta$	1000	46.7	13.2
$L + FeSi \Leftrightarrow \tau_1 + FeSi_2\text{-L}$	940	37.9	32.0
$L + \tau_1 + \theta \Leftrightarrow \tau_{23}$	921	33.7	20.1
$L + \tau_1 + FeSi_2\text{-L} \Leftrightarrow \tau$	899	33.8	32.2
$L + \tau_1 \Leftrightarrow \tau + \tau_{23}$	884	30.8	26.0
$L + FeSi_2\text{-L} \Leftrightarrow \tau + Si$	877	29.5	35.2
$L + \theta + \tau_{23} \Leftrightarrow \gamma$	851	23.3	21.6
$L + Si + \tau \Leftrightarrow \delta$	834	22.2	31.7
$L + \tau \Leftrightarrow \tau_{23} + \delta$	825	22.1	25.7
$L + \tau_{23} \Leftrightarrow \gamma + \delta$	823	21.8	25.4
$L + \theta + \gamma \Leftrightarrow \alpha$	715	6.64	10.8
$L + \gamma + \delta \Leftrightarrow \beta$	694	6.11	17.1
$L + \gamma \Leftrightarrow \alpha + \beta$	680	4.68	11.6
$L + \theta \Leftrightarrow fcc + \alpha$	630	2.11	4.10
$L + \alpha \Leftrightarrow fcc + \beta$	616	1.76	6.56
$L + \delta \Leftrightarrow \beta + Si$	598	1.22	14.3
$L \Leftrightarrow fcc + \beta + Si$	575	0.73	12.7

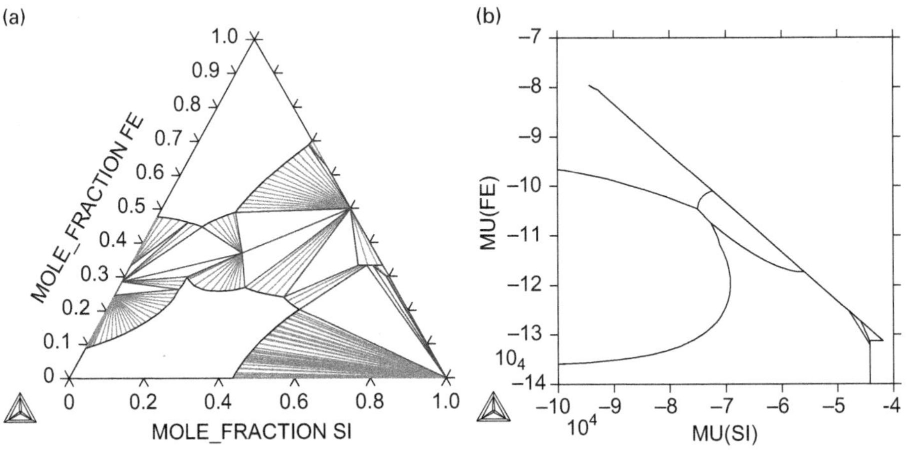

Figure 3.15 *(a) Ternary isothermal section of the Al–Fe–Si ternary system at* T = 1273 K *and* P = 1 atm *and (b) the corresponding potential phase diagram, from* [6] *with permission from Springer.*

of pure Fe, shown in Figure 3.2 and Figure 3.11, respectively, with one-, two-, and three-phase regions and corresponding phase boundaries.

On the other hand, when the phase diagram is sectioned along a molar quantity, usually it would not follow a tie-line because phases in equilibrium usually have different values for the same molar quantity. Consequently, there are no tie-lines inside such phase diagrams in general, and any phase regions only show which phases are in equilibrium with each other without any information on the values of molar quantities of individual phases.

This type of sectioning reduces the dimensionalities of both the phase diagram and phase boundary by the same number, but does not alter the number of phases in the adjacent phase regions. The contact rule, i.e. Eq. 3.66, thus remains valid and is applicable to phase regions with the same dimensionality as that of the sectioned phase diagram. Similarly, Schreinemakers' rule shown in Figure 3.12 is valid under the same conditions.

As an example, the two-dimensional phase diagram of the Mg–Al–Zn ternary system sectioned with one atmospheric pressure and the weight fraction of Zn fixed at 0.01 is shown in Figure 3.16, in a plot of temperature versus mole fraction of Al [7]. This phase diagram is commonly called an isopleth and is generated by fixing one potential, the pressure, P, changing the chemical potentials of Al and Zn to their conjugate molar quantities represented by weight fractions of Al and Zn, and sectioning at $w_{Zn} = 0.01$. From the discussions in Section 3.4.2, the phase regions with number of phases equal to three or fewer, i.e. $p \leq n_m + 1 = 3$, have the same dimensionality as the phase diagram, i.e. they are two-dimensional in the present case, and the phase

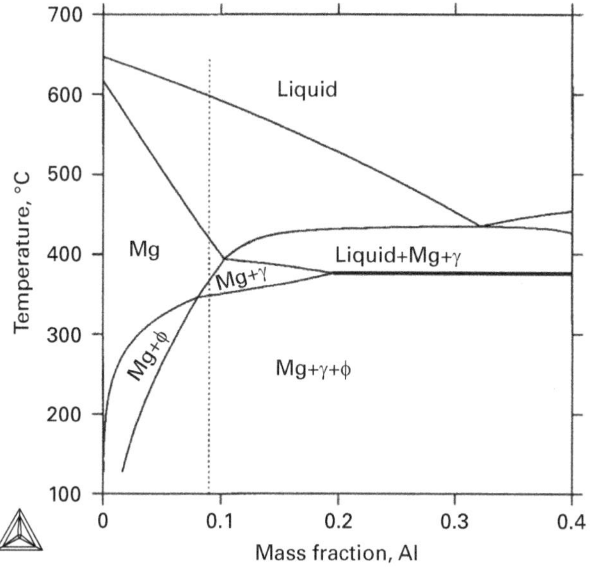

Figure 3.16 *Isopleth with the weight fraction of Zn fixed at* 0.01 *of the Mg–Al–Zn ternary system, from* [7] *with permission from TMS.*

boundary rule is applicable. The maximum number of phases co-existing at equilibrium is given by Eq. 3.41 as the following, for the present case:

$$p_{max} = (c - n_s) + 2 = 3 - 1 + 2 = 4 \qquad 3.72$$

This is so because introducing molar quantities only increases the dimensionality of phase regions and does not change the maximum number of co-existing phases.

The dimensionality of a four-phase region is calculated from Eq. 3.54 as

$$v_m = (c - n_s) + 2 - p + n_m - n_{ms} = 3 - 1 + 2 - 4 + 2 - 1 = 1 \qquad 3.73$$

where n_{ms} is the number of sectioned molar quantities. Since the dimensionality of a four-phase region is lower than that of the phase diagram, the phase boundary rule cannot be applied directly. Such a four-phase region, liquid + Mg + γ + ϕ, is shown in Figure 3.16 between three three-phase regions of liquid + Mg + γ, liquid + Mg + ϕ, and Mg + γ + ϕ.

Figure 3.16 also displays information on what phases are in equilibrium for a given alloy at various temperatures. One example is shown by the dotted vertical line marking a weight fraction of Al of 0.09, a widely used Mg alloy called AZ91. Various phases are present at different temperature ranges, but the equilibrium phase fractions and phase compositions are not shown in the figure as the tie-lines are not in the plane of the phase diagram and have to be calculated at each temperature individually. Figure 3.17 shows the amount of each phase of the AZ91 alloy as a function of temperature, with the dotted lines depicting the values under equilibrium conditions and the solid lines depicting the values under the so-called Scheil condition assuming no diffusion in solid phases and infinitely fast diffusion in the liquid. Similarly, the composition of each phase can also be plotted as shown in Figure 3.18.

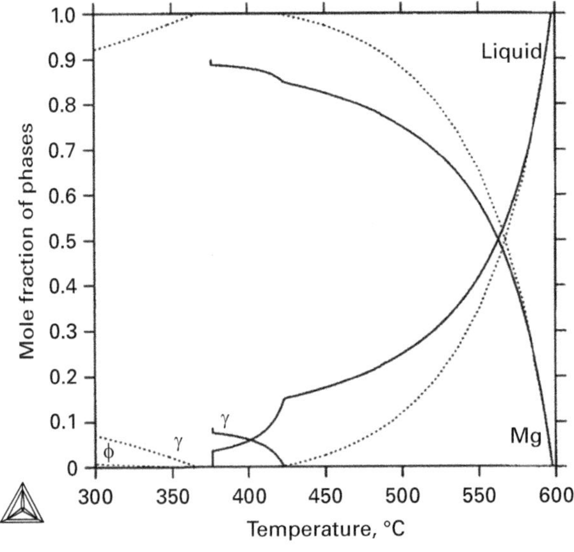

Figure 3.17 *Mole fraction of individual phases under equilibrium (dotted curves) and Scheil (solid curves) conditions in the AZ91 alloy, from* [7] *with permission from TMS.*

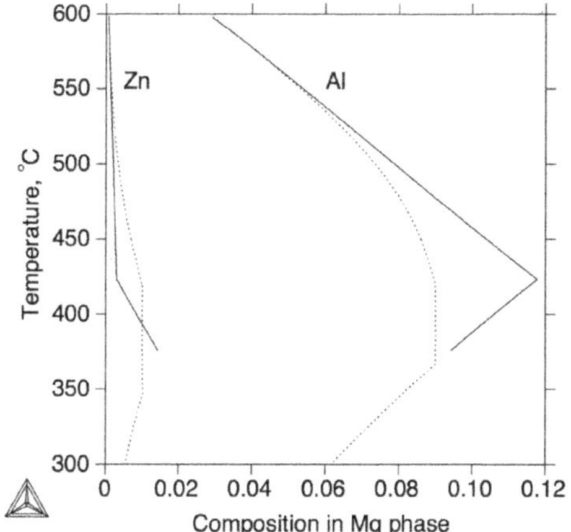

Figure 3.18 *Mass fraction of Al and Zn in the Mg solid solution phase under equilibrium (dotted curves) and Scheil (solid curves) conditions in AZ91, from* [7] *with permission from TMS.*

Exercises

1. In this problem, you need to first write down your steps in solving the problems using equations and thermodynamic variables such as C_P, H_m, S_m, x_i etc. and then obtain the numerical values using any available software. When using software, take screen snapshots as necessary to show your procedure.
    - 1 kg of a steel (Fe + 0.8 mass% C) is heated from a state of equilibrium at 500 °C to a new state of equilibrium at 800 °C. The pressure is kept at 1 atm. How much heat is needed for this operation? (*Hint:* for constant pressure, the enthalpy change is equal to the heat exchange.)
    - A mixture of 2 mol of H_2 and 0.1 mol of O_2 is kept in a very strong cylinder at 25 °C. The cylinder has a moveable piston, working against an outside atmosphere of 1 atm. The mixture is ignited and reacts quickly to reach a state of equilibrium, without giving time for any exchange of heat with the surroundings. Calculate the new temperature. (*Hint:* $dQ = 0$ and $dP = 0$).
2. Calculate the *P*–*T* diagram for Ca with *T* ranging from 600 to 1800 K and *P* ranging from 1 to 1000 Pa. Answer the following questions.
    a. There should be two three-phase equilibrium points (triple points). Complete the phase diagram to lower pressures by hand so that another three-phase equilibrium can be seen.
    b. List the phases in all one-, two- and three-phase regions.

(*cont.*)

c. Draw G–T diagrams at $P = 10^5$, 10^2, and 10^{-4} Pa by hand for all phases, as accurately as possible in terms of transition temperatures.

d. Draw G–P diagrams for $T = 2500$ and 1500 K by hand for all phases, as accurately as possible in terms of transition pressures.

e. Draw the extension of metastable two-phase equilibrium from triple (three-phase) points.

3. Shown below is the potential phase diagram of Fe–C under one atmosphere. There are four phases in the system, i.e. liquid (L), fcc (γ), bcc (a), and graphite (C). Do the following.

a. Derive a Clausius–Clapeyron-like equation for the slope of the lines in the diagram.

b. Discuss the reasons why the lines are curved.

c. Label all phase regions, including one-, two- and three-phase regions.

d. Draw the metastable extensions of two-phase equilibria in the potential phase diagram.

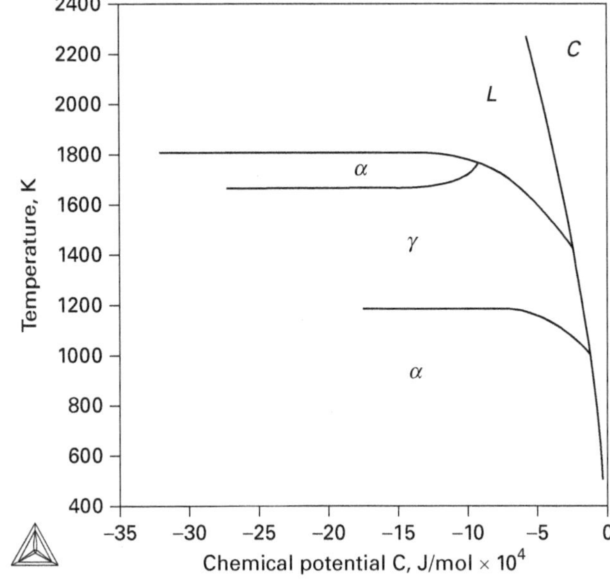

4. Consider the equilibrium between the liquid and solid phases of a pure metal at its melting temperature. It is found that the solid phase has 10^{-3} vacancies per atom. Assuming the entropy of melting is approximately equal to R, calculate the melting temperature if there were no vacancies in the solid phase.

5. Derive the Gibbs phase rule in detail using both words and equations.

6. The P–T potential phase diagram of water is shown. Note the critical point between liquid and vapor, beyond which these is only one phase. Answer the following questions.

(cont.)

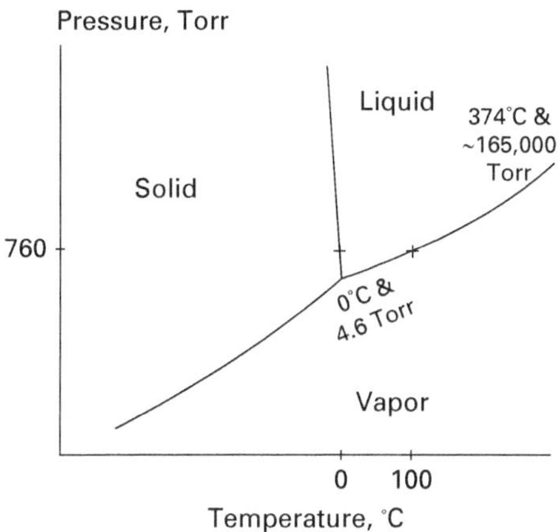

a. The single-phase regions are already labeled. Label all remaining two-phase equilibrium and three-phase equilibrium regions.
b. Derive the Clausius–Clapeyron equation.
c. By comparing the Clausius–Clapeyron equation and the slopes of phase regions, discuss the relative magnitudes of molar volumes and the relative magnitudes of molar entropies of phases in equilibrium.
d. Draw pressure–molar-entropy, molar-volume–temperature, and molar-volume–molar-entropy phase diagrams as accurately as possible with scales of pressure and temperature properly marked wherever needed.
e. In your molar-volume–temperature phase diagram, select the molar volume of the system in the vapor–solid two-phase region and calculate the mole fraction of each phase using the lever rule.
f. In your molar-volume–molar-entropy phase diagram, select the molar volume and molar entropy of the system in the vapor–solid–liquid three-phase region and calculate the mole fraction of each phase.

7. A *T–P* phase diagram for a unary system (pure *A*) is given in the figure. It shows four phases. Construct a reasonable T–μ_1 property diagram at P_1. It should show all the stable and metastable two-phase equilibria at P_1.

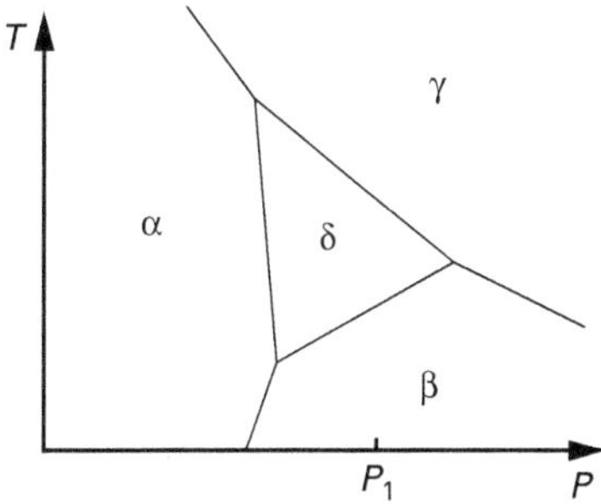

(cont.)

8. The Al–Ni binary phase diagram is shown below. The γ and γ' phases have the disordered fcc and ordered $L1_2$ structures, respectively, which are represented by one Gibbs energy similar to that of a miscibility gap.
   a. Draw molar Gibbs energy curves of ALL phases including both stable and metastable ones at 1200 K and 1800 K as accurately as possible.
   b. Calculate the phase fractions of an alloy with 30%at. Ni as a function of temperature using the lever rule.

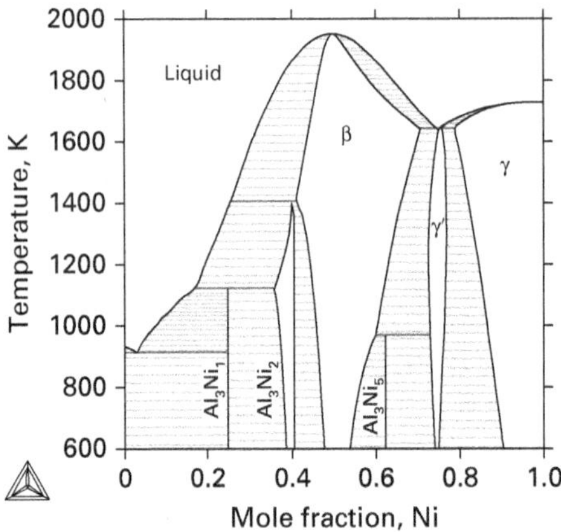

   c. For the most part, the two-phase regions between β and liquid are two-dimensional. However, at the highest temperature, we have a point, and the transformation between these two phases is called a congruent transformation. Draw a Gibbs energy diagram at this temperature for these two phases and discuss whether this point is an invariant point.
   d. Thirty grams of Ni and 14 grams of Al are mixed to form a single-phase solid solution. How many moles of solution are there? What are the mole fractions of Al and Ni? Which phase region is this solution in, at 1500 K?
   e. Convert the phase diagram into a potential phase diagram.
   f. Write down all invariant reactions and types of reactions in the binary system with the high temperature phase(s) on the left and low temperature phase(s) on the right of the reactions.

9. Refer to the enlarged Al–Ni binary phase diagram shown below. For an alloy with $x_{Ni} = 0.8$, calculate and plot the phase fractions of $\gamma'(L1_2)$ and γ (fcc) as a function of temperature. To obtain a microstructure with a

(cont.)

two-phase mixture of 70 volume% γ phase and 30 volume% γ' phase, assuming the volume fraction is equal to the mole fraction, what temperature should one choose? Discuss whether the assumption that the volume fraction is equal to the mole fraction is a good one.

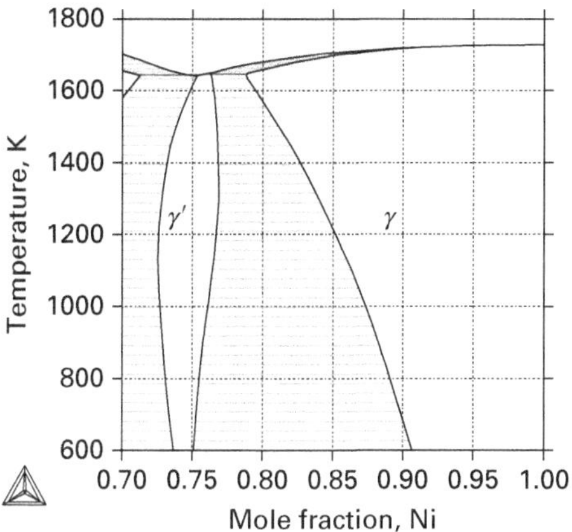

10. Below are shown the results of increasing the pressure at constant temperature and increasing the temperature at constant pressure of a one-component system. The latter type of experiment is routinely done using differential thermal analysis (DTA) or differential scanning calorimetry (DSC). There are four phases: two solid (S_1 and S_2), liquid (L) and vapour (V).
   - Based on these plots, construct the potential phase diagram. You may want to work out the phase boundary between S_1 and the vapor phase first and then the boundary between S_1 and S_2. Assume that phase boundaries are straight lines.
   - Plot a molar–potential phase diagram by changing the pressure axis to its conjugate variable, molar volume.
   - The plots contain steps. Assume a constant heating rate (dQ/dt) and a constant rate of pressurization (dV/dt). Do the durations of the horizontal portions of the plots mean anything? If so, what?
   - Do the slopes of the curves mean anything? If so, what?

(cont.)

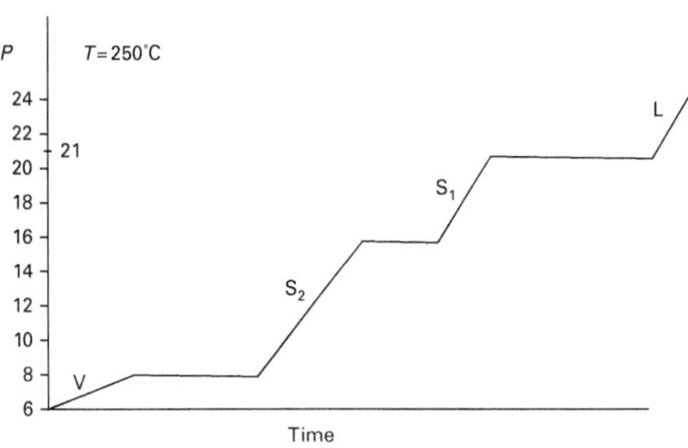

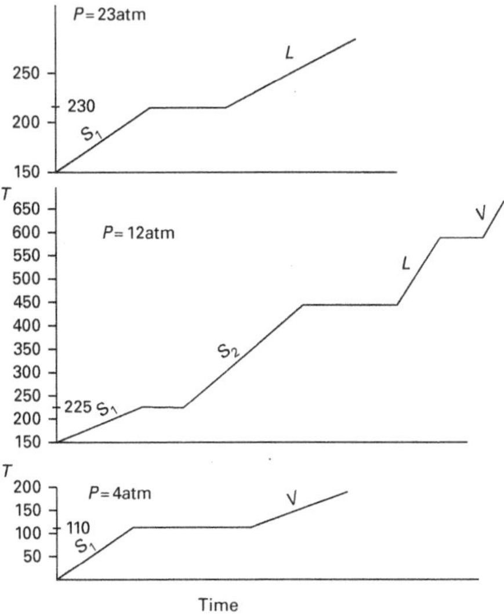

11. The specific volume of poly(tetrafluoroethylene) can be found on page 209 and 210 in *Standard Pressure Volume Temperature Data for Polymers,* by P. Zoller and D. J. Walsh (CRC Press, 1995). Do the following as accurately as possible.
    a. Convert the specific volume to molar volume assuming a constant molecular weight of 340000.
    b. Construct a molar volume–temperature phase diagram for this one-component system. Draw schematically the spinodal curve, i.e. the limit of stability of the single phase in your phase, diagram.

(*cont.*)

c. Convert your molar volume–temperature phase diagram to a pressure–temperature phase diagram.

d. For a system at 40 MPa pressure and 0.55 cm^3/g specific volume, use the lever rule to calculate the volume fractions of the two phases in equilibrium based on your phase diagram.

12. Use the following phase diagrams to answer the questions. The polystyrenes-cyclohexane phase diagram is taken from the reference, M. Tsuyumoto, Y. Einaga, and H. Fujita, "Phase equilibrium of the ternary system consisting of two monodisperse polystyrenes and cyclohexane", *Polym. J.* 16 (1984) 229–240, with permission from Nature.

a. Draw the Gibbs energy of stable phases as a function of concentration for the following systems at given temperatures as accurately as possible.

Polystyrene-cyclohexane (first diagram with $\xi_4 = 0$)	26 °C		
Fe–C (second and third diagrams)	1300 °C	800 °C	
CaO–SiO_2 (fourth diagram)	2400 °C	2000 °C	1800 °C

b. Write down all invariant reactions and the type of reactions in the above systems with the high temperature phase(s) on the left and low temperature phase(s) on the right of the reactions.

c. Label all single-phase and two-phase regions.

d. Convert the mixed phase diagrams to potential phase diagrams.

e. Calculate the phase fractions at the temperatures given in (a) with the following compositions.

Polystyrene-cyclohexane (first diagram with $\xi_4 = 0$)	$\phi = 0.05$		
Fe–C (second and third diagrams)	$x_C = 0.10$	$x_C = 0.15$	
CaO–SiO_2 (fourth diagram)	$x_{SiO2} = 0.10$	$x_{SiO2} = 0.30$	$x_{SiO2} = 0.90$

(cont.)

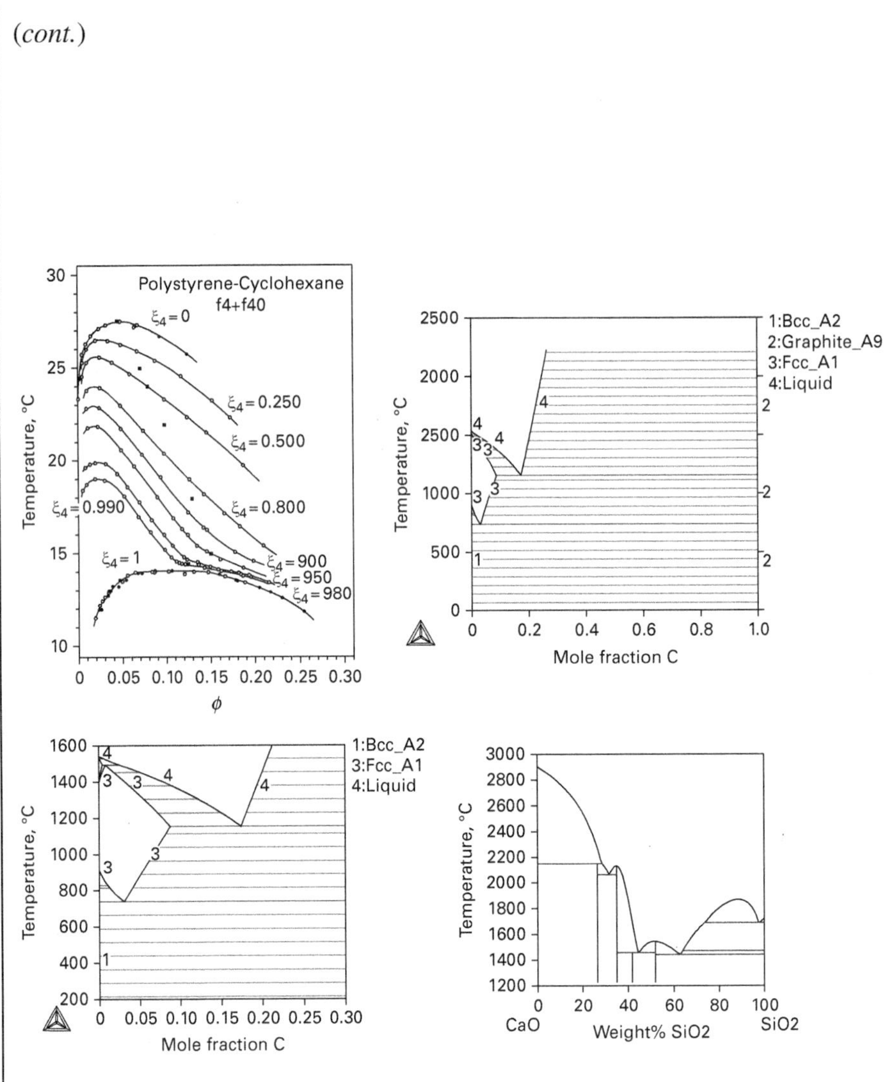

13. The liquidus projection of the Cu–Sn–Ti ternary system under constant pressure is shown here with the arrows showing the directions of decreasing temperature, taken from the reference J. Wang *et al.*, "Experimental investigation and thermodynamic assessment of the Cu-Sn-Ti ternary system,". *CALPHAD* 35 (2011) 82–94, with permission from Elsevier. Assume that all intermetallic compounds are stoichiometric. Answer the following questions.

(*cont.*)

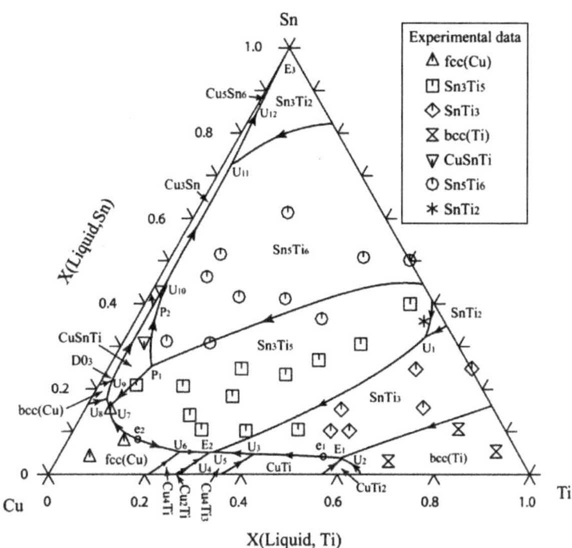

a. Mark accurately the alloy with 0.2 mole fraction of Sn and 0.3 mole fraction of Ti in the composition triangle.

b. Invariant reactions are labeled in the diagram with E, P, and U. Write all invariant reactions with the high temperature phases on the left side and the low temperature phases on the right side and list whether they are eutectic or peritectic.

c. Write down and explain the solidification path of the alloy from part a above.

d. Write down the Gibbs phase rule for this system. Explain whether the Gibbs phase rule can be applied to the liquid projection.

14. The isopleth section at 15 at.% Sn in the Cu–Sn–Ti ternary system under constant pressure is shown here, taken from the reference J. Wang *et al.*, "Experimental investigation and thermodynamic assessment of the Cu–Sn–Ti ternary system," *CALPHAD* 35 (2011) 82–94, with permission from Elsevier. Answer following questions.

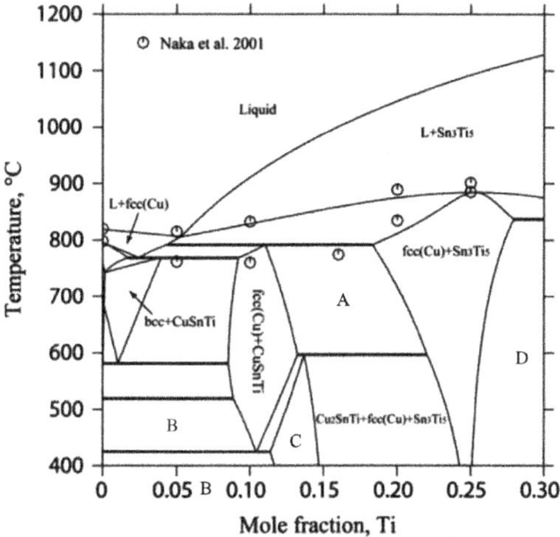

(cont.)

a. There are seven four-phase equilibria in the isopleth. Mark all of them on the diagram with numbers. For those that you know all four phases, list the four phases.

b. Explain whether the phase boundary rule ($D_{+} + D_{-} = r - b$) can be applied to the four-phase equilibria.

c. List the phases in the A, B, C, D phase regions.

d. Explain whether one can use the lever rule to calculate the phase fractions from the phase diagram.

15. The isopleth of the multi-component Fe–1.5Cr–3Ni–0.5Mn–0.3Si–C (all in weight percent) system is shown here. The numbers indicate the stability limits of various phases (zero phase fraction lines). It is important to realize that the tie-lines are not in the plane of the isopleth. The single fcc phase region is labeled.

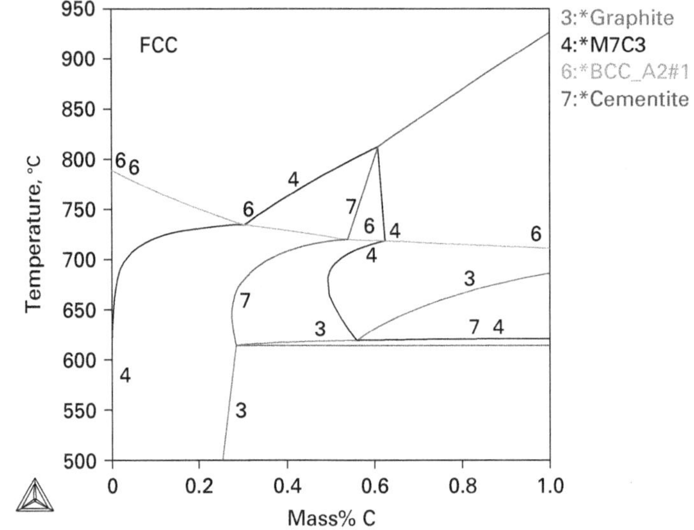

a. List the phases in all the phase regions.

b. Explain whether it is possible to draw a complete molar phase diagram from the given phase diagram.

c. Explain whether one can use the lever rule to calculate the phase fraction from the given phase diagram.

d. Draw schematic diagrams of phase fractions as a function of temperature for 0.1%, 0.2%, 0.4%, and 0.8% C, respectively, as accurately as possible with scales for both axes of your diagrams.

e. Explain whether there any invariant reactions in the phase diagram.

16. The liquidus projection and the isothermal section at 1500 °C of the quasi-ternary system CaO–$A_{12}O_3$–SiO_2 under constant pressure are shown below, taken from

(cont.)

the reference H. Hao *et al.*, "Thermodynamic assessment of the CaO–Al 203-SiO2 system, *J. Am. Ceram. Soc.* 89 (2006) 298–308, with permission from John Wiley & Sons. The numbered lines in the liquidus projection are isotherms at 100 °C intervals with the smallest number corresponding to the lowest temperature. The tie-lines in two-phase regions in the isothermal section are shown. Answer the following questions.

a. What are the maximum numbers of phases possible in the two diagrams?

b. Mark the alloy with 30% Al_2O_3 and 30% SiO_2 on both the liquidus projection and the isothermal section.

c. When this alloy is cooled from its liquid state, which solid phase will form first? Explain.

d. At 1500 °C, what are the equilibrium phases present in the alloy in part a? Use the isothermal section to identify the approximate compositions of the phases.

e. Use the lever rule to estimate the composition change of the liquid phase of the alloy in part a as a function of temperature during solidification. Explain.

f. Based on your results in part d, which phase will be the second solid phase formed during solidification? Explain.

g. As the temperature continues to decrease, which phase will be the third solid phase to form? Is this an invariant reaction? Explain. If so, write down the reaction and its type, with the high temperature phase(s) on the left and low temperature phase(s) on the right.

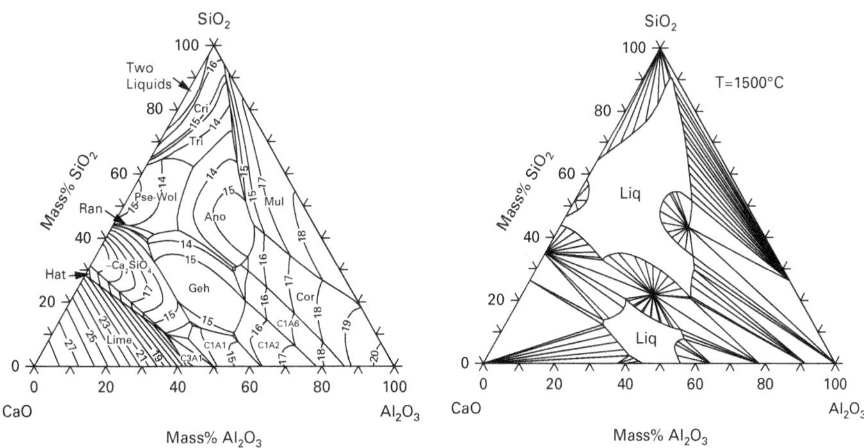

17. The magnetic contribution to entropy can be represented by $\Delta S = R\ln(\beta + 1)$ with β the number of unpaired electrons. The Gibbs energy difference between the fcc and bcc phases is shown in the figure below. Assume that the fcc phase is non-magnetic. Use the data in the figure to estimate the values of ΔS and β of the bcc phase.

(*cont.*)

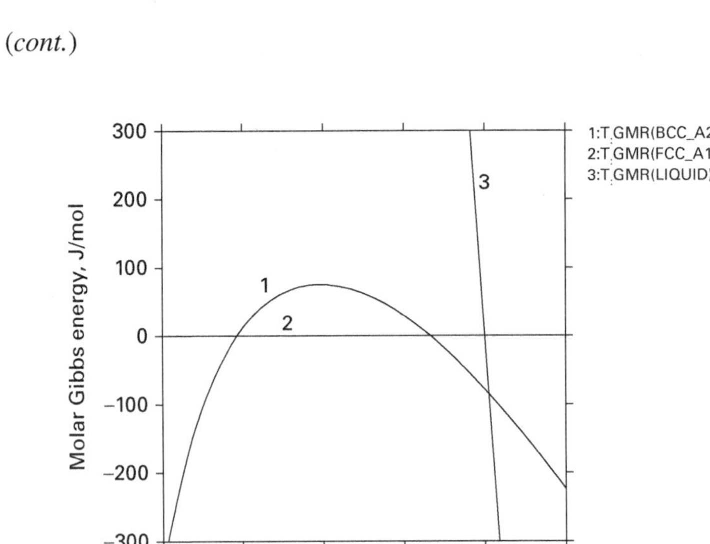

18. Find the Gibbs energy, molar volume, and the interfacial energy between liquid and solid pure Al from the literature. Plot the critical nucleus size and barrier for homogeneous nucleation as a function of temperature. At what temperature would you expect that nucleation can be observed? You may assume that the strain energy equals zero and both liquid and solid have the same molar volume.

19. When a liquid phase is cooled fast enough below its glass transition temperature (T_g), crystallization can be suppressed and instead an amorphous phase is formed. Estimate the difference of enthalpy between the two states in terms of the melting temperature, assuming T_g is one third of the melting temperature and assuming the following.
    a. Both the amorphous and crystalline states have the same entropy at 0 K.
    b. For temperatures below T_g, both states have the same heat capacity.
    c. For temperatures above T_g, the difference of the heat capacity for both phases is a linear function of T.
    d. At the melting temperature, both phases have the same heat capacity again, and the entropy of melting is approximately equal to R.

20. The polystyrene-cyclohexane phase diagram, from problem 12 enlarged as shown below, with $\xi_4 = 1$ representing the miscibility gap of a binary phase diagram with the following properties.
    - Cyclohexane: solvent C_6H_{12}, density 0.779×10^6 g/m^3, molar volume 1.078×10^{-4} m^3/mol.
    - Polystyrene: homopolymer of styrene monomer CH_2–CH–C_6H_5, density (ρ) 1.04×10^6 g/m^3, molar volume 0.0420 m^3/mol.

(*cont.*)

a. Apply the following Gibbs energy per lattice site to this binary system using the above data and derive expressions for the chemical potentials of cyclohexane and polystyrene, respectively:

$$\frac{G}{n} = G_n = \sum \frac{\phi_i}{m_i} {}^0G_{im} + kT\left(\sum \frac{\phi_i}{m_i} \ln\phi_i + \sum \phi_i\phi_j\chi_{ij}\right)$$

where ϕ_i and m_i are the volume fraction and number of lattice sites per molecule i, and n is the total number of lattice sites; ${}^0G_{im}$ is the Gibbs energy of pure molecule i per molecule, which can be set to zero at any temperatures a the reference state, k is the Boltzmann constant, and χ_{ij} is the dimensionless interaction parameter between molecule i and molecule j.

b. Estimate the interaction parameter, χ, for $\xi_4 = 1$ at $T = 12$ °C using the data from the figure.

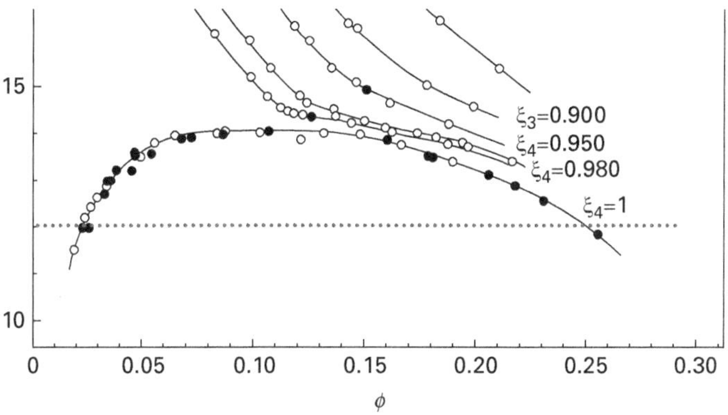

4 Experimental data for thermodynamic modeling

The most widely used thermodynamic modeling technique is the CALPHAD (CALculation of PHAse Diagram) method to be discussed in detail in Chapter 6. The input data in the evaluations of thermodynamic model parameters came primarily from experiments and estimations until first-principles calculations based on the density functional theory [8] became a user tool in the later 1990s. Experimental data include both thermodynamic and phase equilibrium data; the first-principles calculations, which provide thermodynamic data for individual phases, are discussed more extensively in Chapter 5.

Three fairly recently published books summarize the methods commonly used for experimental measurements of the thermodynamic properties of single [9] and multiple phases [10] and phase diagrams [11]. The methods are briefly discussed here, and readers are referred to these books for details. The main techniques for crystal structure analysis include X-ray diffraction, electron backscatter diffraction (EBSD), electron diffraction in transmission electron microscopy, neutron scattering, and synchrotron scattering, which are not discussed in this book.

4.1 Phase equilibrium data

4.1.1 Equilibrated materials

The most common method to determine phase equilibria is to use equilibrated materials. This method typically involves material preparation through high temperature melting or powder metallurgy, homogenization heat treatment, isothermal or cooling/heating procedures, and identification of crystal structures and phase compositions. It is important to avoid macro-inhomogeneity as it can be difficult to remove the inhomogeneity in subsequent treatments. It is also important to use starting materials of the highest purity and to minimize the loss and contamination of materials during the entire experiment using a protective atmosphere of inert gas or vacuum. Typical melting techniques include high temperature furnaces with crucibles, arc melting, and induction melting. Attention needs to be paid to possible reactions between materials and crucibles/containers, which can be avoided by levitating the materials by electromagnetic fields or other means. In addition to using pure elements as raw materials, master alloys with well-controlled compositions are often utilized because the compositions and melting properties of master alloys are usually much closer to those of the final materials than the pure elements. For materials with very high melting temperature or volatile components, the powder metallurgy method can be used where compacts are made, capsulated, and sintered.

Homogenization during subsequent heat treatment is achieved through diffusion, in which time and temperature are two important parameters. To accelerate the homogenization process, the heat treatment temperature should be as close to the solidus temperature as possible, taking into account the composition inhomogeneity with variable solidus temperatures. When there are phase transformations taking place during the heat treatment, extra time is needed for the heat treatment.

The phase boundaries are then determined through measurements of either the compositions of individual phases that are in equilibrium under constant temperature, pressure/stress/strain, and electric/magnetic field or the discontinuity in some physical properties of the material due to a phase transition from the continuous change of temperature or pressure/stress/strain or electric/magnetic fields. The measurement of compositions is usually carried out under ambient conditions so it is necessary that the phases are fully equilibrated under experimental conditions, which requires rigorous verification, and can be "quenched" to ambient conditions to remain unaltered during "quenching" at least in terms of composition. The compositions are typically measured by scanning electron microscopy (SEM) equipped with energy dispersive spectrometry (EDS) or wavelength dispersive spectroscopy (WDS), which has micrometer spatial resolution and better compositional resolution than EDS. A dedicated SEM with WDS gives another important, widely used composition measurement technique called electron probe microanalysis (EMPA). For submicrometer sized phases, analytical transmission electron microscopy equipped with EDS can be used, though care must be taken to avoid interference from neighboring phases.

To accurately identify phases in equilibrium under experimental conditions, in situ characterizations are necessary, which complicates the experimentation. An alternative indirect method is to measure a physical property that changes discontinuously or dramatically for a first- or second-order phase transition, such as heat, volume, electric conductivity, or magnetization. There are two widely used techniques for measuring heat: differential thermal analysis (DTA) and differential scanning calorimetry (DSC). Both measure the difference in temperature with the same amount of power supplied between a sample and an inert standard during heating or cooling; DSC may alternatively measure the difference in the amount of power supplied to keep the temperatures identical. The deviation of this difference from a baseline indicates a phase transition in the sample and is plotted as a function of time or temperature of either the sample or the inert standard. It is evident that the major challenges in both DTA and DSC techniques are to reach thermal equilibrium between the sample or standard and the instrument and thermodynamic equilibrium within the sample due to continuous heating or cooling. The thermal equilibrium can be improved or mitigated through altering various aspects of the experiment such as sample shape and size, type of crucible, mixture with the material used for the inert standard and by using thermocouples in direct contact with the sample and the inert standard. However, thermodynamic equilibria within the sample can only be reached when the heating/cooling rate is comparable with the rate of the phase transition in the sample, which is almost impossible if the phase transition typically involves diffusion in solid phases. Therefore, extreme care is needed in interpreting the temperature determination and the amount of heat associated with the DTA/DSC curves as discussed in detail in reference [11].

4.1.2 Diffusion couples/multiples

The major challenge in the equilibrated material approach is to ensure that the whole sample reaches equilibrium. On the other hand, the diffusion couple/multiple technique does not require the whole sample to be in equilibrium and is based on the assumption that any two phases in contact are in equilibrium with each other at the phase interface, and that the phase compositions can be obtained by extrapolation of concentration profiles in the two phases to the phase interface. Since the total system of a diffusion couple is not at equilibrium, many kinetic phenomena related to diffusion can be studied in a diffusion couple, such as the interdiffusion coefficients, the parabolic growth kinetics of product layer thickness, the diffusion path (represented by the local overall compositions) in ternary and multicomponent systems for visualizing the microstructure of reaction zones, and other properties, all as a function of composition; these are beyond the scope of the present book.

Typical diffusion couples are in the solid state with two materials brought into intimate contact to allow diffusion of elements between the two materials, though solid–liquid diffusion couples are also used. The contacting faces are commonly ground and polished flat, clamped together using mechanical mechanisms, and annealed at high temperatures where diffusion can take place to a significant degree in a matter of days, weeks, or months. The samples are then quenched to retain the high temperature equilibrium features. For metallic systems, diffusion couples can also be prepared by eletrolytical and electroless plating techniques. It is important to avoid the formation of liquid during annealing as it ruins the sample geometry. Furthermore, good adherence at the interfaces is critical for reliable data.

Since diffusion couples are not in a fully equilibrium state, the tie-lines between two phases at the phase interfaces need to be obtained by extrapolation of concentration profiles in neighboring phases. The electron propagation in quantitative EPMA is typically in the range 1–2 μm, yielding an excitation volume of approximately 2–4 μm diameter and a requirement of reasonable layer widths of phases on both sides of the interface for accurate extrapolation. Therefore, the reliable composition of a single phase must be taken several micrometers away from the interface. When steep concentration gradients exist near the interfaces, the extrapolation may lead to large errors, and analytical electron microscopy is then needed. Furthermore, fluorescence effects, where the primary excitation can be powerful enough to excite other elements in the sample, result in enhanced X-ray production and the need for proper corrections. For a new phase to become observable experimentally, it must nucleate and grow to reach the resolution limit of analytical tools. It is thus not uncommon that some known equilibrium phases are not found, and some non-equilibrium phases form. One way to circumvent this issue is to use next to the phase of interest incremental diffusion couples with narrow concentrations so that only this phase may be formed.

For ternary and higher-order systems, a more efficient approach can be used by placing a thin layer of a third alloy between two other alloys. The thin central layer is eventually consumed, and the diffusion path is not fixed as in the semi-infinite diffusion couples. The phase compositions change continuously with time as a result of the overlapping of two quasi-equilibrated diffusion zones.

A diffusion multiple contains three or more pure elements or alloys of different compositions and is a sample with multiple diffusion couples and diffusion triples in it. It is more efficient in terms of both materials and time in comparison with equilibrated alloys and diffusion couples. All alloy blocks are prepared individually and sealed in a vacuum in a cylinder, which is also used as one alloy for the diffusion multiple. The sealed cylinder with a vacuum inside also serves as the can for subsequent hot isostatic pressing to achieve intimate interfacial contacts. The cylinder can then be cut into disks for further annealing treatments. A broad range of design strategies is needed for complex diffusion multiples, along with automated plotting procedures, due to the large amounts of EMPA data. The major source of error lies in the extraction of tie-lines from EPMA results, due to very condensed information relating to a small area and the deviation of scanned lines from those perpendicular to the interface.

In terms of the local equilibrium characteristics of diffusion couples/multiples, it is evident that in the equilibrated materials approach it may not be necessary to reach full equilibrium for the whole sample if one is only interested in the local equilibrium compositions between two neighboring phases. This can even provide information on metastable extensions of two phases if the two phases are in a metastable equilibrium at the annealing temperature.

4.1.3 Additional methods

The electrical resistivities of different phases are usually different. A change of slope of electric resistivity as a function of composition or temperature or pressure reflects a phase transformation. This technique is simple and reliable.

Magnetic transitions can be measured using a vibrating sample or superconducting quantum interference device (SQUID) magnetometer by determining the magnetic moment of a sample in the presence of an applied magnetic field. Magnetic field lines form closed loops, resulting in an external dipolar and demagnetizing field in a finite-sized sample. The effective field sensed by the sample is thus the difference between the applied field and the demagnetizing field. The magnetic transition temperature is evaluated from Arrott plots, where the ratio of magnetic field over magnetization with a proper exponent is plotted with respect to the magnetization with another proper exponent for a series of temperatures. These proper exponents result in parallel isotherm lines, and the isotherm line intersecting the origin corresponds to the magnetic transition temperature.

Thin films with composition gradients, commonly referred to as combinatorial libraries, can be used to study the phase relations similar to diffusion couples/multiples, though the results may differ due to the effects of surface and potential interactions with the substrate.

Phase relations at high pressures are measured by equipment using diamond anvil cells (DAC) or multi-anvil devices. The high pressure is realized by decreasing the area, i.e. the anvil culet size. Pressures up to 100 GPa can be created in a DAC with a culet size of 0.3 mm for small samples of the order of 0.2 to 0.4 mm. For large samples, the large-volume press (LVP) technique has been developed, typically using WC with the pressure mostly limited to 30 GPa and the sample size ranging from 1 mm^3 to 1 cm^3. The pressure can be measured either using the ruby (Cr^{3+} doped Al_2O_3) fluorescence line shift or monitoring the molar volume of a pressure marker by X-ray diffraction. The samples in DAC apparatus are heated by lasers or resistive wires or a small heater around the samples, while high temperatures in LVPs are achieved by resistive heaters. Crystal structures are detected by in situ X-ray or synchrotron diffraction. Attention needs to be paid to temperature and pressure homogeneity and the non-hydrostatic stresses, which are both better controlled in LVP equipment.

4.2 Thermodynamic data

Broadly speaking, thermodynamic data represent the values of the Gibbs free energy and its first and second derivatives. For thermochemical properties, the main topic of this book, calorimetric, electrochemical, and vapor pressure methods are the primary techniques used, with the first used for accurate measurement of enthalpy and entropy and the latter two used for direct determination of Gibbs energy and activity. The electrochemical method is discussed in Chapter 8. The calorimetric method is divided into solution, combustion, direct reaction, and heat capacity calorimetry, respectively, which all involve chemical reactions to be discussed in detail in Chapter 7. The vapor pressure method involves the equilibrium of volatile species between gas and samples and is divided into Knudsen effusion and equilibration methods, respectively.

4.2.1 Solution calorimetry

The book edited by Marsh and O'Hare [12] documents the detailed experimental techniques used for solution calorimetry. In one experiment, the enthalpy of solution of a single phase is measured in a particular solvent. To convert this enthalpy of solution into enthalpy of formation of the phase, a thermodynamic cycle is set up for a chemical reaction to form this phase from either constitutive pure elements or compounds. Therefore, in another experiment, the enthalpy of solution of the constitutive pure elements or compounds is measured in a solvent which is as identical as possible to that used in the first experiment. The difference of the two enthalpies of solution thus gives the enthalpy of formation of the single phase from its constitutive elements or compounds at the temperature of the samples, usually at room temperature, before they are dropped into the solvent.

The solvent can be aqueous solutions at ambient temperatures and pressures or metallic/salt/oxide melts at high temperatures under either adiabatic or isoperibol conditions. Adiabatic calorimetry measures the temperature change of the solvent and is usually more accurate than isoperibol calorimetry, which measures the heat generated during the dissolution, though adiabatic calorimetry requires more complex instruments. It is important that the choices of solvent and temperature ensure the complete dissolution of all substances into the solvent to form a homogeneous solution. Furthermore, the effects of dilution and of changes in solvent composition need to be considered in the calculation of the enthalpy of solution.

A large number of solvents are used. For aqueous solvents, hydrofluoric acid or mixtures of HF and HCl are often used. For oxides, buffer-type systems are typical such as lead and alkali borates and alkali tungstates or molybdates. For metallic phases, low melting metals such as Sn, Bi, In, Pb, and Cd, or sometimes Al and Cu, are used. Factors such as solubility, dissolution kinetics, thermal history, stirring, heat flow, particle size, and system size are optimized for accurate measurements. To enhance the solution kinetics, the compound to be studied can be mixed with another element or compound so that the mixture can form liquid in the solvent at a higher reaction rate. In all cases, it is important to calibrate the system with pure elements and compounds of known enthalpy of formation.

4.2.2 Combustion, direct reaction, and heat capacity calorimetry

In combustion calorimetry, the sample is ignited and reacts with reactive gases like oxygen or fluorine. To accurately calculate the enthalpy of formation from the enthalpy of combustion, reliable characterization of the reactants and reaction products is critical, in order to identify problems such as incomplete combustion, impurities in the reactants, which are often ill defined, and more than one reaction of the gaseous species and condensed phase. Combustion calorimeters are usually of the isoperibol type around room temperature in a water bath.

Direct reaction calorimetry is similar to combustion calorimetry, though it is carried out at high temperatures in heat-flux or adiabatic environments. The partial enthalpy of reaction can also be measured if the partial pressure of volatile species can be controlled and measured. The key factor for accurate results is that both the reactants and reaction products are well characterized and the reaction goes to completion quickly, like in combustion calorimetry. For reactive reactants, special procedures are needed to avoid loss of the reactants before the reaction takes place.

Heat capacity is defined as the amount of heat needed to increase the temperature by 1 K, as shown by Eq. 2.7, and its integration with respect to temperature from 0 K gives the entropy as shown by Eq. 2.32. At low temperatures, adiabatic calorimetry gives more accurate data regarding heat capacity though it is time consuming and requires complex instruments. At high temperatures, the efficient but less accurate DSC method is more widely used.

4.2.3 Vapor pressure method

In the Knudsen effusion method, a small amount of volatile species in the gas phase effuses through a small orifice of 0.1 to 1 mm with negligible influence on the equilibrium in the Knudsen cell. The vapor is ionized and analyzed with a mass spectrometer. The partial pressure of a species can be calculated from its ionization area and intensity using a calibration factor determined by a reference material with known partial pressure. For high temperature measurements, care must be taken to avoid reactions between the cell and sample and the fragmentation of gas species on ionization. The typical vapor pressure range is between 10^{-7} and 10 Pa.

In the various equilibration methods, the total vapor pressure is usually measured directly using pressure gauges in the range of 10^{-7} and 100 kPa. Other direct or indirect methods include the following:

- the thermogravimetric method for measuring the vapor mass;
- atomic absorption spectroscopy for measuring gas composition;
- measurement of sample composition equilibrated with a gas of well-defined activity of the volatile species;
- the dew point method in which the condensation temperature of the volatile component is measured from the vapor equilibrated with the sample at a higher temperature;
- the chemical transport method, to be discussed in Section 7.4.

The main error in all these methods is often due to inadequate equilibration between vapor and sample.

Exercises

1. A container of liquid lead is to be used as a calorimeter to determine the enthalpy of formation of Mg_2Si. It has been determined by experiment that the heat capacity of the bath is 4000 J/K at 300 °C. With the bath originally at 300 °C, the following experiments are performed.
   a. A mixture of 0.2 mol of pure Mg (4.861 grams) and 0.1 mol of pure Si (2.808 grams) at 25 °C is dropped into the calorimeter. When the mixture has dissolved completely, the temperature of the bath is found to have increased by 0.37 °C from 300 °C.
   b. One tenth of a mol of formula of Mg_2Si (7.669 grams) at 25 °C is dropped into the calorimeter. When the compound is dissolved completely, the temperature of the bath decreases by 1.60 °C from 300 °C.
   Answer the following questions.
      i. What is the enthalpy of formation of Mg_2Si per mole of formula and per mole of atoms?
      ii. To what temperature does the result for the enthalpy of formation apply? Explain.
2. Design an experiment similar to Exercise 1 to measure the enthalpy of mixing of a liquid solution of Pb–Sn at 250 °C with $x_{Sn} = 0.7$.
   a. Derive an expression to calculate the enthalpy of mixing, representing your measurements by symbols.
   b. Evaluate the interaction parameter, assuming the solution is a regular solution.
3. The melting point of gold is 1336 K. The vapor pressure of liquid gold is given by

$$\ln P = 23.716 - \frac{43552}{T} - 1.222 \ln T$$

where the units of P and T are atmospheres and degrees kelvin.

(*cont.*)

a. Calculate the heat of vaporization of gold at its melting point.

Answer parts b, c, and d numerically *only* if the data given in this problem statement are sufficient to support the calculations. If there are not enough data, write "solution not possible."

b. What is the vapor pressure of solid gold at its melting temperature?

c. What is the vapor pressure of solid gold at 1200 K?

d. What is the heat of fusion of solid gold?

4. The heat capacity of CuO is shown in the figure below. Do the following.

   a. Evaluate the parameters in the following expression for the heat capacity using the data in the figure: $C_P = c + dT + e/T^2 + fT^2$.

   b. Derive an expression for the entropy, with $S_{298} = 42.74$ J/K per mole of CuO. What is the reference state of this entropy?

   c. Derive an expression for the enthalpy, with $H_{298} = -155800$ J/K per mole of CuO. Discuss the reference state.

   d. Derive an expression for the Gibbs energy. State your reference state.

   e. Plot the heat capacity, entropy, enthalpy, and Gibbs energy as functions of temperature using your expressions.

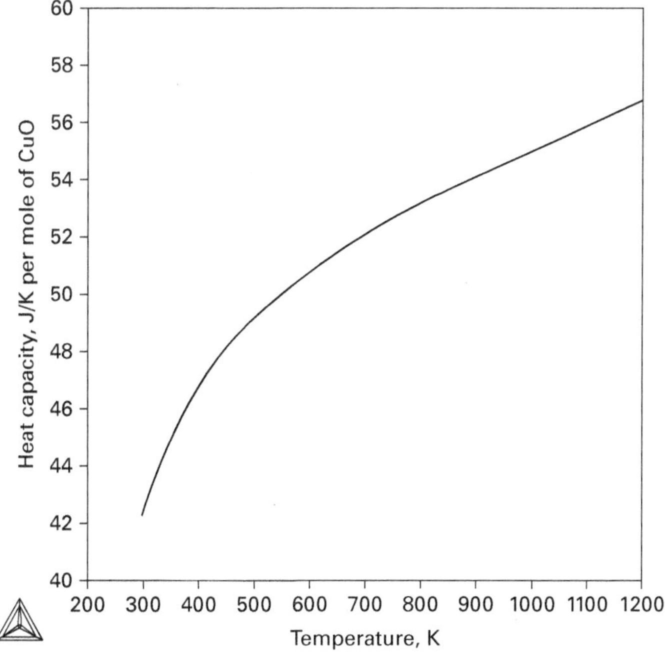

5. Calcium has two solid structures in addition to liquid and gas phases: fcc at low temperatures and bcc at high temperatures. Their heat capacity data are shown in the figure below. Their transition temperatures and enthalpies of transition are as follows:

(*cont.*)

fcc → bcc	716 K	930 J/mol
bcc → liquid	1115 K	8542 J/mol
liquid → gas	1762 K	148282 J/mol

a. Evaluate the parameters in the following expression for the heat capacity by measuring four data points for the fcc and bcc phases in the figure and using least squares fitting or any of your favored programs: $C_P = c + dT + e/T^2 + fT^2$. You can assume constant heat capacities for liquid and gas.

b. Derive expressions for the enthalpies of the solid structures. Specify your reference state.

c. Derive the expression for their entropies using $S_{298}^{fcc-Ca} = 42$ J/(K mol). What is the reference state for the entropy?

d. Derive the expressions for their Gibbs energies. State your reference state.

e. Plot the heat capacity, entropy, enthalpy, and Gibbs energy of each phase as functions of temperature using your expressions.

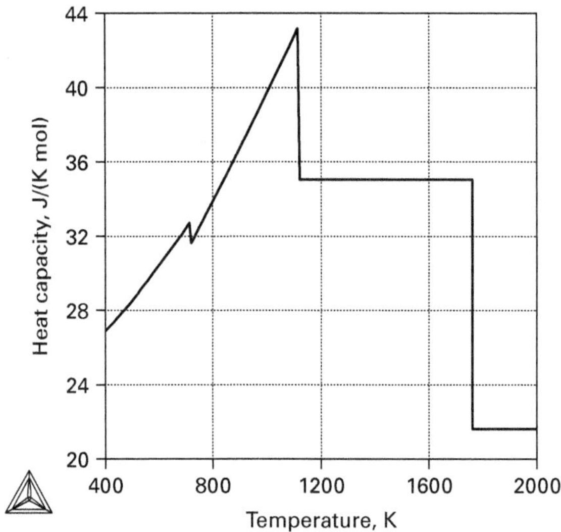

6. Use the following thermodynamic data for Ca to answer the questions below.

Molar weight Ca = 40.078 g/mol

$V(\mathrm{Ca}, \alpha) = 26.025$ cc/mol, $V(\mathrm{Ca}, \beta) = 26.200$ cc/mol, $V(\mathrm{Ca}, \mathrm{l}) = 28.361$ cc/mol

$S_{298\mathrm{K}}(\mathrm{Ca}, \alpha) = 41.6$ J/(K mol), $S_{298\mathrm{K}}(\mathrm{Ca}, \mathrm{g}) = 154.8$ J/(K mol)

(*cont.*)

Heat capacity (J/mol-K) valid between 298–2500 K at 1 atm

$C_p(\text{Ca}, \alpha) = 16.0 + 12.0 \times 10^{-3}T$

$C_p(\text{Ca}, \beta) = 14.0 + 15.0 \times 10^{-3}T \qquad T(\alpha \to \beta) = 716\text{ K}$

$C_p(\text{Ca}, \text{l}) = 33.5 \qquad T(\text{fusion}) = 1115\text{ K}$

$C_p(\text{Ca}, \text{g}) = 20.8$

$\alpha \to \beta$: $\Delta\text{H}[\text{Ca}(\alpha) = \text{Ca}(\beta)] = 930\text{ J/mol at }716\text{ K}$

Fusion: $\Delta H[\text{Ca}(\beta) = \text{Ca}(\text{l})] = 8542\text{ J/mol at }1115\text{ K}$

Sublimation: $\Delta H[\text{Ca}(\alpha) = \text{Ca}(\text{g})] = 177800\text{ J/mol at }298\text{ K}$

a. In the pressure–temperature potential phase diagram, calculate the slopes of the phase boundaries between α and β and between β and liquid phases.

b. The pressure of Ca(g) in equilibrium with the stable condensed calcium phase is 1.0×10^{-9} bar at 627 K, and 1.0×10^{-2} bar at 1228 K. Using the results from Problem a and other information given above, draw a pressure–temperature potential phase diagram of Ca and make sure that all boundaries have correct slopes with all metastable extensions included.

c. Label all phase regions and apply the Gibbs phase rule to all phase regions in the potential phase diagram you create.

d. Draw two Gibbs energy diagrams: one as a function of temperature for a given pressure and one as a function of pressure for a given temperature. Include all stable and metastable phase equilibria.

e Assume Ca(g) is in equilibrium with Ca(α) at 298 K and calculate the equilibrium vapor pressure of the gas at this temperature.

f. What is the entropy, $S[\text{Ca}(\beta)]$, at 500 K and 1 atm?

g. Determine a general equation for the entropy of $S[\text{Ca}(\alpha)]$ at 1 atm (in other words, determine an equation for $S[\text{Ca}(\alpha)] = f(T)$).

h. A sample of 100 g of Ca(α) at 298 K is added to 100 g of Ca(l) at 1200 K, and then allowed to equilibrate under adiabatic conditions in a closed system, i.e. the enthalpy of the system is constant. The system volume is small so that no appreciable amount of vapor can form; only the condensed phases need to be considered. Calculate the final equilibrium state of the system.

5 First-principles calculations and theory

In the previous chapter, the experimental techniques used to obtain the thermochemical and phase equilibrium data that were the inputs for the thermodynamic modeling of a system were summarized. However, experimental data are not always available. This is due to the fact that (i) the experiments are expensive, especially when they involve developing new materials, and (ii) the experiments cannot reliably access the non-stable phases in most cases. The alternative approach is to predict the thermochemical data by first-principles calculations. The prediction of material properties, without using phenomenological parameters, is the basic spirit of first-principles calculations. In particular, the steady increase of both computer power and the efficiency of computational methods have made the first-principles predictions of most thermodynamic properties possible, including both enthalpy and entropy as a function of temperature, volume, and/or pressure.

By definition, the term "first-principles" represents a philosophy that the prediction is to be based on a basic, fundamental proposition or assumption that cannot be deduced from any other proposition or assumption. This implies that the computational formulations are based on the most fundamental theory of quantum mechanics, the Schrödinger equation or density functional theory, and the inputs to the calculations must be based on well-defined physical constants – the nuclear and electronic masses and charges. In other words, once the atomic species of an assigned material are known, the theory should predict the energies of all possible crystalline structures, without invoking any phenomenological fitting parameters.

This chapter is organized in sequence from thermodynamic calculations to fundamental theory, to help those readers who are more interested in realistic calculations using existing computer codes. Detailed theoretical discussions follow the subsections on thermodynamic calculations for those readers who are also interested in the derivation of the formulations used in the thermodynamic calculations. The subsections are arranged accordingly in the order: (i) examples of the commonly adopted calculation procedures for thermodynamic properties using the elemental metal nickel as the main prototype; (ii) derivation of the Helmholtz energy expression under the first-principles framework; (iii) introduction of the solution to the electronic Schrödinger equation within two well-developed frameworks – the quantum chemistry approach and the density functional theory; (iv) detailed description of the procedure on how to solve the Schrödinger equation for the motions of atomic nuclei by means of lattice dynamics; and (v) First-principles approaches to disordered alloys.

5.1 Nickel as the prototype

This section exemplifies the step-by-step procedures for calculating the thermal properties within the framework of the first-principles phonon approach, using the elemental metal Ni as the prototype. The calculation of the formation enthalpy of Ni_3Al is given at the end. The calculation in this section is limited to the ferromagnetic phase, i.e. a single microstate, implying that no configurational mixtures or magnetic phase transitions are considered. Those are discussed in Section 5.2.5 and Chapter 6.

The Vienna Ab-Initio Simulation Package (VASP) [13, 14] has been employed for electronic calculations, and the YPHON code [15] has been employed for phonon calculations. VASP is a code based on the pseudopotential approach to density functional theory using plane wave functions as the basis set, by which only the valence electrons are handled explicitly and the core electrons are approximated by an effective pseudopotential. The same energy cutoff value, which determine the number of plane waves in the expansion of the electronic wave function, has been used for Ni, Al, and Ni_3Al. The rationale for the derivations of the formulations used in this section is given in Section 5.2 for readers who wish to have an in-depth understanding of the physics behind the formulations used.

5.1.1 Helmholtz energy and quasi-harmonic approximation

At present, the most rigorous method for predicting the thermodynamic properties of a material at finite temperatures is the phonon approach. In such an approach, the microscopic Hamiltonian is expanded up to the second order. All the thermodynamic quantities are calculated using formulations derived from statistical physics without further approximation. The great advantage of phonon theory is that all the input parameters can be obtained by means of first-principles calculations without using any phenomenological parameters.

Let us consider a system with an average atomic volume V. Neglecting the electron phonon coupling, it is a well-demonstrated procedure [16] to decompose the Helmholtz energy $F(V,T)$ of the system at temperature T into three additive contributions as follows:

$$F(V,T) = E_c(V) + F_{vib}(V,T) + F_{el}(V,T) \tag{5.1}$$

where E_c is the static total energy, which is the total energy of the system at 0 K without any atomic vibrations, F_{vib} is the vibrational contribution due to the lattice ions, and F_{el} is the electronic contribution, due to thermal electronic excitation at finite temperature, which can become important for metals at high temperature.

The terminology "quasi-harmonic approximation" arises from the fact that for a given volume, $F_{vib}(V,T)$ is calculated under the harmonic approximation and the anharmonic effects are included solely through the volume dependence of the phonon frequency. The easiest computational implementation of Eq. 5.1 is to first independently calculate the Helmholtz energies at several selected volumes near the equilibrium volume and then use numerical interpolation to find the Helmholtz energy at an arbitrary volume. The volume interval is usually at the scale of 3%~5% of the equilibrium volume. Too small a volume interval can result in numerical instability due to the numerical uncertainties in the static total energy calculation, in particular, when one numerically computes the first- and especially the second-order derivatives of the Helmholtz energy in deriving the thermodynamic quantities. It should be noted that, whenever available, analytic formulas should be used instead of a numerical second-order derivative to avoid numerical errors. For instance, when the phonon approach is employed, the constant volume heat capacity has an analytic expression in terms of the phonon density of states.

Nickel metal adopts the fcc structure at ambient conditions and the primitive unit cell contains one atom. Almost all the existing first-principles codes have a function for calculating the static total energy. The static total energy E_c in Eq. 5.1 should be calculated using the primitive unit cell. As the Helmholtz energy is to be calculated at several volumes, it is good practice to plot the calculated static total energy points together with an interpolated energy curve to examine the convergence of the static total energy calculation. Since the first-principles method often employs the self-consistent technique, it could occur that calculations at certain volumes may not be convergent, which should be fixed by trying the various algebraic schemes provided in most of the existing codes. Furthermore, since certain calculations involve the second-order derivative of the Helmholtz energy, a minor uncertainty along the static total energy curve can result in a large deviation for calculated properties such as the thermal expansion coefficient and bulk modulus. In that case, a reasonable solution is to smooth the static total energy using the modified Birch–Murnaghan equation of state (EOS) [17, 18]

$$E_c = a + bV^{-2/3} + cV^{-4/3} + dV^{-2} + eV^{-8/3} \tag{5.2}$$

The calculated static total energy of elemental metal Ni is plotted in Figure 5.1, with the circles representing the calculated values and the curve representing the EOS fitting.

The vibrational contribution to the Helmholtz energy from phonon theory can be computed by [19]

$$F_{vib}(V,T) = k_B T \int_0^\infty \ln\left(2 \sinh \frac{\hbar\omega}{2k_B T}\right) g(\omega, V) d\omega \tag{5.3}$$

where k_B is Boltzmann's constant, ω represents the phonon frequency, and $g(\omega, V)$ is the phonon density of states. It is recommended that $g(\omega, V)$ is calculated at the same volume set at which the static total static energies are calculated.

For the present prototype of Ni, the supercell method for the calculation of $g(\omega, V)$ was employed. The procedure is follows.

i. Make a supercell by enlarging the primitive unit cell according to the defined neighbor interaction distance; employ the first-principles code (VASP [13, 14] in this work) to calculate the interatomic force constants.

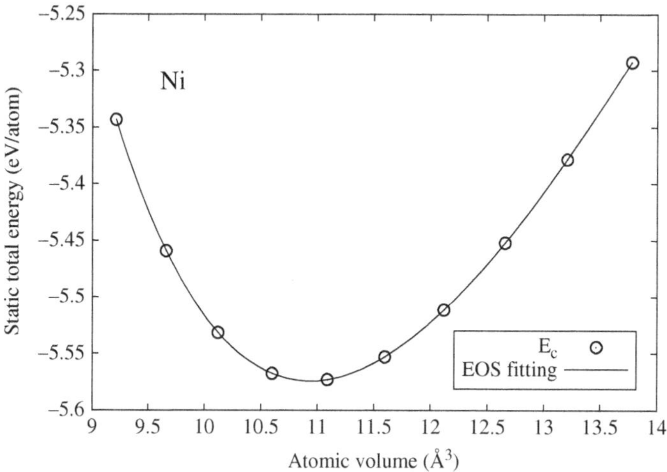

Figure 5.1 *Static total energy of nickel.*

ii. Assign the mesh in the wave vector (**q**) space; make the dynamical matrix at each **q** point; diagonalize the dynamical matrix to find the phonon frequencies at each **q** point; and finally collect all the phonon frequencies for all **q** points. The detailed formulation for phonon calculations is given in Section 5.2.

For the phonon calculations, one can use the open source code YPHON [15] developed by the present authors. Other choices are the free ATAT code [20] or the free PHON code [21]. For calculation of the phonon density of states, we used a supercell containing 64 atoms, which is a 4×4×4 supercell of the primitive unit cell. Figure 5.2 is a plot of the calculated phonon density of states using YPHON code at the calculated static equilibrium volume compared with the measured data at 10 K [22] (symbols).

For a first-principles thermodynamic calculation, an important step to avoid possible calculation errors is to examine the phonon dispersions first. Phonon dispersion [23] means the evolution of phonon frequencies along the designated direction for a crystal. Phonon dispersion can be measured rather accurately by inelastic neutron scattering [24–26] or inelastic X-ray scattering [27] experiments. Figure 5.3 shows the calculated phonon dispersions (curves) along the [00ζ], [0ζ1], [0ζζ], and [ζζζ] directions of Ni using the YPHON code and the neutron scattering data at 296 K (symbols).

For the calculation of F_{el} in Eq. 5.1, the most computationally convenient approach is to use the Mermin statistics, as follows:

$$F_{el}(V,T) = E_{el}(V,T) - TS_{el}(V,T), \tag{5.4}$$

where E_{el} is the thermal electronic energy, and S_{el} is the bare electronic entropy. Both the calculations of E_{el} and S_{el} need the electronic density of states (EDOS) as input. The electronic density of states can be obtained during the calculation of the static total energy. The detailed formulations for E_{el} and S_{el} are given in Section 5.2.2. Since Ni is

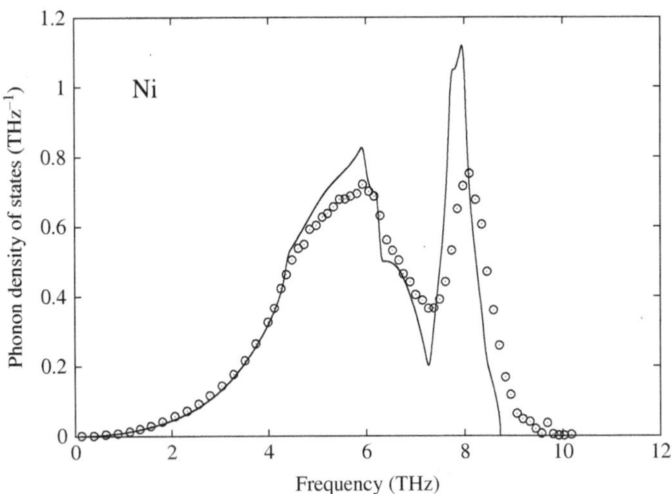

Figure 5.2 *Phonon density of states of nickel.*

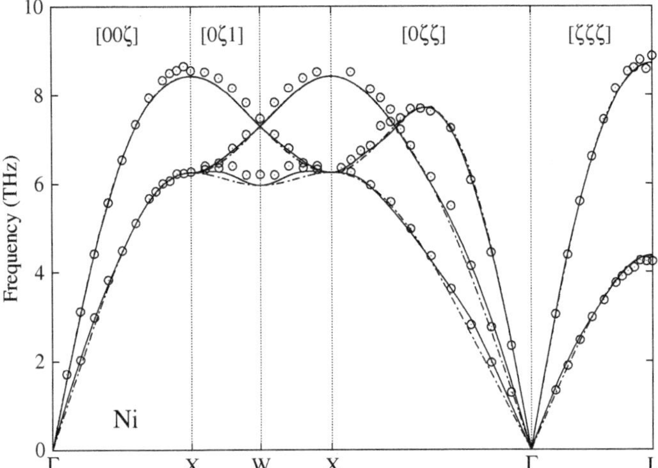

Figure 5.3 *Phonon dispersions of nickel. The solid lines represent results calculated using a supercell containing* 256 *atoms which is a* 4×4×4 *supercell of the conventional cubic unit cell. The dot-dashed lines represent results calculated using a supercell containing* 64 *atoms which is a* 4×4×4 *supercell of the primitive unit cell.*

magnetic, the EDOS of Ni can be partitioned into those of spin up and spin down, due to the spin freedom of electrons. The calculated EDOS for Ni is shown in Figure 5.4 where the solid, dot-dashed, and dashed lines represent the total, spin up, and spin down EDOS with the Fermi energy set to zero.

The calculated temperature evolution of the Helmholtz energy as a function of volume for Ni is illustrated in Figure 5.5. The circles represent the calculated static

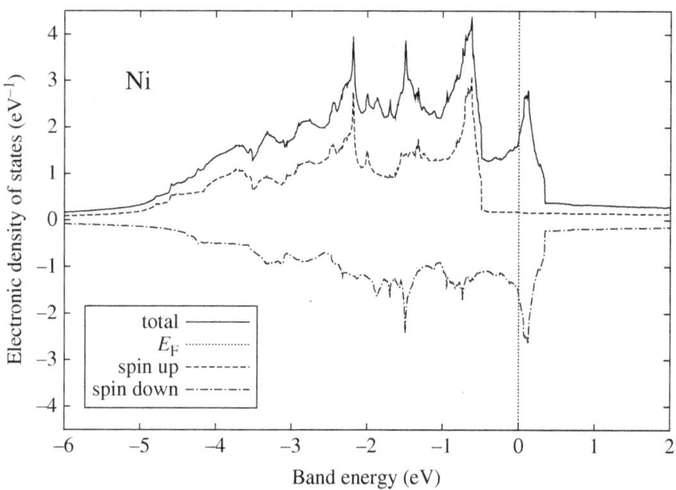

Figure 5.4 *Electronic density of states of nickel. That due to spin up is plotted as a positive value and that due to spin down is plotted as a negative value purely for the clarity of the figure. The "total" is the sum of the absolute values of those for spin up and spin down.*

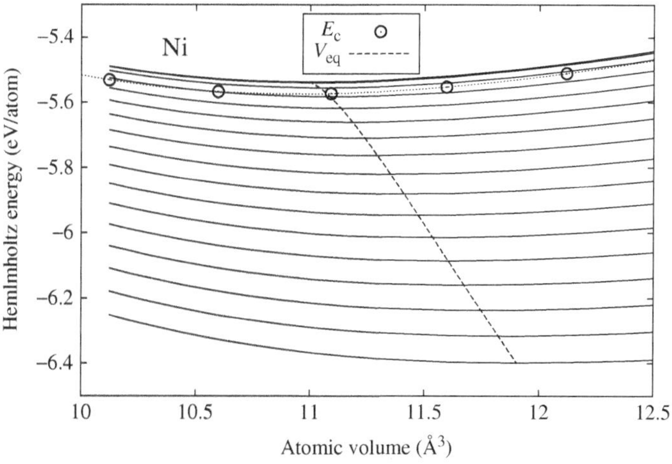

Figure 5.5 *Temperature evolution of the Helmholtz energy for nickel.*

total energies. The solid curves represent the Helmholtz energy curves from 0 to 1600 K at a temperature increment of 100 K as displayed from top to bottom in the figure. The dashed line marks the evolution of the equilibrium volume at $P = 0$ with increasing temperature. It is to be noted that the Helmholtz energy always decreases with increasing temperature due to the entropy term $-TS$. Note that at 0 K the Helmholtz energy is higher than the static total energy due to the zero point vibrational energy as can be seen when $T \rightarrow 0$ which reduces Eq. 5.3 to

$$F_{vib}(V,T)|_{T=0} = \int_0^{\infty} \frac{\hbar\omega}{2} g(\omega, V)d\omega \quad 5.5$$

which is positive.

5.1.2 Volume, entropy, enthalpy, thermal expansion, bulk modulus, and heat capacity

The equilibrium volume $V_{eq}(P,T)$ at given T and P can be obtained by finding the root of the following equation:

$$-\left(\frac{\partial F(V,T)}{\partial V}\right)_T = P \quad 5.6$$

As mentioned above, the dashed line in Figure 5.5 illustrates $V_{eq}(P,T)$ as a function of T from 0 to 1600 K at $P = 0$ for Ni.

The entropy can be calculated through F by

$$S(V,T) = -\left(\frac{\partial F(V,T)}{\partial T}\right)_V \quad 5.7$$

Figure 5.6 is a plot of the calculated entropy (curve) of Ni as a function of temperature from 0 to 1600 K at $P = 0$ and the recommended data (symbols) with details in reference [16].

Based on F and S, the enthalpy at given P and T can be computed as

$$H(V,T) = F(V,T) + TS(V,T) + PV \quad 5.8$$

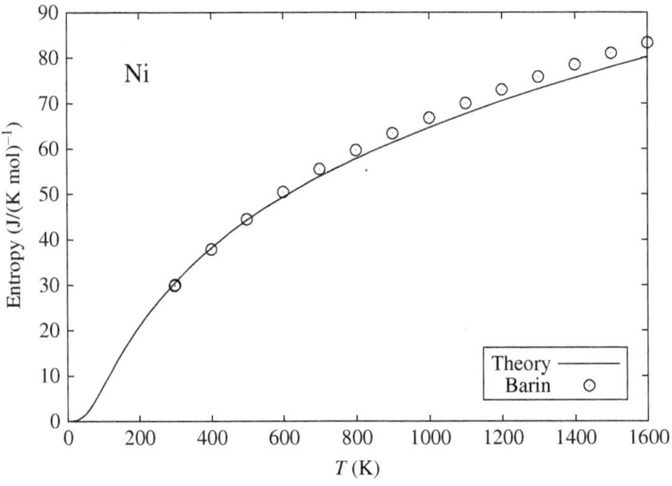

Figure 5.6 *Entropy of nickel as a function of temperature.*

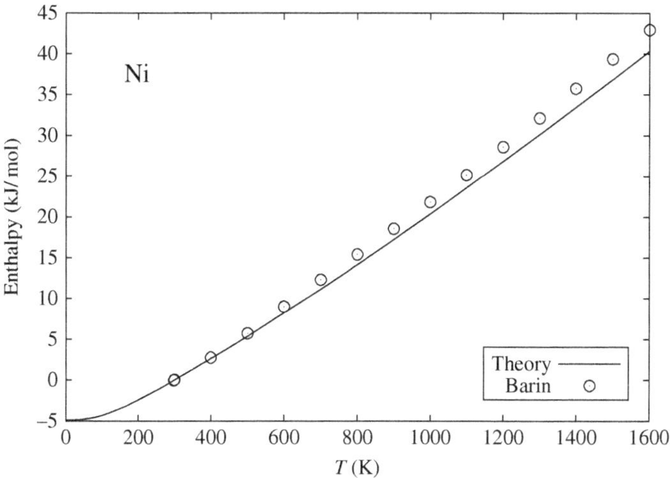

Figure 5.7 *Enthalpy of nickel as a function of temperature.*

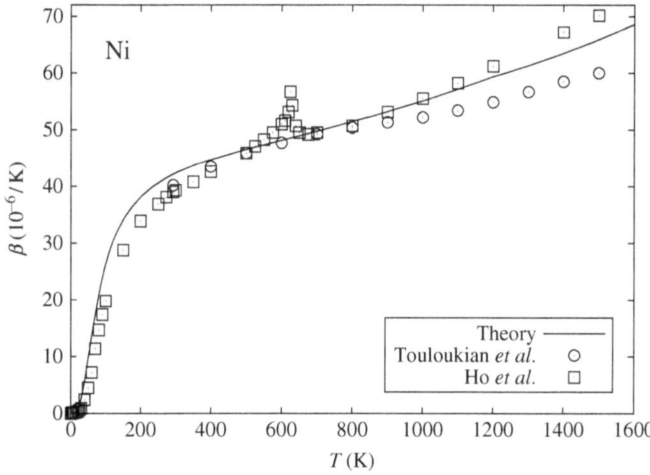

Figure 5.8 *Volume thermal expansion coefficient of nickel as a function of temperature.*

Figure 5.7 is a plot of the calculated enthalpy (curve) of Ni as a function of temperature from 0 to 1600 K at $P = 0$ and the recommended data (open circles) with details in reference [16].

With the equilibrium volume $V_{eq}(P,T)$ calculated by Eq. 5.6, the volume thermal expansion coefficient defined by Eq. 2.8 can be calculated using

$$\beta_P(P,T) = \frac{1}{V_{eq}} \left(\frac{\partial V_{eq}(P,T)}{\partial T} \right)_P \qquad 5.9$$

Figure 5.8 is a plot of the calculated thermal expansion coefficient (curve) of nickel as a function of temperature from 0 to 1600 K at $P = 0$ compared with experimental data (symbols) with details in reference [16].

The bulk modulus of a material represents the substance's resistance to uniform compression. Depending on how the temperature varies during compression, a distinction should be made between the isothermal bulk modulus (constant temperature) and the adiabatic bulk modulus (constant entropy or no heat transfer). As a matter of fact, most of the available experimental data are adiabatic whereas most of the published theoretical data are isothermal.

The isothermal bulk modulus, as defined in Eq. 2.9 in terms of the Gibbs energy, can be calculated from

$$B_T(V,T) = V\left(\frac{\partial F^2(V,T)}{\partial V^2}\right)_T \qquad 5.10$$

Based on the isothermal bulk modulus, the adiabatic bulk modulus can be calculated from

$$B_S(V,T) = (C_P/C_V)B_T(V,T) \qquad 5.11$$

where C_P and C_V represent the constant pressure heat capacity and constant volume heat capacity, respectively. Figure 5.9 is a plot of the calculated bulk moduli (curves) of Ni as a function of temperature from 0 to 1600 K at $P = 0$. The experimental data are from ultrasonic measurements (symbols, see reference [28] for more details) and are therefore adiabatic bulk moduli calculated based on the measured adiabatic elastic constants, using the relation

$$B_S = \left(C_{11}^S + 2C_{12}^S\right)/3 \qquad 5.12$$

The heat capacity at constant volume, as defined in Eq. 2.28, can be calculated from

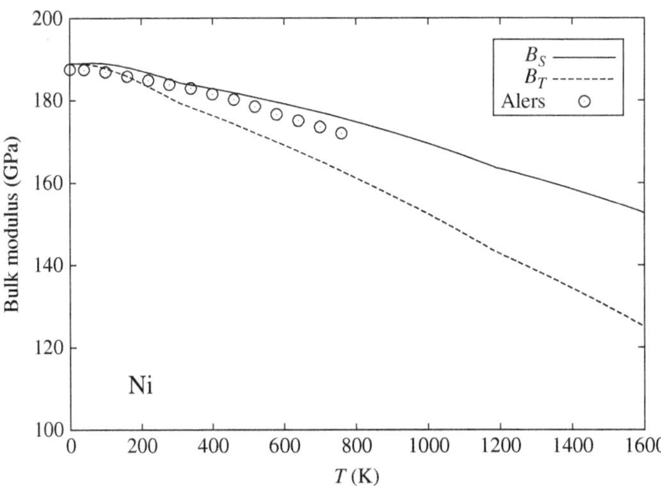

Figure 5.9 *Bulk moduli of nickel as a function of temperature. Solid line, adiabatic; dashed line, isothermal.*

$$C_V(V,T) = \left(\frac{\partial U(V,T)}{\partial T}\right)_V \qquad 5.13$$

where $U(V,T) = F(V,T) + TS(V,T)$ represents the internal energy. The heat capacity at constant pressure (see Eq. 2.31) can then be calculated as

$$C_P(P,T) = C_V(V,T) + VTB_T(V,T)(\beta_P(P,T))^2 \qquad 5.14$$

utilizing the calculated thermal expansion coefficient in Eq. 5.9 and bulk modulus in Eq. 5.10.

It can be seen that thermal expansion makes the difference between the heat capacity at constant volume and the heat capacity at constant pressure. The calculated contributions to the heat capacity of Ni as a function of temperature from 0 to 1600 K at $P = 0$ are illustrated in Figure 5.10, where the lattice vibration and the thermal electron contributions have been separated out. From Figure 5.10, it is observed that there is a large difference between the calculated C_P (solid line) and the experimental data (symbols, see reference [18] for more details) at 600 K due to the magnetic phase transition which has not been considered in the calculation. It should be pointed out that for Ni the thermal electronic contribution to the heat capacity (dotted line in Figure 5.10) is substantial at high temperatures.

5.1.3 Formation enthalpy of Ni_3Al

One can do similar calculations for elemental metal Al and the compound Ni_3Al, which has the $L1_2$ structure, by following the same steps as in the calculations for Ni. The formation enthalpy in units of per mole atom can be calculated as

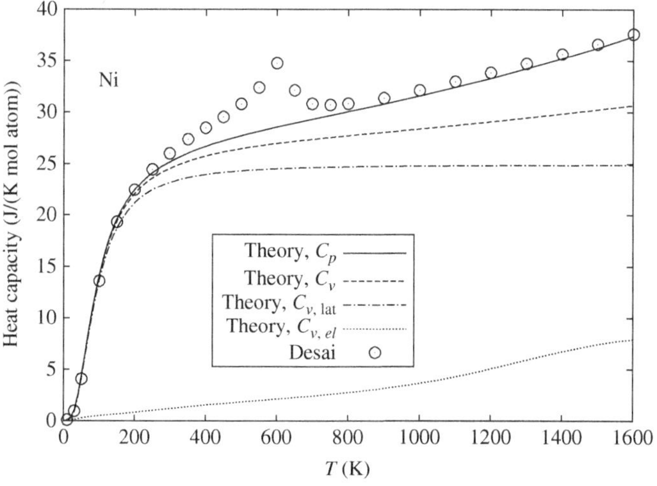

Figure 5.10 *Heat capacity of nickel as a function of temperature; C_V^{vib} represents the calculated lattice vibration contribution and C_V^{el} represents the calculated thermal electronic contribution.*

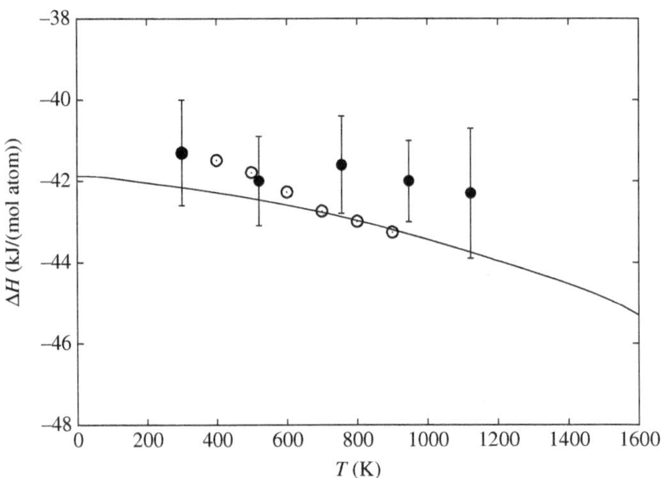

Figure 5.11 *Formation enthalpy of $L1_2$–Ni_3Al with respect to pure fcc Ni and Al.*

$$\Delta_f H^{Ll_2-Ni_3Al} = H_m^{Ll_2-Ni_3Al} - \frac{3}{4} H_m^{fcc-Ni} - \frac{1}{4} H_m^{fcc-Al} \tag{5.15}$$

where $H_m^{Ll_2-Ni_3Al}$, H_m^{fcc-Ni}, and H_m^{fcc-Al} represent the enthalpies of Ni_3Al, Ni, and Al in energy units per mole of atoms, respectively. Figure 5.11 is a plot of the calculated formation enthalpy of Ni_3Al (curve) as a function of temperature from 0 to 1600 K at $P = 0$ compared with experimental data (symbols) with details in reference [16].

5.2 First-principles formulation of thermodynamics

5.2.1 Helmholtz energy

In this chapter so far, all our discussions have been limited to the case of a system which consists of a single microstate (microscopic state). Here, the terminology "microstate" refers to the microscopic structure that is distinguished by the crystal structure, atom distributions in the lattice sites, and the arrangements of the local atomic spin and electronic angular momentum distributions among the lattice sites. From this section on, the index σ is employed to label the microstate. For a solid material at finite temperatures, a phase can be formed by a single microstate or a mixture of multiple microstates.

Let us consider a canonical system made of N atoms with an average atomic volume V. The study is limited to the motions of the atomic nuclei and electrons. For such a system, one can use $\varepsilon_{\mathbf{g},\mathbf{n}}(N, V; \sigma)$ to denote the energy eigenvalues of the corresponding microscopic Hamiltonian associated with microstate σ. The subscript $\mathbf{g}$ symbolically labels the different vibrational states for the motions of the atomic nuclei and the subscript $\mathbf{n}$ symbolically labels the electronic states distinguished by the different distributions of the electrons between the electronic valence and conduction bands. Neglecting electron–phonon coupling, one can assume that the contributions to $\varepsilon_{\mathbf{g},\mathbf{n}}(N, V; \sigma)$ between the vibrational and electronic states are additive, so that

$$\varepsilon_{\mathbf{g},\mathbf{n}}(N,V;\sigma) = E_c(N,V;\sigma) + \varepsilon_{\mathbf{g}}(N,V;\sigma) + \varepsilon_{\mathbf{n}}(N,V;\sigma) \qquad 5.16$$

where E_c is the static total energy of the microstate σ. Note that in Eq. 5.16 $\varepsilon_{\mathbf{g}}(N,V;\sigma)$ and $\varepsilon_{\mathbf{n}}(N,V;\sigma)$ represent the energies of the vibrational state and the electronic state respectively.

One then can formulate the canonical partition function of the microstate σ at the given temperature T and volume V as

$$\begin{aligned} Z(N,V,T;\sigma) &= \sum_{\mathbf{g},\mathbf{n}} \exp\left[-\beta\varepsilon_{\mathbf{g},\mathbf{n}}(N,V;\sigma)\right] \\ &= \exp[-\beta E_c(N,V;\sigma)]\sum_{\mathbf{g}}\exp\left[-\beta\varepsilon_{\mathbf{g}}(N,V;\sigma)\right]\sum_{\mathbf{n}}\exp[-\beta\varepsilon_{\mathbf{n}}(N,V;\sigma)] \end{aligned} \qquad 5.17$$

where $\beta = 1/k_BT$, k_B being Boltzmann's constant. As a result, with $F = -k_BT\ln Z$, the Helmholtz energy F per atom for the microstate σ is derived as follows:

$$F(V,T;\sigma) = E_c(V;\sigma) + F_{vib}(V,T;\sigma) + F_{el}(V,T;\sigma) \qquad 5.18$$

where the variable N has been abbreviated using

$$E_c(V;\sigma) = \frac{E_c(N,V;\sigma)}{N}. \qquad 5.19$$

We have

$$F_{vib}(V,T;\sigma) = -\frac{k_BT}{N}\ln\sum_{\mathbf{g}}\exp\left[-\beta\varepsilon_{\mathbf{g}}(N,V;\sigma)\right] \qquad 5.20$$

$$F_{el}(V,T;\sigma) = -\frac{k_BT}{N}\ln\sum_{\mathbf{n}}\exp[-\beta\varepsilon_{\mathbf{n}}(N,V;\sigma)] \qquad 5.21$$

The calculation of E_c is straightforward in most of the existing first-principles codes as discussed earlier.

5.2.2 Mermin statistics for the thermal electronic contribution

For the calculation of F_{el} in Eq. 5.21, the most computationally flexible approach is to use the Mermin statistics [28] by which

$$F_{el}(V,T;\sigma) = E_{el}(V,T;\sigma) - TS_{el}(V,T;\sigma) \qquad 5.22$$

where the bare electronic entropy S_{el} takes the form, after replacing the summation in Eq. 5.21 over the electronic states with integration,

$$S_{el}(V,T;\sigma) = -k_B\int n(\varepsilon,V;\sigma)\left\{\begin{array}{l} f(\varepsilon,V,T;\sigma)\ln f(\varepsilon,V,T;\sigma) \\ +[1-f(\varepsilon,V,T;\sigma)]\ln[1-f(\varepsilon,V,T;\sigma)] \end{array}\right\}d\varepsilon \qquad 5.23$$

by utilizing $n(\varepsilon,V;\sigma)$, the electronic density of states. Here f in Eq. 5.23 is the Fermi distribution and takes the form

$$f(\varepsilon, V, T; \sigma) = \frac{1}{\exp\left[\frac{\varepsilon - \mu(V, T; \sigma)}{k_B T}\right] + 1} \quad 5.24$$

Note that $\mu(V, T; \sigma)$ in Eq. 5.24 is the electronic chemical potential, which is strongly temperature dependent. At each given T, $\mu(V, T; \sigma)$ should be carefully calculated keeping the number of electrons unchanged in satisfying the following equation:

$$\int n(\varepsilon, V; \sigma) f(\varepsilon, V, T; \sigma) d\varepsilon = \int^{\varepsilon_F} n(\varepsilon, V; \sigma) d\varepsilon \quad 5.25$$

noting that ε_F is the Fermi energy calculated at 0 K. With respect to Eq. 5.22, the thermal electronic energy E_{el}, due to the thermal electron excitations, can be calculated using

$$E_{el}(V, T; \sigma) = \int n(\varepsilon, V; \sigma) f(\varepsilon, V, T; \sigma) \varepsilon d\varepsilon - \int^{\varepsilon_F} n(\varepsilon, V; \sigma) \varepsilon d\varepsilon \quad 5.26$$

At low temperatures, Eq. 5.22, Eq. 5.23, and Eq. 5.26 are reduced to

$$F_{el}(V, T; \sigma) = -\frac{1}{2}\lambda(V; \sigma) T^2 \quad 5.27$$

where λ is the so-called electronic specific heat coefficient, calculated as

$$\lambda(V; \sigma) = \frac{\pi^2}{3}(k_B)^2 n(\varepsilon_F, V; \sigma) \quad 5.28$$

where $n(\varepsilon_F, V; \sigma)$ is the electronic density of states at the Fermi level, and

$$E_{el}(V, T; \sigma) = \frac{1}{2}\lambda(V; \sigma) T^2 \quad 5.29$$

$$S_{el}(V, T; \sigma) = \lambda(V; \sigma) T \quad 5.30$$

From Eq. 5.29, one can easily derive the electronic contribution to the specific heat at low temperature as

$$C_{el}(V, T; \sigma) = \lambda(V; \sigma) T \quad 5.31$$

Usually, the dependence of $\lambda(V; \sigma)$ on V is weak. Therefore for a normal conductor (except for heavy fermion metals or superconductor related materials), at low temperatures, the electronic contribution to the heat capacity is linear with T. Equation 5.31 also indicates that, for insulators, from Eq. 5.28 the electronic contribution to the heat capacity is zero since for insulators $n(\varepsilon_F, V; \sigma) = 0$.

5.2.3 Vibrational contribution by phonon theory

Under the harmonic/quasi-harmonic approximation, lattice dynamics or phonon theory is currently the most established method. It truncates the interaction potential up to the second order. In such a case, $\varepsilon_{\mathbf{g}}(N, V; \sigma)$ in Eq. 5.16 can be expressed in terms of phonon frequency $\omega_j(V; \sigma)$ as follows:

$$\varepsilon_{\mathbf{g}}(N, V; \sigma) = \sum_{j=1}^{3N} (g_j + 1/2)\hbar\omega_j(V; \sigma) \quad 5.32$$

where the label **g** has the meaning $(g_1, g_2, \ldots, g_{3N})$ and the g_j can take any integer values from zero to infinity.

As a result, Eq. 5.20 is reduced to

$$F_{vib}(V, T; \sigma) = \frac{k_B T}{N} \sum_{j=1}^{3N} \ln\left[2\sinh\frac{\hbar\omega_j(V; \sigma)}{2k_B T}\right] \quad 5.33$$

or equivalently,

$$F_{vib}(V, T; \sigma) = k_B T \int_0^{\infty} \ln\left[2\sinh\frac{\hbar\omega}{2k_B T}\right] g(\omega, V; \sigma) d\omega \quad 5.34$$

where an integration has been used to replace the summation in Eq. 5.20 by means of introducing a function, $g(\omega, V; \sigma)$, named the phonon density of states (PDOS) whose integration over ω is equal to three per atom.

Accordingly, the formula to calculate the entropy becomes

$$S(V, T; \sigma) = S_{el}(V, T; \sigma) + k_B \int_0^{\infty} \left[\frac{(\hbar\omega/k_B T)}{e^{\hbar\omega/k_B T} - 1} - \ln\left(1 - e^{-\hbar\omega/k_B T}\right)\right] g(\omega, V; \sigma) d\omega \quad 5.35$$

the formula to calculate the internal energy becomes

$$U(V, T; \sigma) = E_{el}(V, T; \sigma) + k_B \int_0^{\infty} \left(\frac{\hbar\omega}{2} + \frac{\hbar\omega}{e^{\hbar\omega/k_B T} - 1}\right) g(\omega, V; \sigma) d\omega \quad 5.36$$

and the formula to calculate the heat capacity at constant volume becomes

$$C_V(V, T; \sigma) = \left(\frac{\partial E_{el}(V, T; \sigma)}{\partial T}\right)_V + k_B \int_0^{\infty} \frac{(\hbar\omega/k_B T)^2 e^{\hbar\omega/k_B T}}{\left(e^{\hbar\omega/k_B T} - 1\right)^2} g(\omega, V; \sigma) d\omega \quad 5.37$$

5.2.4 Debye–Grüneisen approximation to the vibrational contribution

Strictly speaking, the Debye theory is only accurate at very low temperatures. It assumes a parabolic type of dependence of the PDOS on the phonon frequency. This assumption is only correct at the scale of tens of degrees kelvin because at low temperatures only the low frequency acoustic phonons are activated, and so they play the major role in the parabolic type of PDOS found in the low frequency range, as shown in Figure 5.2. That is why there are two kinds of Debye temperature: the low temperature

Debye temperature and the high temperature Debye temperature. The low temperature Debye temperature can be strictly derived by fitting low temperature heat capacity data. The high temperature Debye temperature is usually a phenomenological fitting parameter.

The Debye model approximates the PDOS in Eq. 5.34 by

$$g(\omega, V;\sigma) = \begin{cases} 9\omega^2/\omega_D^3(V;\sigma), & \text{if } \omega \le \omega_D(V;\sigma) \\ 0, & \text{if } \omega > \omega_D(V;\sigma) \end{cases} \quad 5.38$$

where ω_D is the so-called Debye cutoff frequency, related to the Debye temperature Θ_D by

$$\Theta_D(V;\sigma) = \frac{\hbar\omega_D(V;\sigma)}{k_B} \quad 5.39$$

As the result, the vibrational contribution to the Helmholtz energy under the Debye approximation becomes

$$F_{vib}(V,T;\sigma) = \frac{9}{8}k_B\Theta_D(V;\sigma) + k_BT\left\{3\ln\left[1-\exp\left(-\frac{\Theta_D(V;\sigma)}{T}\right)\right] - D\left(\frac{\Theta_D(V;\sigma)}{T}\right)\right\} \quad 5.40$$

where $D(\Theta_D/T)$ is the Debye function given by $D(x) = 3/x^3\int_0^x\left\{t^3/[\exp(t)-1]\right\}dt$.

Under the Debye approximation, the formula for calculating the entropy becomes

$$S(V,T;\sigma) = S_{el}(V,T;\sigma) + k_B\left\{-3\ln\left[1-\exp\left(-\frac{\Theta_D(V;\sigma)}{T}\right)\right] + 4D\left(\frac{\Theta_D(V;\sigma)}{T}\right)\right\} \quad 5.41$$

the formula for calculating the internal energy becomes

$$U(V,T;\sigma) = E_c(V;\sigma) + E_{el}(V,T;\sigma) + U_{vib}(V,T;\sigma) \quad 5.42$$

where

$$U_{vib}(V,T;\sigma) = \frac{9}{8}k_B\Theta_D(V;\sigma) + 3k_BTD\left(\frac{\Theta_D(V;\sigma)}{T}\right) \quad 5.43$$

and the formula for calculating the heat capacity at constant volume becomes

$$C_V(V,T;\sigma) = \left(\frac{\partial E_{el}(V,T;\sigma)}{\partial T}\right)_V + 9k_B\left(\frac{T}{\Theta_D(V;\sigma)}\right)^3\int_0^{\Theta_D(V;\sigma)/T}\frac{x^4e^x}{(e^x-1)^2}dx \quad 5.44$$

Here it is noted that $\Theta_D(V;\sigma)$ is volume dependent, and this dependence is often written in terms of the Grüneisen constant:

$$\gamma_{vib}(V;\sigma) = -\frac{\partial\ln\Theta_D(V;\sigma)}{\partial\ln V} \quad 5.45$$

It has been found that the dependence of $\gamma_{vib}(V;\sigma)$ on V is usually weak and hence $\gamma_{vib}(V;\sigma)$ is often approximated as a constant. With $\gamma_{vib}(V;\sigma)$, the formula for calculating the pressure becomes

$$P(V,T;\sigma) = -\frac{\partial E_c(V;\sigma)}{\partial V} - \left(\frac{\partial F_{el}(V,T;\sigma)}{\partial V}\right)_T + \frac{\gamma_{vib}(V;\sigma)}{V} U_{vib}(V,T;\sigma) \quad 5.46$$

An important result of the Debye approximation is that when $T \to 0$, together with Eq. 5.31, the heat capacity in Eq. 5.44 is reduced to

$$C_V(V,T;\sigma) = \lambda(V;\sigma)T + \frac{12\pi^4}{5} k_B \left(\frac{T}{\Theta_D(V;\sigma)}\right)^3 \quad 5.47$$

This gives the Debye T^3 law when the thermal electron contribution $\lambda(V;\sigma)T$ is neglected. In the analysis of superconductor and heavy fermion materials, Eq. 5.47 is often rewritten as

$$\frac{C_V(V,T;\sigma)}{T} = \lambda(V;\sigma) + \frac{12\pi^4}{5} k_B \frac{T^2}{\Theta_D(V;\sigma)^3} \quad 5.48$$

which is more convenient for examining the heat capacity measured at extremely low temperatures.

5.2.5 System with multiple microstates (MMS model)

For a system consisting of multiple microstates, the total partition function is the summation over the partition functions of all microstates, Eq. 5.17:

$$Z(N,V,T) = \sum_\sigma w^\sigma Z(N,V,T;\sigma) \quad 5.49$$

where w^σ is the multiplicity of the microstate σ. It is immediately apparent that $x^\sigma = w^\sigma Z(N,V,T;\sigma)/Z(N,V,T)$ is the thermal population of the microstate σ. Furthermore, with $F = -k_B T \ln Z$, one obtains

$$\begin{aligned} F(N,V,T) &= -k_B T \sum_\sigma x^\sigma \ln Z(N,V,T;\sigma) + k_B T \left[\sum_\sigma x^\sigma \ln Z(N,V,T;\sigma) - x^\sigma \ln Z(N,V,T)\right] \\ &= \sum_\sigma x^\sigma N F(V,T;\sigma) + k_B T \sum_\sigma x^\sigma \ln(x^\sigma / w^\sigma) \end{aligned} \quad 5.50$$

Equation 5.50 relates the total Helmholtz energy, *F(N,V,T)*, of a system with mixing among multiple microstates and the Helmholtz energy, $F(V,T;\sigma)$, of the individual microstates. An important result of Eq. 5.50 is the configurational entropy due to the mixing among multiple microstates, called the microstate configurational entropy) (MCE) in this book,

$$\Delta S_f(N,V,T) = -k_B \sum_\sigma w^\sigma \left[x^\sigma / w^\sigma \ln(x^\sigma / w^\sigma)\right] \quad 5.51$$

which makes the entropy of a system with mixing among multiple microstates equal to

$$S(N,V,T) = \Delta S_f(N,V,T) + \sum_\sigma x^\sigma N S(V,T;\sigma) \quad 5.52$$

Similarly, one can obtain the heat capacity at constant volume of a system with mixing among multiple microstates as

$$C_V(V,T) = \Delta C_f(V,T) + \sum_\sigma x^\sigma C_V(V,T;\sigma) \tag{5.53}$$

where the configurational contribution to the heat capacity due to the mixing among multiple microstates is

$$\Delta C_f(V,T) = \frac{1}{k_B T^2}\left\{\sum_\sigma x^\sigma [E(V,T;\sigma)]^2 - \left[\sum_\sigma x^\sigma E(V,T;\sigma)\right]^2\right\} \tag{5.54}$$

Moreover, the isothermal bulk modulus of a system with mixing among multiple microstates can be also computed similarly as

$$B_T(V,T) = \Delta B_f(V,T) + \sum_\sigma x^\sigma B_T(V,T;\sigma) \tag{5.55}$$

with

$$\Delta B_f(V,T) = \frac{V}{k_B T}\left\{\left[\sum_\sigma w^\sigma x^\sigma P(V,T;\sigma)\right]^2 - \sum_\sigma w^\sigma x^\sigma [P(V,T;\sigma)]^2\right\} \tag{5.56}$$

This multiple microstate model (MMS model) is used in Chapter 9 to quantitatively predict thermal expansion anomalies.

5.3 Quantum theory for the motion of electrons

5.3.1 Schrödinger equation

The Schrödinger equation is typically written as follows:

$$i\hbar \frac{\partial}{\partial t}\Psi(\mathbf{X},t) = \widehat{H}\Psi(\mathbf{X},t) \tag{5.57}$$

where Ψ is the wave function of the system, $\hbar$ is the reduced Planck constant, $\mathbf{X}$ is an abbreviation of the space coordinates and spin states of the multiple particle system, $\widehat{H}$ is the energy operator the called, Hamiltonian. When $\widehat{H}$ is independent of time t, one can separate the coordinate $\mathbf{X}$ from the time t in finding the solution of Eq. 5.57 by writing

$$\Psi(\mathbf{X},t) = \Psi(\mathbf{X})\Psi(t) \tag{5.58}$$

by which the stationary solutions of Eq. 5.57 can be expressed through setting

$$\frac{i\hbar \frac{\partial}{\partial t}\Psi(t)}{\Psi(t)} = \frac{\widehat{H}\Psi(\mathbf{X})}{\Psi(\mathbf{X})} = E \tag{5.59}$$

resulting in

$$\widehat{H}\Psi(\mathbf{X}) = E\Psi(\mathbf{X}) \tag{5.60}$$

$$\Psi(t) = \exp\left(-i\frac{E}{\hbar}t\right) \tag{5.61}$$

Note that $E/\hbar$ is the frequency of the de Broglie matter wave.

For any trial function $\Lambda(\mathbf{X})$ (in the Hilbert space) for $\Psi(\mathbf{X})$, the variational principle tells us that the energy of the system always has a lower bound corresponding to a ground state with energy E_0, as

$$E = \frac{\displaystyle\int \Lambda^*(\mathbf{X})\widehat{H}\Lambda(\mathbf{X})d\mathbf{X}}{\displaystyle\int \Lambda^*(\mathbf{X})\Lambda(\mathbf{X})d\mathbf{X}} \geq E_0 \tag{5.62}$$

where $\Lambda^*(\mathbf{X})$ represents the complex conjugate of $\Lambda(\mathbf{X})$, resulting in

$$\frac{\delta E}{\delta \Psi} = 0 \tag{5.63}$$

which is known as the Rayleigh–Ritz variational principle.

5.3.2 Born–Oppenheimer approximation

For a time independent atomic system, it is often accurate enough to write $\widehat{H}$ in Eq. 5.57 or Eq. 5.60 in terms of the electron coordinates $\mathbf{r}$ and nuclei coordinates $\mathbf{R}$,

$$\begin{aligned}\widehat{H}(\mathbf{r},\mathbf{R}) = &\sum_i -\frac{\hbar^2}{2m_e}\nabla_i^2 + \sum_I -\frac{\hbar^2}{2M_I}\nabla_I^2 \\ &-\sum_{I,i}\frac{Z_I e^2}{|\mathbf{R}_I - \mathbf{r}_i|} + \frac{1}{2}\sum_{i\neq j}\frac{e^2}{|\mathbf{r}_i - \mathbf{r}_j|} + \frac{1}{2}\sum_{I,J}\frac{Z_I Z_J e^2}{|\mathbf{R}_I - \mathbf{R}_J|}\end{aligned} \tag{5.64}$$

where e represents the electron charge and i and j label the electrons, I and J the atomic nuclei, Z_I the atomic nuclear charge number of atom I, m_e the electron mass, M_I the mass of atomic nuclei I, ∇_i^2 the Laplace operator for electron i, and ∇_I^2 the Laplace operator for atomic nuclei I, noting that

$$\nabla^2 = \frac{\partial^2}{\partial x^2} + \frac{\partial^2}{\partial y^2} + \frac{\partial^2}{\partial z^2} \tag{5.65}$$

with respect to the Cartesian axes x, y, and z.

Considering the fact that the electron mass is two thousand times smaller than the mass of the atomic nuclei, implying that the motions of the electrons are much faster than those of the atomic nuclei, Born and Oppenheimer proposed that the wave function of the whole system can be simply approximated as the product of the electron wave function $\Psi(\mathbf{r})$ and the atomic nuclei wave function $\Psi(\mathbf{R})$ as

$$\Psi(\mathbf{r},\mathbf{R}) = \Psi(\mathbf{r}) \times \Psi(\mathbf{R}) \qquad 5.66$$

With the auxiliary approximation of neglecting the dynamic coupling between the motions of electrons and atomic nuclei, the Schrödinger equation for the motion of the electrons becomes

$$\widehat{H}_e\Psi(\mathbf{r}) = E(\mathbf{R})\Psi(\mathbf{r}) \qquad 5.67$$

where

$$\widehat{H}_e = \sum_i -\frac{\hbar^2}{2m_e}\nabla_i^2 - \sum_{I,i}\frac{Z_I e^2}{|\mathbf{R}_I - \mathbf{r}_i|} + \frac{1}{2}\sum_{i\neq j}\frac{e^2}{|\mathbf{r}_i - \mathbf{r}_j|} \qquad 5.68$$

and the Schrödinger equation for the motion of the atomic nuclei becomes

$$\widehat{H}_N\Psi(\mathbf{R}) = \varepsilon_{\mathbf{n}}\Psi(\mathbf{R}) \qquad 5.69$$

where

$$\widehat{H}_N = \sum_I -\frac{\hbar^2}{2M_I}\nabla_I^2 + \frac{1}{2}\sum_{I,J}\frac{Z_I Z_J e^2}{|\mathbf{R}_I - \mathbf{R}_J|} + E(\mathbf{R}) \qquad 5.70$$

with

$$E(\mathbf{R}) = \frac{\int \Psi^*(\mathbf{r})\widehat{H}_e\Psi(\mathbf{r})d\mathbf{r}}{\int \Psi^*(\mathbf{r})\Psi(\mathbf{r})d\mathbf{r}} \qquad 5.71$$

where $\Psi^*(\mathbf{r})$ represents the complex conjugate of $\Psi(\mathbf{r})$.

5.3.3 Hartree–Fock approximation to solve the Schrödinger equation

It was Hartree who first assumed that the electron wave function in Eq. 5.67 can be expressed as a product of a collection of N independent one-electron wave functions, $\Phi_i(\mathbf{r},\mathbf{s})$ where $i = 1, 2, \ldots, N$, N being the number of electrons in a system, in terms of its space coordinate $\mathbf{r}$ and spin state $\mathbf{s}$. After that, Fock modified the Hartree approximation by considering the fact that the wave function of a multi-fermionic system should satisfy anti-symmetry requirements and thus the Pauli exclusion principle that the total wave function changes sign upon the exchange of fermions. Accordingly, the wave function of an N- electron system under the Hartree–Fock approximation is expressed as the Slater determinant [29]

$$\Psi^{Slater}(\mathbf{r},\mathbf{s}) = \frac{1}{\sqrt{N!}}\begin{vmatrix} \Phi_1(\mathbf{r}_1,\mathbf{s}_1) & \Phi_1(\mathbf{r}_2,\mathbf{s}_2) & \ldots & \Phi_1(\mathbf{r}_N,\mathbf{s}_N) \\ \Phi_2(\mathbf{r}_1,\mathbf{s}_1) & \Phi_2(\mathbf{r}_2,\mathbf{s}_2) & \ldots & \Phi_2(\mathbf{r}_N,\mathbf{s}_N) \\ \vdots & \vdots & \vdots & \vdots \\ \Phi_N(\mathbf{r}_1,\mathbf{s}_1) & \Phi_N(\mathbf{r}_2,\mathbf{s}_2) & \ldots & \Phi_N(\mathbf{r}_N,\mathbf{s}_N) \end{vmatrix} \qquad 5.72$$

For brevity, one can use atomic units, thus setting $\hbar = 1$, $e = 1$, and $m_e = 1$ in Eq. 5.68, so that

$$\widehat{H}_e = \sum_i h(i) + \frac{1}{2}\sum_{i \neq j} \frac{1}{r_{ij}} \tag{5.73}$$

where $r_{ij} = |\mathbf{r}_i - \mathbf{r}_j|$, and

$$h(i) = -\frac{1}{2}\nabla_i^2 - \sum_n \frac{Z_n}{|\mathbf{r}_i - \mathbf{R}_n|} \tag{5.74}$$

Accordingly, the total energy of the system is expressed as

$$E = \sum_i \langle \Phi_i | h(i) | \Phi_i \rangle + \frac{1}{2}\sum_{ij} [J_{ij} - K_{ij}] \tag{5.75}$$

where J_{ij} is called the Coulomb–Hartree term:

$$\begin{aligned} J_{ij} &= \left\langle \Phi_i(\mathbf{r}_1, \mathbf{s}_i)\Phi_i(\mathbf{r}_1, \mathbf{s}_i)\frac{1}{r_{12}}\Phi_j(\mathbf{r}_2, \mathbf{s}_j)\Phi_j(\mathbf{r}_2, \mathbf{s}_j) \right\rangle \\ &= \iint \Phi_i^*(\mathbf{r}_1, \mathbf{s}_i)\Phi_i(\mathbf{r}_1, \mathbf{s}_i)\frac{1}{r_{12}}\Phi_j^*(\mathbf{r}_2, \mathbf{s}_j)\Phi_j(\mathbf{r}_2, \mathbf{s}_j)d\mathbf{r}_1 d\mathbf{r}_2 \end{aligned} \tag{5.76}$$

and where K_{ij} is called the exchange term:

$$\begin{aligned} K_{ij} &= \left\langle \Phi_i(\mathbf{r}_1, \mathbf{s}_i)\Phi_j(\mathbf{r}_1, \mathbf{s}_j)\frac{1}{r_{12}}\Phi_j(\mathbf{r}_2, \mathbf{s}_j)\Phi_i(\mathbf{r}_2, \mathbf{s}_i) \right\rangle \\ &= \delta_{\mathbf{s}_i, \mathbf{s}_j} \iint \Phi_i^*(\mathbf{r}_1, \mathbf{s}_i)\Phi_j(\mathbf{r}_1, \mathbf{s}_j)\frac{1}{r_{12}}\Phi_j^*(\mathbf{r}_2, \mathbf{s}_j)\Phi_i(\mathbf{r}_2, \mathbf{s}_i)d\mathbf{r}_1 d\mathbf{r}_2 \end{aligned} \tag{5.77}$$

Here $\delta_{\mathbf{s}_i, \mathbf{s}_j = 1}$ if spins $\mathbf{s}_i$ and $\mathbf{s}_j$ point in the same direction and $\delta_{\mathbf{s}_i, \mathbf{s}_j} = 0$ if spins $\mathbf{s}_i$ and $\mathbf{s}_j$ point in the opposite direction.

By utilizing the variational condition $\delta E/\delta\Phi_i = 0$, one obtains

$$[h(\mathbf{r}_1) + J(\mathbf{r}_1) - K(\mathbf{r}_1)]\Phi_i(\mathbf{r}_1, \mathbf{s}_i) = \varepsilon_i(\mathbf{s}_i)\Phi_i(\mathbf{r}_1, \mathbf{s}_i) \tag{5.78}$$

where $\varepsilon_i(\mathbf{s}_i)$ is called the one-electron energy, and

$$J(\mathbf{r}_1) = \int \frac{1}{r_{12}} \sum_j \Phi_j^*(\mathbf{r}_2, \mathbf{s}_j)\Phi_j(\mathbf{r}_2, \mathbf{s}_j)d\,\mathbf{r}_2 = \int \frac{\rho(\mathbf{r}_2)}{r_{12}} d\mathbf{r}_2 \tag{5.79}$$

with ρ the electronic charge density, whose expression is

$$\rho(\mathbf{r}) = \sum_j \Phi_j^*(\mathbf{r}, \mathbf{s}_j)\Phi_j(\mathbf{r}, \mathbf{s}_j) \tag{5.80}$$

The third term on the left in Eq. 5.78 is given by

$$K(\mathbf{r}_1)\Phi_i(\mathbf{r}_1, \mathbf{s}_i) = \delta_{\mathbf{s}_i, \mathbf{s}_j} \int \frac{1}{r_{12}} \sum_j \Phi_j^*(\mathbf{r}_2, \mathbf{s}_j)\Phi_i(\mathbf{r}_2, \mathbf{s}_i)\Phi_j(\mathbf{r}_1, \mathbf{s}_j)d\mathbf{r}_2 \tag{5.81}$$

It should be especially noted here that to solve the Hartree–Fock equation, Eq. 5.78, the most time consuming part is due to the non-local exchange term $K(\mathbf{r}_1)$, because the one-electron wave function $\Phi_i(\mathbf{r}_1, \mathbf{s}_i)$ being evaluated is also contained in the expression on the left-hand side of Eq. 5.78 in view of Eq. 5.81.

The configurational interaction method is a generalization of the Hartree–Fock approximation. In such a case, Y, the number of one-electron wave functions, can be larger than the number of electrons, N, in the system. Accordingly, from the Y one-electron wave functions Φ_y, $y = 1, 2, \ldots, Y$, one can build a number M of Stater determinants by combinatorial mathematics such that the maximum of M can be

$$M = \binom{Y}{N} \tag{5.82}$$

As a result, the wave function of an N electron system becomes a combination of the M Stater determinants:

$$\Psi(\mathbf{r}, \mathbf{S}) = \sum_{\sigma=1}^{M} C^{\sigma} \Psi_{\sigma}^{Slater}(\mathbf{r}, \mathbf{s}_{\sigma}) \tag{5.83}$$

where the coefficients are found from the multiple linear equation

$$\begin{Bmatrix} H_{1,1} & H_{1,2} & \ldots & H_{1,N} \\ H_{2,1} & H_{2,2} & \ldots & H_{2,N} \\ \vdots & \vdots & \vdots & \vdots \\ H_{N,1} & H_{N,2} & \ldots & H_{N,N} \end{Bmatrix} \begin{Bmatrix} C_1^j \\ C_2^j \\ \vdots \\ C_N^j \end{Bmatrix} = E_j \begin{Bmatrix} C_1^j \\ C_2^j \\ \vdots \\ C_N^j \end{Bmatrix} \tag{5.84}$$

The matrix elements in Eq. 5.84 are determined by the integral

$$H_{\mu\nu} = \left\langle \Psi_{\mu}^{Slater}(\mathbf{r}, \mathbf{s}_{\mu}) | \widehat{H}_e | \Psi_{\nu}^{Slater}(\mathbf{r}, \mathbf{s}_{\nu}) \right\rangle \tag{5.85}$$

5.3.4 Density functional theory (DFT) and zero temperature Kohn–Sham equations

The density functional theory assumes that the properties of a system are solely dictated by its electronic density distribution (or, equally, its charge density), $\rho(\vec{r})$, in real space. This is to say that for an arbitrary $\rho(\vec{r})$, the total energy of the system, E, is always larger or equal to a value, E_0, called the ground state energy:

$$E[\rho] \geq E_0 \tag{5.86}$$

In terms of variational principle, Eq. 5.86 is equivalent to

$$\frac{\delta E[\rho]}{\delta \rho} = 0 \tag{5.87}$$

Kohn and Sham [8] proposed writing the total energy $E[\rho]$ as

$$E[\rho] = T[\rho] + V_{ext}[\rho] + V_H[\rho] + V_{xc}[\rho_{\downarrow}, \rho_{\uparrow}] \tag{5.88}$$

where $T[\rho]$ represents the kinetic energy of the system, $V_{ext}[\rho] = \int V_{ext}(\vec{r})\rho(\vec{r})d\vec{r}$ is the external potential acting on the system, $V_H[\rho] = \frac{e^2}{2}\iint \left(\rho(\vec{r})\rho(\vec{r}')/|\vec{r} - \vec{r}'|\right)d\vec{r}\,d\vec{r}'$ is the Hartree energy, and $V_{xc}[\rho_\downarrow, \rho_\uparrow]$ is the so-called exchange-correlation energy, with $\rho = \rho_\downarrow + \rho_\uparrow$ where $\rho_\downarrow$ and $\rho_\uparrow$represent the charge densities of electrons with spin down and spin up, respectively. Using

$$\rho(\vec{r}) = \sum_{i=1}^{N}\sum_{s=1}^{2}\left|\Phi_i(\vec{r}, s)\right|^2 \tag{5.89}$$

together with the variational principle of Eq. 5.87, one can obtain the one-electron Schrödinger equation

$$\left[-\frac{\hbar^2}{2m_e}\nabla^2 + v_{eff}\left(\vec{r};\rho_\downarrow,\rho_\uparrow\right)\right]\Phi_i\left(\vec{r}, s\right) = \varepsilon_i\Phi_i\left(\vec{r}, s\right) \tag{5.90}$$

$$v_{eff}\left(\vec{r};\rho_\downarrow,\rho_\uparrow\right) = V_{ext}\left(\vec{r}\right) + e^2\int\frac{\rho\left(\vec{r}'\right)}{|\vec{r} - \vec{r}'|}d\vec{r}' + \frac{\delta V_{xc}[\rho_\downarrow,\rho_\uparrow]}{\delta\rho} \tag{5.91}$$

so that the total energy is obtained as

$$E[\rho] = \sum_{i=1}^{N}\varepsilon_i - V_H[\rho] + V_{xc}[\rho] - \int\frac{\delta V_{xc}[\rho_\downarrow,\rho_\uparrow]}{\delta\rho}\rho\left(\vec{r}\right)d\vec{r} \tag{5.92}$$

The major challenge within DFT is that an accurate formulation of the exchange-correlation energy is unknown. Except for a uniform electron gas, no exact analytical form for the exchange-correlation energy has yet been obtained. Therefore approximations must be made for the exchange-correlation energy in calculating a realistic system. At present, the two most popular approximations are the local density approximation (LDA) [30] and the generalized gradient approximation (GGA) [31, 32].

The local density approximation (LDA)) takes the exchange-correlation energy to be the same as that for a locally uniform electron gas. In this case one can write V_{xc} as

$$V_{xc}[\rho_\downarrow,\rho_\uparrow] = \int\varepsilon_{xc}(\rho_\downarrow,\rho_\uparrow)\rho\left(\vec{r}\right)d\vec{r} \tag{5.93}$$

Although this approximation is extremely simple, it works reasonably well for many systems. The only remaining problem is to find an approximate solution to $\varepsilon_{xc}(\rho_\downarrow,\rho_\uparrow)$. One of the most commonly employed parameterized expressions for $\varepsilon_{xc}(\rho)$ is that of Perdew and Zunger [30].

Many modern DFT codes use the more advanced generalized gradient approximation (GGA) [31, 32] to the exchange-correlation energy to improve accuracy for certain physical properties. As the LDA approximates the energy of the true density by the energy of a local constant density, it fails in situations where the density undergoes rapid changes, such as in molecules. An improvement to this can be made by considering the gradient of the electron density. Symbolically, this can be written as

$$V_{xc}[\rho_\downarrow,\rho_\uparrow] = \int \varepsilon_{xc}(\rho_\downarrow,\rho_\uparrow,\nabla\rho_\downarrow,\nabla\rho_\uparrow)\rho(\vec{r})d\vec{r} \qquad 5.94$$

The most commonly used GGA is that due to Perdew *et al.* [31, 32].

5.3.4.1 Solving the Kohn–Sham equations for a solid

For a solid, Eq. 5.90 is still a mathematical challenge, with an infinite number of one-electron wave functions to be solved, and therefore cannot be solved directly in real space. To reduce the dimension of the problem, one can choose to solve the equation at a specific $\vec{k}$ point in the reciprocal space. According to Blöch's theorem, the wave function for a solid can be written as the product of a wave-like part, $\exp(i\vec{k}\cdot\vec{r})$, and a cell-periodic part, $u_j(\vec{r};\vec{k})$:

$$\Phi_i(\vec{r}, s; \vec{k}) = \exp(i\vec{k}\cdot\vec{r})u_i(\vec{r};\vec{k})|s|\rangle \qquad 5.95$$

where $u_j(\vec{r};\vec{k})$ can be expressed as a sum of a finite number of plane waves whose wave vectors are reciprocal lattice vectors of the crystal:

$$u_j(\vec{r};\vec{k}) = \sum_G C_j(\vec{k}+\vec{G})\exp(i\vec{G}\cdot\vec{r}) \qquad 5.96$$

so that

$$\sum_{\vec{G}'}\left[\frac{\hbar^2}{2m_e}, |\vec{k}+\vec{G}|^2\delta_{\vec{G}\vec{G}'} + v_{eff}(\vec{G}-\vec{G}')\right]C_j(\vec{k}+\vec{G}') = \varepsilon_j(\vec{k})C_j(\vec{k}+\vec{G}) \qquad 5.97$$

where the band index j is used to number the eigenenergies $\varepsilon_j(\vec{k})$ and the eigenvectors $C_j(\vec{k}+\vec{G})$ at a given $\vec{k}$. The number of plane waves is determined by the following equation:

$$\frac{\hbar^2}{2m_e}|\vec{k}+\vec{G}|^2 \le E_{cut} \qquad 5.98$$

where E_{cut} is the energy cutoff.

Utilizing the obtained wave functions, the charge density can be calculated using Brillouin zone integration:

$$\rho(\vec{r}) = \frac{\Omega}{(2\pi)^3}\int_{BZ}\sum_j \Phi_j^*(\vec{r};\varepsilon_j(\vec{k}))\Phi_j(\vec{r};\varepsilon_j(\vec{k}))\overline{\Theta}(\varepsilon_j(\vec{k})-\varepsilon_F)d\vec{k} \qquad 5.99$$

where

$$\overline{\Theta}\left(\varepsilon_j\left(\vec{k}\right)-\varepsilon_F\right)=\left\{\begin{array}{ll}1, & \text{if} \quad \varepsilon_j\left(\vec{k}\right)\leq\varepsilon_F; \\ 0, & \text{if} \quad \varepsilon_j\left(\vec{k}\right)>\varepsilon_F\end{array}\right\}, \qquad \int_{\Omega}\rho\left(\vec{r}\right)d\vec{r}=N \tag{5.100}$$

where the parameter ε_F is needed to make the integration over the charge density within the primitive unit cell equal to the number of electrons, N, in the primitive unit cell. Numerically, the integration can be approximated by summation over a discrete k-mesh as

$$\rho\left(\vec{r}\right)=\sum_{i=1}^{N_{\mathrm{BZ}}}\sum_{j}\Phi_j^*\left(\vec{r};\varepsilon_j\left(\vec{k}_i\right)\right)\Phi_j\left(\vec{r};\varepsilon_j\left(\vec{k}_i\right)\right)\overline{\Theta}\left(\varepsilon_j\left(\vec{k}_i\right)-\varepsilon_F\right) \tag{5.101}$$

where N_{BZ} represents the number of $\vec{k}$ points in the first Brillouin zone in the k-mesh. When a solid possesses symmetry, the summation in the above equation can be further reduced to a summation over the irreducible Brillouin zone (IBZ)):

$$\rho\left(\vec{r}\right)=\sum_{i=1}^{N_{\mathrm{IBZ}}}w\left(\vec{k}_i\right)\sum_{j}\Phi_j^*\left(\vec{r};\varepsilon_j\left(\vec{k}_i\right)\right)\Phi_j\left(\vec{r};\varepsilon_j\left(\vec{k}_i\right)\right)\overline{\Theta}\left(\varepsilon_j\left(\vec{k}_i\right)-\varepsilon_F\right) \tag{5.102}$$

where $w\left(\vec{k}_i\right)$ is a weight factor that represents the number of points that are equivalent to $\vec{k}_i$ by space group symmetry.

5.4 Lattice dynamics

5.4.1 Quantum theory for motion of atomic nuclei

For convenience of discussion, the following notation convention is used: α and β label the Cartesian axes x, y, and z, j and k label atoms in the primitive unit cell, m_j is the atomic mass of the jth atom in the primitive unit cell, $\mathbf{r}(j)$ is the position of the jth atom in the primitive unit cell, P and Q are the indices of the primitive unit cell in the system, $\mathbf{R}(P)$ is the position of the Pth primitive unit cell in the system, and V is the average volume of the primitive unit cell.

The quantum theory for the motion of atomic nuclei replicates closely the quantum theory for the motion of electrons. That is, the wave function $\Phi(\mathbf{R})$is solved for the motions of the atomic nuclei for a Schrödinger equation whose potential is the total electronic energy $E(\mathbf{R})$ derived from Eq. 5.71. The symbol $E(\mathbf{R})$ is replaced by $E(\mathbf{R}+\mathbf{u};\sigma)$ to represent the static total electronic energy, with $\mathbf{R}$ representing the static equilibrium positions of the atomic nuclei, $\mathbf{u}$ the displacements of atomic nuclei away from their static equilibrium positions, and σ the additional degrees of freedom such as electronic states. The Schrödinger equation for the motion of atomic nuclei is then

$$H_N\Psi(\mathbf{R}+\mathbf{u})=\varepsilon_{\mathbf{g}}(N,V;\sigma)\Psi(\mathbf{R}+\mathbf{u}) \tag{5.103}$$

where

$$H_N=K_N+E(\mathbf{R}+\mathbf{u};\sigma) \tag{5.104}$$

with K_N representing the kinetic energy operator:

$$K_N = -\frac{1}{2}\sum_{P}^{N_c}\sum_{j}^{N_a}\sum_{\alpha}^{3}\frac{\hbar^2}{m_j}\frac{\partial^2}{\partial u_\alpha^2(j;P)} \quad 5.105$$

where N_c is the number of primitive unit cells in the system, N_a is the number of atoms in the primitive unit cell, $\hbar$ is the Planck constant, $u_\alpha(j;P)$ represents the αth Cartesian component of $\mathbf{u}$ for the atom at the jth lattice site in the Pth primitive unit cell in the system.

The harmonic approximation [23, 33] truncates the term $E(\mathbf{R}+\mathbf{u};\sigma)$ to the second order in its Taylor series:

$$E(\mathbf{R}+\mathbf{u};\sigma) = \frac{1}{2}\sum_{P,Q}^{N_c}\sum_{j,k}^{N_a}\sum_{\alpha,\beta}^{3}\Phi_{\alpha\beta}^{jk}(P,Q)u_\alpha(j;P)u_\beta(k;Q) \quad 5.106$$

where $\Phi_{\alpha\beta}^{jk}$ is the real-space interatomic force constant. With the approximation of Eq. 5.106, it can be demonstrated that finding the solution of Eq. 5.103 is equivalent to finding the vibrational frequencies of a classical system with N=N_cN_a particles for small mechanical oscillations.

Let us rewrite Eq. 5.106 as

$$E(\mathbf{R}+\mathbf{u};\sigma) = \frac{1}{2}\sum_{P,Q}^{N_c}\sum_{j,k}^{N_a}\sum_{\alpha,\beta}^{3}C_{\alpha\beta}^{jk}(P,Q)w_\alpha(j;P)w_\beta(k;Q) \quad 5.107$$

where

$$C_{\alpha\beta}^{jk}(P,Q) = \frac{\Phi_{\alpha\beta}^{jk}(P,Q)}{\sqrt{m_j m_k}} \quad 5.108$$

$$w_\alpha(j;P) = \sqrt{m_j}u_\alpha(j;P) \quad 5.109$$

Accordingly, the kinetic energy operator in Eq. 5.105 becomes

$$K_N = -\frac{1}{2}\sum_{P}^{N_c}\sum_{j}^{N_a}\sum_{\alpha}^{3}\hbar^2\frac{\partial^2}{\partial w_\alpha^2(j;P)} \quad 5.110$$

5.4.2 Normal coordinates, eigenenergies, and phonons

One way to simplify the solution to the Schrödinger equation for the motion of atomic nuclei is to follow the study of the vibrations of atoms in crystal lattice dynamics. The essential step in lattice dynamics is to transform the problem of the correlated motions of $3N$ particles into a problem of $3N$ independent harmonics. For this purpose, one can introduce a set of new coordinates ζ_i (i = 1, 2, . . ., $3N$), known as normal coordinates, by the transformation

$$\zeta_i = \sum_{P}^{N_c}\sum_{j}^{N_a}\sum_{\alpha}^{3}e_\alpha^i(j;P)w_\alpha(j;P) \quad 5.111$$

where $e^i_\alpha(j;P)$ is the transformation coefficient, which can be determined by solving $3N$ simultaneous equations

$$\omega^2 e_\alpha(j;P) = \sum_Q^{N_c}\sum_k^{N_a}\sum_\beta^{3} C^{jk}_{\alpha\beta}(P,Q)e_\beta(k;Q) \qquad 5.112$$

Here ω^2 is to be determined such that one can find $3N$ $e_\alpha(j;P)$ which are not all zero. The equations are linear and homogeneous. Following the basic theorem in linear algebra that, to find the non-zero solutions of the equations, the determinant formed by $C^{jk}_{\alpha\beta}(P,Q)$ must equal zero, we obtain

$$\left|C^{jk}_{\alpha\beta}(P,Q) - \delta_{\alpha\beta}\delta_{jk}\delta_{PQ}\omega^2\right| = 0 \qquad 5.113$$

where δ is the Kronecker delta symbol. Since Eq. 5.113 is an equation with $3N$ degrees, one can always find $3N$ values of ω_i^2 ($i = 1, \ldots, 3N$). Each of the ω_i^2 yields a set of $e^i_\alpha(j;P)$ which can be chosen such that

$$\sum_Q^{N_c}\sum_k^{N_a}\sum_\beta^{3} e^i_\beta(k;Q)e^l_\beta(k;Q) = \delta_{il} \qquad 5.114$$

where δ_{il} represents the Kronecker delta symbol and

$$\sum_{i=1}^{3N} e^i_\alpha(j;P)e^i_\beta(k;Q) = \delta_{\alpha\beta}\delta_{jk}\delta_{PQ} \qquad 5.115$$

Then with the normal coordinates defined in Eq. 5.111 and utilizing Eq. 5.114, $w_\alpha(j;P)$ defined in Eq. 5.109 is obtained as

$$w_\alpha(j;P) = \sum_{i=1}^{3N} e^i_\alpha(j;P)\zeta_i \qquad 5.116$$

With this equation, Eq. 5.107 is simplified by the following process:

$$\begin{aligned} E(\mathbf{R}+\mathbf{u};V,\sigma) &= \frac{1}{2}\sum_{P,Q}^{N_c}\sum_{j,k}^{N_a}\sum_{\alpha,\beta}^{3} C^{jk}_{\alpha\beta}(P,Q)w_\alpha(j;P)w_\beta(k;Q) \\ &= \frac{1}{2}\sum_{i=1}^{3N}\zeta_i\sum_P^{N_c}\sum_j^{N_a}\sum_\alpha^{3} e^i_\alpha(j;P)\sum_{i'=1}^{3N}\zeta_{i'}\sum_Q^{N_c}\sum_k^{N_a}\sum_\beta^{3} C^{jk}_{\alpha\beta}(P,Q)e^{i'}_\beta(k;Q) \\ &= \frac{1}{2}\sum_{i=1}^{3N}\zeta_i\sum_P^{N_c}\sum_j^{N_a}\sum_\alpha^{3} e^i_\alpha(j;P)\sum_{i'=1}^{3N}\zeta_{i'}\omega_{i'}^2 e^{i'}_\alpha(j;P) \\ &= \frac{1}{2}\sum_{i=1}^{3N}\sum_{i'=1}^{3N}\zeta_i\zeta_{i'}\omega_{i'}^2\sum_P^{N_c}\sum_j^{N_a}\sum_\alpha^{3} e^i_\alpha(j;P)e^{i'}_\alpha(j;P) \\ &= \frac{1}{2}\sum_{i=1}^{3N}\sum_{i'=1}^{3N}\zeta_i\zeta_{i'}\omega_{i'}^2\delta_{ii'} \\ &= \frac{1}{2}\sum_{i=1}^{3N}\zeta_i^2\omega_i^2 \end{aligned} \qquad 5.117$$

noting that in the above process Eq. 5.112 and Eq. 5.114 have been utilized.

Furthermore, using Eq. 5.116, the kinetic energy operator in Eq. 5.110 can be transformed as follows:

$$
\begin{aligned}
K_N &= -\frac{1}{2}\sum_P^{N_c}\sum_j^{N_a}\sum_\alpha^{3}\hbar^2\frac{\partial^2}{\partial w_\alpha^2(j;P)} \\
&= -\frac{1}{2}\sum_P^{N_c}\sum_j^{N_a}\sum_\alpha^{3}\hbar^2\frac{\partial}{\partial w_\alpha(j;P)}\sum_{i=1}^{3N}\frac{\partial \zeta_i}{\partial w_\alpha(j;P)}\frac{\partial}{\partial \zeta_i} \\
&= -\frac{1}{2}\sum_P^{N_c}\sum_j^{N_a}\sum_\alpha^{3}\hbar^2\frac{\partial}{\partial w_\alpha(j;P)}\sum_{i=1}^{3N}\frac{\partial}{\partial \zeta_i}e_\alpha^i(j;P) \\
&= -\frac{1}{2}\sum_P^{N_c}\sum_j^{N_a}\sum_\alpha^{3}\hbar^2\sum_{i'=1}^{3N}\frac{\partial}{\partial \zeta_{i'}}e_\alpha^{i'}(j;P)\sum_{i=1}^{3N}\frac{\partial}{\partial \zeta_i}e_\alpha^i(j;P) \\
&= -\frac{1}{2}\sum_{i'=1}^{3N}\hbar^2\frac{\partial}{\partial \zeta_{i'}}\sum_{i=1}^{3N}\frac{\partial}{\partial \zeta_i}\sum_P^{N_c}\sum_j^{N_a}\sum_\alpha^{3}e_\alpha^{i'}(j;P)e_\alpha^i(j;P) \\
&= -\frac{1}{2}\sum_{i'=1}^{3N}\hbar^2\frac{\partial}{\partial \zeta_{i'}}\sum_{i=1}^{3N}\frac{\partial}{\partial \zeta_i}\delta_{ii'} \\
&= -\frac{1}{2}\sum_{i=1}^{3N}\hbar^2\frac{\partial^2}{\partial \zeta_i^2}
\end{aligned}
\qquad 5.118
$$

noting that in the above process Eq. 5.112 and Eq. 5.114 are utilized again.

As a result, under the harmonic approximation, the Hamiltonian in Eq. 5.104 is simplified as

$$H_N = \sum_{i=1}^{3N}\frac{1}{2}\left(\frac{\partial^2}{\partial \zeta_i^2} + \omega_i^2\zeta_i^2\right) \qquad 5.119$$

which represents a quantum system containing $3N$ independent harmonics. Corresponding to each of the ω_i, quantum theory tells us that the eigenenergy of a harmonic has the form

$$\varepsilon_i(N,V;\sigma) = (g_i + 1/2)\hbar\omega_i(V;\sigma) \qquad 5.120$$

with g_i = 0, 1, ..., ∞. Such a harmonic behaves like a boson particle with energy $\hbar\omega_i(V;\sigma)$ and forms the concept of the phonon.

Furthermore, a state of the whole system is specified by the set of $3N$ independent quantum numbers $\mathbf{g} = (g_1, g_1, \ldots, g_{3N},)$. Finally, the energy of a state of the system formed by $3N$ independent harmonics, $\varepsilon_{\mathbf{g}}(N,V;\sigma)$, introduced in Eq. 5.16, is obtained from the summation of the energies of the $3N$ independent harmonics as

$$\varepsilon_{\mathbf{g}}(N,V;\sigma) = \sum_{i=1}^{3N}\varepsilon_i(N,V;\sigma) \qquad 5.121$$

This concludes the rationale by which Eq. 5.32 is derived.

5.4.3 Dynamical matrix and phonon mode

Because of the periodicity of a crystal, one can make an initial guess that the solutions of Eq. 5.103 are elastic plane waves based on collective atomic vibrations [23, 33], from the harmonic approximation of Eq. 5.107,

$$u_\alpha(t;\mathbf{R}) = u_\alpha(j;P)\exp(-i\omega t) = u_\alpha(j)\exp\{i\mathbf{q}_t.[\mathbf{R}(P)+\mathbf{r}(j)] - i\omega t\} \qquad 5.122$$

where ω represents the frequency of the plane wave, and $\mathbf{q}_t$ is a wave vector designating the wave number and direction along which the plane wave propagates. It should be pointed out that $u_\alpha(j)$ in Eq. 5.122 is now independent of the index P. That is, except for a phase factor, atoms that are equivalent by translational symmetry among different primitive unit cells will experience the same type of atomic motion, independently of the positions of these primitive unit cells in the system. This is equivalent to applying the periodic condition, so that $w_\alpha(j;P)$ in Eq. 5.107 obeys

$$w_\alpha(j;P) = \exp\{i\mathbf{q}_t\cdot[\mathbf{R}(P)+\mathbf{r}(j)]\}w_\alpha(j) \qquad 5.123$$

Note that $w_\alpha(j)$ is now independent of the index P.

Furthermore, one wants to limit the $\mathbf{q}_t$ in Eq. 5.125 and Eq. 5.126 to those known as exact wave vectors, which represent a special set of points in the reciprocal space that satisfy the condition

$$\frac{1}{N_c}\sum_P^{N_c}\exp\{i\mathbf{q}_t\cdot[\mathbf{R}(P)-\mathbf{R}(0)]\} = \delta(\mathbf{q}_t) \qquad 5.124$$

where $\delta(\mathbf{q}_t)$ is the Kronecker delta function. In fact, the number of $\mathbf{q}_t$ equals to the number of primitive unit cells contained in the system.

Utilizing the translational invariance by which $C^{jk}_{\alpha\beta}$ in Eq. 5.108 (or $\Phi^{jk}_{\alpha\beta}(P,Q)$ in Eq. 5.106) depends on P and Q only through the difference $\mathbf{R}(P)-\mathbf{R}(Q)$, the following Fourier transformation can be employed to simplify Eq. 5.113:

$$D^{jk}_{\alpha\beta}(\mathbf{q}_t) = \frac{1}{N_c}\sum_{P,Q}^{N_c} C^{jk}_{\alpha\beta}(P,Q)\exp\{i\mathbf{q}_t\cdot[\mathbf{R}(P)+\mathbf{r}(j)-\mathbf{R}(Q)-\mathbf{r}(k)]\} \qquad 5.125$$

and one obtains

$$\left|D^{jk}_{\alpha\beta}(\mathbf{q}_t) - \delta_{\alpha\beta}\delta_{jk}\omega^2(\mathbf{q}_t,V;\sigma)\right| = 0 \qquad 5.126$$

The counterpart of Eq. 5.126 with respect to Eq. 5.112 is

$$\omega^2(\mathbf{q}_t,V;\sigma)e_\alpha(j;\mathbf{q}_t) = \sum_k^{N_a}\sum_\beta^{3} D^{jk}_{\alpha\beta}(\mathbf{q}_t)e_\beta(k;\mathbf{q}_t) \qquad 5.127$$

Equation 5.126 is now an equation with $3N_a$ degrees of freedom. At each $\mathbf{q}_t$, one can always find $3N_a$ eigenvalues of $\omega_i^2(\mathbf{q}_t,V;\sigma)$ (i = 1, . . ., $3N_a$). The $3N_a$ vibrations are often known as phonon modes, noting again that N_a is the number of atoms in the primitive unit cell. Each of the $\omega_i^2(\mathbf{q}_t, V; \sigma)$ yields a set of $e_\alpha^i(j; \mathbf{q}_t)$ which can be chosen such that

$$\sum_k^{N_a} \sum_\beta^3 \left(e_\beta^i(k; \mathbf{q}_t)\right)^* e_\beta^{i'}(k; \mathbf{q}_t) = \delta_{ii'} \quad 5.128$$

where $\left(e_\beta^i(k; \mathbf{q}_t)\right)^*$ represents the complex conjugate of $e_\beta^i(k; \mathbf{q}_t)$, and

$$\sum_i \left(e_\alpha^i(j; \mathbf{q}_t)\right)^* e_\beta^i(k; \mathbf{q}_t) = \delta_{\alpha\beta}\delta_{jk} \quad 5.129$$

Finally, for a solid, the summation over Q in Eq. 5.125 can be factorized out, resulting in, after transforming $C_{\alpha\beta}^{jk}(P, Q)$ back to $\Phi_{\alpha\beta}^{jk}(P, Q)$ together with using Eq. 5.108,

$$D_{\alpha\beta}^{jk}(\mathbf{q}_t) = \frac{1}{\sqrt{m_j m_k}} \sum_P^{N_c} \Phi_{\alpha\beta}^{jk}(P, 0)\exp\{i\mathbf{q}_t \cdot [\mathbf{R}(P) + \mathbf{r}(j) - \mathbf{R}(0) - \mathbf{r}(k)]\} \quad 5.130$$

Note that, because of translational invariance, $\Phi_{\alpha\beta}^{jk}$ depends on P and Q only through the difference $\mathbf{R}(P) - \mathbf{R}(Q)$.

In the case where a system is built from a parallelepiped multiplication of the primitive unit cell with lattice vectors $\mathbf{a}_\alpha$ ($\alpha = x$, y, and z) in the form $N_c^x\mathbf{a}_x \times N_c^y\mathbf{a}_y \times N_c^z\mathbf{a}_z$, $\mathbf{q}_t$ in Eq. 5.126 and Eq. 5.130 takes the form

$$\mathbf{q}_t^\alpha = \frac{m^\alpha}{N_c^\alpha}\mathbf{b}^\alpha \quad 5.131$$

where $m^\alpha = 0, 1, \ldots, N_c^\alpha - 1$, and $\mathbf{b}^\alpha$ is the primitive lattice vector in the reciprocal space, given by

$$\begin{aligned} \mathbf{b}^x &= (2\pi/V_a)(\mathbf{a}_y \times \mathbf{a}_z) \\ \mathbf{b}^y &= (2\pi/V_a)(\mathbf{a}_z \times \mathbf{a}_x) \\ \mathbf{b}^z &= (2\pi/V_a)(\mathbf{a}_x \times \mathbf{a}_y) \end{aligned} \quad 5.132$$

with V_a, representing the volume of the primitive unit cell, given as

$$V_a = \mathbf{a}_x \cdot (\mathbf{a}_y \times \mathbf{a}_z) \quad 5.133$$

Combining Eq. 5.132 and Eq. 5.133, it is easy to demonstrate that

$$\mathbf{b}^\alpha \cdot \mathbf{a}_\beta = 2\pi\delta_{\alpha\beta} \quad 5.134$$

where $\delta_{\alpha\beta}$ is the Kronecker delta symbol.

It can be shown that the $\mathbf{q}$ points defined in Eq. 5.131 represent the exact wave vector points. First, $\mathbf{R}(P) - \mathbf{R}(0)$ can be written as

$$\mathbf{R}(P) - \mathbf{R}(0) = \sum_\alpha p_\alpha \mathbf{a}_\alpha \quad 5.135$$

where $p_\alpha = 0, 1, \ldots, N_c^\alpha - 1$. Then, the left-hand side of Eq. 5.124 for the definition of an exact wave vector becomes

$$\frac{1}{N_c}\sum_P \exp\{i\mathbf{q}_t\cdot[\mathbf{R}(P)-\mathbf{R}(0)]\}$$
$$=\frac{1}{N_c^xN_c^yN_c^z}\prod_\alpha\sum_{p_\alpha=0}^{N_c^\alpha-1}\exp\left(2\pi ip_\alpha\frac{m^\alpha}{N_c^\alpha}\right)$$
$$=\frac{1}{N_c^xN_c^yN_c^z}\prod_\alpha\frac{1-\exp(2\pi im^\alpha)}{1-\exp\left(2\pi i\frac{m^\alpha}{N_c^\alpha}\right)} \quad 5.136$$

Knowing the fact that

$$\frac{1-\exp(2\pi im^\alpha)}{1-\exp\left(2\pi i\frac{m^\alpha}{N_c^\alpha}\right)}=\begin{cases}0, & \text{if} \quad m^\alpha\neq 0\\ N_c^\alpha, & \text{if} \quad m^\alpha=0\end{cases} \quad 5.137$$

Hence, the $\mathbf{q}$ points defined in Eq. 5.131 are the exact wave vector points.

5.4.4 Linear-response method versus supercell method

The problem of lattice vibrations in a solid is now transformed into computing the dynamical matrix in Eq. 5.126. The first-principles solution of the problem is currently divided into two categories: the linear-response method [34] and the supercell method [35]. In the linear-response method, utilizing the electronic linear response of the undistorted crystals [36], evaluation of the dynamical matrix can be performed through the density functional perturbation theory [34] without the approximation of the cutoff in neighboring interactions.

Compared with the linear-response method, the supercell method is conceptually simple and is easy to implement computationally. The supercell method adopts the frozen phonon approximation, in which the changes in total energy or forces are calculated in the real space by displacing the atoms from their equilibrium positions. The advantage of the supercell method is that the phonon frequencies at the exact wave vectors, which are commensurable with the supercell, are calculated exactly with no further approximation [37]. The shortcoming of the supercell method is that it is often limited by the size of the supercell that can be handled with current computing resources.

In the supercell approach, inaccuracies are thought to arise from the truncation of the force constants [34, 35]. This is only partially true. In the supercell method, the calculated $\varphi_{\alpha\beta}^{jk}$ represents the cumulative contributions of the atom indexed by k and P in the supercell and all its images resulting from translational transformation of the supercell in the whole space. Let $\mathbf{L}_\alpha$ represent the lattice vectors of the supercell, then

$$\varphi_{\alpha\beta}^{jk}(P,0)=\sum_{n_x=-\infty,\,n_y=-\infty,\,n_z=-\infty}^{\infty}\Phi_{\alpha\beta}^{jk}\left[R(P)+n_x\mathbf{L}_x+n_y\mathbf{L}_y+n_z\mathbf{L}_z,0\right] \quad 5.138$$

For the exact wave vectors in Eq. 5.131, one has

$$\mathbf{q}_t^{\alpha} \cdot L_{\alpha} = 2\pi k_l \qquad 5.139$$

where k_l is an integer. Replacing $\Phi_{\alpha\beta}^{jk}(P,0)$ from Eq. 5.130 with $\varphi_{\alpha\beta}^{jk}(P,0)$ in Eq. 5.138, one obtains

$$\begin{aligned} D_{\alpha\beta}^{jk}(\mathbf{q}_t) &= \frac{1}{\sqrt{m_j m_k}} \sum_{P}^{N_c} \varphi_{\alpha\beta}^{jk}(P,0) \exp\{i\mathbf{q}_t \cdot [\mathbf{R}(P) + \mathbf{r}(j) - \mathbf{R}(0) - \mathbf{r}(k)]\} \\ &= \frac{1}{\sqrt{m_j m_k}} \sum_{P=-\infty}^{\infty} \Phi_{\alpha\beta}^{jk}(P,0) \exp\{i\mathbf{q}_t \cdot [\mathbf{R}(P) + \mathbf{r}(j) - \mathbf{R}(0) - \mathbf{r}(k)]\} \qquad 5.140 \end{aligned}$$

Therefore, the phonon frequencies calculated at the exact wave vectors by the cumulative force constants approach are exact, and the supercell size will not lead to errors in the calculated phonon frequencies [37].

Generally speaking, if a supercell contains N_c primitive unit cells, one can always find N_c corresponding exact wave vectors in the $\mathbf{q}$ space. In most linear response calculations, the common choice of a 4×4×4 $\mathbf{q}$-mesh is exactly equivalent to a 4×4×4 supercell in real space. Furthermore, since the supercell approach does not impose any approximation on the electronic response to the distortion of the lattice geometry, the effects of electron–phonon interactions can be accounted for by the supercell method.

In the supercell method, due to the imposition of periodic conditions, the calculated force constant in real space cannot account for the effects of the vibration-induced electric field for polar materials. It has been demonstrated that such an effect adds an additional term to the dynamical matrix in reciprocal space in the form

$$d_{\alpha\beta}^{jk}(na) = \frac{4\pi e^2}{V_a} \frac{[\mathbf{q} \cdot \mathbf{Z}^*(j)]_{\alpha} [\mathbf{q} \cdot \mathbf{Z}^*(k)]_{\beta}}{\mathbf{q} \cdot \boldsymbol{\varepsilon}_{\infty} \cdot \mathbf{q}} \bigg|_{\mathbf{q} \to 0} \qquad 5.141$$

where $\mathbf{Z}^*(j)$ represents the Born effective charge tensor of the jth atom in the primitive unit cell and $\boldsymbol{\varepsilon}_{\infty}$ the high frequency static dielectric tensor, i.e. the contribution to the dielectric permittivity tensor from electronic polarization. As a result, for polar materials, the matrix element at wave vector $\mathbf{q} \to 0$ of the dynamical matrix in Eq. 5.130, derived by means of Eq. 5.138, should have the form

$$D_{\alpha\beta}^{jk}(0) = \frac{1}{\sqrt{m_j m_k}} \left[d_{\alpha\beta}^{jk}(na) + \sum_{P}^{N_c} \varphi_{\alpha\beta}^{jk}(P,0) \right] \qquad 5.142$$

It can be demonstrated [38] that this is equivalent to replacing the real-space force constant $\Phi_{\alpha\beta}^{jk}(P,0)$ in Eq. 5.106 as follows:

$$\Phi_{\alpha\beta}^{jk}(P,0) \to \varphi_{\alpha\beta}^{jk}(P,0) + \frac{d_{\alpha\beta}^{jk}(na)}{N_c} \qquad 5.143$$

At present, implementations of the first-principles method for calculating phonon frequencies are mostly limited by the supercell size when using Eq. 5.113 or the number of exact wave vector points when using Eq. 5.126. A 4 × 4 × 4 supercell built on the primitive unit cell or $4 \times 4 \times 4$ exact wave vector mesh is usually the common limit. If only the phonon frequencies derived from Eq. 5.113 or Eq. 5.126 are used, $g(\omega, V; \sigma)$ can be rather unsmooth, which will result in inaccurate thermodynamic properties when it is used with Eq. 5.34. The mixed-space approach [38] can circumvent this difficulty by use of the Fourier interpolation to calculate the dynamical matrix for an arbitrary wave vector $\mathbf{q}$ as

$$D_{\alpha\beta}^{jk}(\mathbf{q}) = \frac{1}{\sqrt{m_j m_k}} \sum_P^{N_c} \left[\varphi_{\alpha\beta}^{jk}(P, 0) + \frac{d_{\alpha\beta}^{jk}(na)}{N_c} \right] \exp\{i\mathbf{q} \cdot [\mathbf{R}(P) + \mathbf{r}(j) - \mathbf{R}(0) - \mathbf{r}(k)]\} \tag{5.144}$$

with the help of Eq. 5.106, Eq. 5.138, and Eq. 5.141 for polar materials. Note that the term involving $d_{\alpha\beta}^{jk}(na)$ is for polar materials only.

The quantity $g(\omega, V; \sigma)$ can be calculated as

$$g(\omega, V; \sigma) = \frac{1}{N_\mathbf{q}} \sum_\mathbf{q} \sum_{i=1}^{3N_a} \delta\Big(\omega - \omega_i(\mathbf{q}, V; \sigma)\Big) \tag{5.145}$$

where the function $\delta(x)$ is usually taken as a Gaussian:

$$\delta(x) = \frac{1}{\Delta\sqrt{2\pi}} \exp\left(-\frac{x^2}{2\Delta^2}\right) \tag{5.146}$$

where Δ is an adjustable damping (broadening) parameter whose role is to smear out the $g(\omega, V; \sigma)$ curve; $N_\mathbf{q}$ in Eq. 5.145 is the number of $\mathbf{q}$ points used. Empirically, a $\mathbf{q}$ mesh of $60 \times 60 \times 60$ is accurate enough for most purposes and can be done efficiently with the YPHON code [15]; this is discussed in Appendix A.

5.5 First-principles approaches to disordered alloys

The first-principles calculations discussed so far strictly rely on the exact atomic positions in the unit cells. A brute-force approach for a random solution phase would be to directly construct a large supercell and randomly decorate the host lattice with different types of atoms. Such an approach would necessarily require very large supercells to adequately mimic the statistics of the random solutions. Since first-principles methods are computationally constrained by the number of atoms that one can treat, this brute-force approach is computationally prohibitive. Take a binary $A_{1-x}B_x$ substitutional alloy as illustrated in Figure 5.12 as an example. For a system containing N atoms, there is a possible number 2^N of configurations, which is an astronomically large number when N is large. It is an impossible task to explore such a huge configuration space with the available computing resources.

As a result, approximations must be made to the first-principles calculations. At present, there are three main approaches to calculating disordered solution phases: the coherent potential approximation (CPA)) [41], the cluster expansion) (CE) [42], and the special quasi-random structures (SQS)) [43] approach.

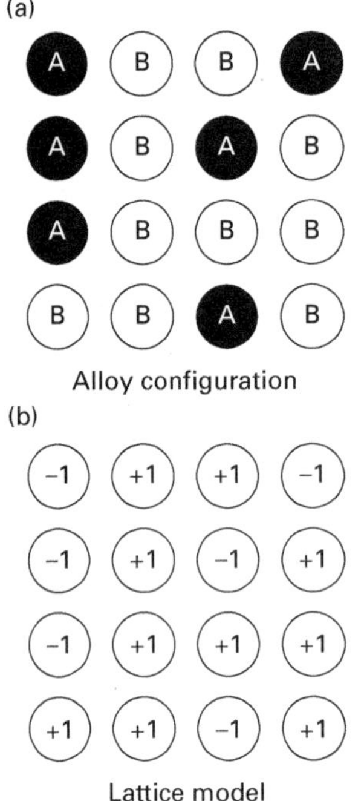

Figure 5.12 *Mapping of a substitutional $A_{1-x}B_x$ alloy into an Ising-like lattice model [39, 40].*

The coherent potential approximation [41] treats random alloys by considering the *average* occupations of lattice sites in solving the Kohn–Sham equation. Since a mean-field approach is employed, the dependence of properties on the local environments surrounding an atom is not treated explicitly in CPA. In a random solution, there exists a distribution of local environments (e.g., in bcc alloys, A or B surrounded by the various $A_X B_{8-X}$ coordination shells with X between 0 and 8), resulting in local environmentally dependent quantities such as charge transfer and local displacements of atoms from their ideal lattice positions. Even in random $A_{1-x}B_x$ solid solutions, the average A–A, A–B, and B–B bond lengths are generally different. These effects are considered in the CE and SQS approaches, on which we focus in the next two subsections. In all the following subsections, unless specifically noted, the formulism for the binary system is discussed for the sake of simplicity.

5.5.1 Cluster expansions

Many properties of a solution phase, such as the energy, are dependent on the *configuration* – the arrangement of atoms on the lattice sites. In a cluster expansion [35, 42], the configuration dependence of properties is formulated efficiently by a "lattice algebra" which maps a substitutional configuration into an Ising-like lattice model. Taking a binary $A_{1-x}B_x$ solution phase for instance, A atoms are represented by the "down" spins ($S_i = -1$) and B atoms are represented by the "up" spins ($S_i = +1$) as illustrated in Figure 5.12. Using the cluster expansion technique, for a system containing N atoms, the total energy of any alloy configuration $\sigma = (S_1, S_2, \ldots, S_N)$ can be conveniently evaluated using the following Ising-like Hamiltonian:

$$E(\sigma) = J_0 + \sum_i J_i S_i(\sigma) + \sum_{i,j} J_{ij} S_{ij}(\sigma) + \sum_{i,j,k} J_{ijk} S_{ijk}(\sigma) + \sum_{i,j,k,l} J_{ijkl} S_{ijkl}(\sigma) + \ldots \quad 5.147$$

where the J's are the effective cluster interactions (ECIs)); $S_i(\sigma)$ is a number representing the atomic occupation at the lattice i under the configuration σ, which takes the values -1 and 1 for binary and -1, 0, and 1 for ternary systems, etc. In Eq. 5.147, the 2-site, 3-site, and 4-site correlations are written as follows,

$$S_{ij}(\sigma) = S_i(\sigma)S_j(\sigma) \quad 5.148$$

$$S_{ijk}(\sigma) = S_i(\sigma)S_j(\sigma)S_k(\sigma) \quad 5.149$$

$$S_{ijkl}(\sigma) = S_i(\sigma)S_j(\sigma)S_k(\sigma)S_l(\sigma) \quad 5.150$$

The expansion in Eq. 5.147 is exact as long as *all* the n-site interactions are included. For a binary system, this can be observed by using the combination law that $\sum_{n=0}^{N} \binom{n}{N} = 2^N$ where $\binom{n}{N} = N!/(n!(N-n)!)$ is the number of n-site interactions. However, in actual calculations, one never does an expansion to order N (containing 2^N terms for a binary system) since this would be too long to be practical. In fact, since the interactions between widely separated atoms are expected to be weaker than the interactions between nearer atoms for most of the important properties, the expansion in Eq. 5.147 is usually truncated at a certain distance to include only a few short-range pair (2-site), triple (3-site), and at the most, quadruple (4-site) interactions.

Once a configuration is assigned, the correlations S's are just geometrical factors. The common practice in cluster expansion is as follows: (i) perform first-principles calculations of a selected set of configurations (around 20–100); (ii) evaluate the values of the interactions J's using Eq. 5.147 with the energies from (i); (iii) use the fitted J's to predict the energy for a desired set of configurations; and (iv) find the ensemble average at a given temperature for the energetics of the random alloys through Monte Carlo simulations.

5.5.2 Special quasi-random structures

Special quasi-random structures (SQSs) [43, 44] are specially designed *small unit cell* periodic structures with minimal number of atoms per unit cell, which can be used to closely mimic the most relevant, near-neighbor pair and multi-site correlation functions of random substitutional alloys. The correlation functions are classified by their n-site component "figures" $f = (n,m)$, where the index n is called the vortex for pair, triple, and quadruple correlations (n = 2, 3, and 4); m measures the correlation distance.

In the SQS approach, a distribution of distinct local environments is maintained and their average corresponds to a random alloy. Thus, a single DFT calculation of an SQS can give many important alloy properties (e.g. equilibrium bond lengths, charge transfer, formation enthalpies, etc.) that depend on the existence of those distinct local environments. The SQS approach has been used extensively to study formation enthalpies, bond length distributions, density of states, band gaps, and optical properties in semiconductor alloys. It is to be noted that the cluster expansion (CE) approach can treat short-range ordering efficiently, while it is not clear how the SQS approach can be used to represent short-range ordering.

The key quantities in the SQS approach are the n-site correlation functions. Specifically, the 2-site correlation function corresponding to the 2-site component figures (2,m) is

$$\overline{\Pi}_{2,m}(\sigma) = \frac{1}{N_{2,m}} \sum_{i \neq j,\, R_{ij}=m}^{N} S_{ij}(\sigma) \tag{5.151}$$

where $N_{2,m}$ represents the total number of possible pairs with correlation distance (neighboring distance) R_{ij} equal to m. The 3-site correlation function corresponding to the 3-site component figures (3,m) is

$$\overline{\Pi}_{3,m}(\sigma) = \frac{1}{N_{3,m}} \sum_{i \neq j \neq k,\, R_{ijk}=m}^{N} S_{ijk}(\sigma) \tag{5.152}$$

where $N_{3,m}$ represents the total number of all possible 3-site figures with the correlation distance (size and shape) R_{ijk} equal to m. The 4-site correlation function corresponding to the 4-site component figures (4,m) is

$$\overline{\Pi}_{4,m}(\sigma) = \frac{1}{N_{4,m}} \sum_{i \neq j \neq k \neq l,\, R_{ijkl}=m}^{N} S_{ijkl}(\sigma) \tag{5.153}$$

where $N_{4,m}$ represents the total number of all possible 4-site figures with correlation distance (size and shape) R_{ijkl} equal to m.

With a given supercell size N, the essential task of the SQS approach is to search through all configurations that approach as closely as possible to the correlation functions of a perfectly random (R) structure, and for the binary system their number is

$$\overline{\Pi}_{n,m}(R) = (2x - 1)^n \tag{5.154}$$

Describing random alloys by small unit cell periodical structures surely introduces erroneous correlations beyond a certain distance. However, since interactions between nearest neighbors are generally more important than interactions between more distant neighbors, SQSs can be constructed in such a way that they exactly reproduce the correlation functions of a random alloy between the first few nearest neighbors, deferring errors due to periodicity to more distant neighbors. The practical procedure could be to find the structures that match the 2-site correlation functions up to a given neighboring distance, and then to add the conditions matching the high order correlation functions up to a certain correlation distance.

Appendix B is a collection of SQSs with a variety of compositions for binary fcc, bcc, hcp, and $L1_2$ structures, for ternary fcc, bcc, and B2 structures, and for perovskite in the cubic ABO_3 structure. The format used is that of VASP.

5.5.3 Phonon calculations for SQSs

A somewhat more theoretically demanding application of the SQS approach is the calculation of the phonon dispersions of a random alloy. Considering the fact that the size of an SQS cell in general is around 8–32, phonon calculations based on SQSs are achievable using either the SQS cell or its supercell, for example, $2 \times 2 \times 2$ of the SQS cell. However, one notes that while the phonon density of states can be calculated straightforwardly, calculations of the phonon dispersions run into a problem. That is, since the phonon calculation treats the SQS as a primitive unit cell made of more atoms than the primitive unit cell of the ideal lattice, the number of phonon dispersions derived from a regular phonon calculation is much greater than that measured for a random alloy. For example, if one uses an SQS containing 16 atoms for an fcc solid solution, the regular phonon calculations would produce $3 \times 16 = 48$ phonon dispersions in comparison to just three phonon dispersions from measurement. By averaging over the force constants of an SQS, the dynamical matrix can be calculated with respect to the wave vector space of the ideal lattice of a random alloy.

One consideration that must be taken into account is that the phonon dispersions measured from inelastic neutron scattering experiments only represent the averaged vibrations of the ideal lattice. For random alloys or phases with minor geometry distortion, it was suggested that one should calculate the dynamical matrix using Eq. 5.155 below instead of Eq. 5.125, obtaining (see [45])

$$D_{\alpha\beta}^{jk}(\mathbf{q}) = \frac{1}{\sqrt{\overline{m}_j \overline{m}_k}} \frac{1}{N_c} \sum_{P,Q}^{N_c} \left[\varphi_{\alpha\beta}^{jk}(P,Q) + \frac{d_{\alpha\beta}^{jk}(na)}{N_c} \right] \exp\{i\mathbf{q}\cdot[\mathbf{R}(P) + \mathbf{r}(j) - \mathbf{R}(Q) - \mathbf{r}(k)]\} \quad 5.155$$

where in the case of a random alloy, $\overline{m}_j$ represents the averaged atomic mass at the jth lattice site. The purpose of the summation over Q is to average the effects of local distortions, making it possible to compare the calculated dispersions with the measured dispersions representing the averaged vibrations of the ideal lattice. As a result, the dimension of the SQS supercell dynamical matrix is thus reduced to match that of the primitive unit cell of the ideal lattice for the calculation of the phonon frequencies. The calculational procedure is as follows.

i. Make an SQS supercell based on the primitive unit cell of the ideal lattice, in order to mimic the correlation functions of the random solution;
ii. relax the SQS supercell with respect to the internal atomic positions while keeping the cell shape and volume fixed;

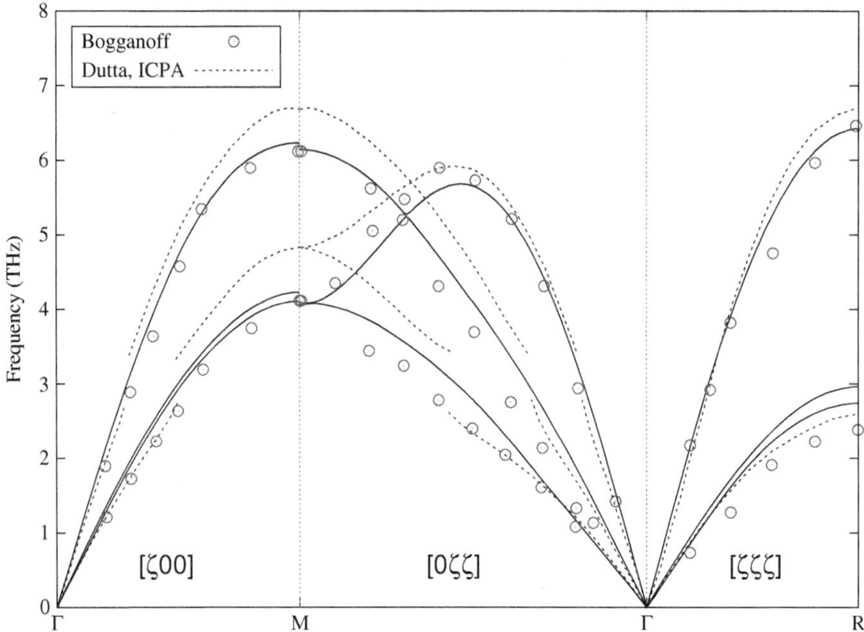

Figure 5.13 *Phonon dispersions for random Cu_3Au. The solid lines represent the present calculation and the open circles represent the inelastic neutron scattering data with details in reference [45]. The dashed lines represent the results calculated using the ab initio transferable force constant model by Dutta et al. [46].*

iii. make the phonon supercell by further enlarging the SQS supercell and calculate the force constants; and

iv. calculate the dynamical matrix $D^{jk}_{\alpha\beta}(\mathbf{q})$, with the wave vector, $\mathbf{q}$, being defined from the primitive unit cell of the ideal lattice, through a Fourier transformation.

The calculated phonon dispersions along the directions (00ξ), (0$\xi\xi$), and ($\xi\xi\xi$) are compared with the inelastic neutron scattering data in Figure 5.13 for Cu_3Au.

Exercises

The exercises are solely for the purpose of practice in using the YPHON package for the calculation of phonon properties.

By default, it is assumed that you have access to VASP.5. Even if you do not have VASP, you can still use the YPHON package since the exercise subfolders contain the force constants already calculated for all demo materials and the dielectric data for the polar materials. If you do not have VASP, you can work on Exercises 1.7, 1.8, 2.8, and 3 to 6.

To start, make an exercise folder named "mYdemo."

(*cont.*)

Exercise 1. Magnesium

Make a folder "Mg" under the folder "Ydemo" and go to the folder Mg. You need to prepare the following files before submitting your VASP.5 job. (*Note*: if you do not have VASP5, you can skip to Exercise 1.7.)

1.1 POSCAR file

First, prepare the VASP input file POSCAR using the primitive unit cell, name the file "POSCAR.prm" and copy/paste the following lines into it

```
Mg HCP
1.00
  1.594090000000   -2.761040000000   0.000000000000
  -3.188170000000  0.000000000000    0.000000000000
  0.000000000000   0.000000000000    -5.186410000000
  Mg
  2
D
0 0 0
0.66666667          0.33333333          0.5
```

Then make the supercell using "Ycell –ss 2 <POSCAR.prm >POSCAR" which builds the supercell POSCAR file containing these lines:

```
Supercell
1.00
  3.1881800000    -5.5220800000    0.0000000000
  -6.3763400000   0.0000000000     0.0000000000
  0.0000000000    0.0000000000     -10.3728200000
Mg
16
D
  0.0000000000    0.0000000000     0.0000000000 Mg
  0.0000000000    0.0000000000     0.5000000000 Mg
  0.0000000000    0.5000000000     0.0000000000 Mg
  0.0000000000    0.5000000000     0.5000000000 Mg
  0.5000000000    0.0000000000     0.0000000000 Mg
  0.5000000000    0.0000000000     0.5000000000 Mg
  0.5000000000    0.5000000000     0.0000000000 Mg
  0.5000000000    0.5000000000     0.5000000000 Mg
  0.3333333350    0.1666666650     0.2500000000 Mg
  0.3333333350    0.1666666650     0.7500000000 Mg
  0.3333333350    0.6666666650     0.2500000000 Mg
  0.3333333350    0.6666666650     0.7500000000 Mg
  0.8333333350    0.1666666650     0.2500000000 Mg
  0.8333333350    0.1666666650     0.7500000000 Mg
  0.8333333350    0.6666666650     0.2500000000 Mg
  0.8333333350    0.6666666650     0.7500000000 Mg
```

(cont.)

1.2 INCAR file

Make the VASP input file INCAR containing the following lines:

```
EDIFF=1.d-6
PREC = A
ISMEAR = 1
SIGMA = 0.2
IBRION = 6
ISIF = 0
NSW=1
```

1.3 KPOINTS file

Make the VASP input file KPOINTS containing the following lines:

```
Magnesium
0
G
3 3 3
0 0 0
```

1.4 POTCAR file

Make the VASP input potential file POTCAR by the Linux command (depending on the VASP pseudopotential file location in your system)

```
zcat /usr/global/msc/vasp/potpaw_PBE/Mg/POTCAR.Z >POTCAR
```

1.5 Run VASP.5

Run VASP interactively (the run can be finished in less than one minute) or make a PBS batch job script containing the following lines and then submit your job (depending on your system environment for batch jobs)

```
#PBS -q debug
#PBS -l nodes=1:ppn=8
#PBS -S /bin/tcsh
#PBS -j oe
#PBS -l walltime=00:30:00
module load vasp/5.3.5
setenv VSPCMD "mpirun -np 8 vasp"
cd $PBS_O_WORKDIR
$VSPCMD
```

1.6 Collect the force constants for input to YPHON

After your job is done, first delete the files not used using "rm CHG DYNMAT IBZKPT WAVECAR CHGCAR DOSCAR EIGENVAL REPORT XDATCAR PCDAT" and then type the following Linux command sequentially, namely to get the force constant by

```
vasp_fij
```

(*cont.*)

You will see a file named "superfij.out" which contains lines like

```
1.5940900000   -2.7610400000   0.0000000000
-3.1881700000  0.0000000000    0.0000000000
0.0000000000   0.0000000000    -5.1864100000
  3.188180     -5.522080       0.000000
  -6.376340    0.000000        0.000000
  0.000000     0.000000        -10.372820
16 8
Direct
  0.00000000                   0.00000000   0.00000000   Mg
  0.00000000                   0.00000000   0.50000000   Mg
  0.00000000                   0.50000000   0.00000000   Mg
  0.00000000                   0.50000000   0.50000000   Mg
  0.50000000                   0.00000000   0.00000000   Mg
  0.50000000                   0.00000000   0.50000000   Mg
  0.50000000                   0.50000000   0.00000000   Mg
  0.50000000                   0.50000000   0.50000000   Mg
  0.33333333                   0.16666667   0.25000000   Mg
  0.33333333                   0.16666667   0.75000000   Mg
  0.33333333                   0.66666667   0.25000000   Mg
  0.33333333                   0.66666667   0.75000000   Mg
  0.83333333                   0.16666667   0.25000000   Mg
  0.83333333                   0.16666667   0.75000000   Mg
  0.83333333                   0.66666667   0.25000000   Mg
  0.83333333                   0.66666667   0.75000000   Mg
  -2.885402    0.000005        0.000000     0.056928     0.000000     0.000000
1.425033       0.000000        0.000000     -0.042757    0.000000     0.000000
0.356201       -0.617095       0.000000     0.016087     0.033974     0.000000
0.356196       0.617091        0.000000     0.016087     -0.033974    0.000000
-0.027400      0.000000        -0.000001    -0.027400    0.000000     0.000001
0.003752       0.000000        0.000000     0.003752     0.000000     0.000000
0.187231       -0.123917       0.312727     0.187231     -0.123917    -0.312727
0.187230       0.123917        -0.312726    0.187230     0.123917     0.312726
...
```

1.7 To calculate the PDOS

Note: if one does not have VASP.5, one can copy the superfij.out file from the YPHON package under the folder "Ydemo/Mg."

To calculate the PDOS, type (if your system has gnuplot installed, you will see the PDOS plotted)

```
Yphon <superfij.out –plot
```

1.8 Calculate the phonon dispersions

Prepare the phonon dispersion instruction file containing the lines (copy and paste works fast)

(cont.)

0.0 0.0 0.0 0.5 0.5 0.0 Gamma K 0 ($1*2) 2
0.5 0.0 0.0 0.0 0.0 0.0 K Gamma 1 ((0.5-$1)*2) 2
0.0 0.0 0.0 0.0 0.0 0.5 Gamma A 2 (($1)*2) 2
0.0 0.0 0.5 0.5 0.5 0.5 A L 3 ($1*2) 2
0.5 0.0 0.5 0.5 0.0 0.0 L M 4 ((0.5-$1)*2) 2

Then execute YPHON as follows to see the phonon dispersion plot.

Yphon <superfij.out –pdis dfile.hcp –expt exp01.dat –plot

Note: the dfile.hcp file contains the instruction on how to plot the experimental data contained in the file "exp01.dat" together with the calculated phonon dispersions. You can get the dfile.hcp and exp01.dat files from the YPHON package under the folder "Ydemo/Mg."

At this step, you will find two useful the files "vdis.plt" and "vdos.plt" which are gnuplot scripts produced by YPHON for usage by gnuplot to make figures of phonon dispersions and PDOS.

Exercise 2. MgO

Make a folder "MgO" under the folder of "mYdemo" and go to the folder MgO.

This exercise just shows how to calculate the phonon dispersions of MgO. You need to prepare the following files and submit your VASP.5 job. (*Note:* if one does not have VASP.5, one can skip to exercise subsection 2.8.))

2.1 POSCAR file

First, prepare the POSCAR file containing the following lines for the primitive unit cell, naming the file "POSCAR.prm."

MgO-Born effective charge
4.212
.0 .5 .5
.5 .0 .5
.5 .5 .0
Mg O
1 1
D
.0 .0 .0 Mg
.5 .5 .5 O

Then type "cp POSCAR.prm POSCAR."

2.2 INCAR file

EDIFF=1.d-6
PREC = High
ISMEAR = -5
IBRION = -1
LEPSILON=.T.
NSW=0

(*cont.*)

2.3 KPOINTS file

```
Magnesium oxide
0
G
9 9 9
 0 0 0
```

2.4 POTCAR file

Get POTCAR by the following Linux commands (depending on the VASP pseudopotential location in your system)

```
zcat /usr/global/msc/vasp/potpaw_LDA/Mg/POTCAR.Z >POTCAR
zcat /usr/global/msc/vasp/potpaw_LDA/O/POTCAR.Z >>POTCAR
```

2.5 Collect the Born effective charge tensor and macroscopic dielectric tensor

Submit your batch job and after the batch job is done, type "vasp_BE" to collect the Born effective charge and macroscopic dielectric tensor and then you will see the file named "dielecfij.out" which contains lines like

```
  0.000000                                2.106000            2.106000
  2.106000                                0.000000            2.106000
  2.106000                                2.106000            0.000000
0.0000000000000000  0.0000000000000000  0.0000000000000000  Mg
0.5000000000000000  0.5000000000000000  0.5000000000000000  O
  3.147             0.000               0.000
  0.000             3.147               0.000
  0.000             0.000               3.147
ion                 1
  1                 1.96085             0.00000             0.00000
  2                 0.00000             1.96085             0.00000
  3                 0.00000             0.00000             1.96085
ion                 2
  1                 -1.96142            0.00000             0.00000
  2                 0.00000             -1.96141            0.00000
  3                 0.00000             0.00000             -1.96141
```

2.6 Modify INCAR for force constant calculation

```
EDIFF=1.d-6
PREC = A
ISMEAR = -5
IBRION = 6
ISIF = 0
NSW=1
```

2.7 Create supercell POSCAR file

```
Ycell –ss 2 <POSCAR.prm >POSCAR
```

You will see a POSCAR file like

(cont.)

```
Supercell
1.00
  0.0000000000    4.2120000000   4.2120000000
  4.2120000000    0.0000000000   4.2120000000
  4.2120000000    4.2120000000   0.0000000000
Mg O
8 8
D
    0.0000000000  0.0000000000  0.0000000000  Mg
    0.0000000000  0.0000000000  0.5000000000  Mg
    0.0000000000  0.5000000000  0.0000000000  Mg
    0.0000000000  0.5000000000  0.5000000000  Mg
    0.5000000000  0.0000000000  0.0000000000  Mg
    0.5000000000  0.0000000000  0.5000000000  Mg
    0.5000000000  0.5000000000  0.0000000000  Mg
    0.5000000000  0.5000000000  0.5000000000  Mg
    0.2500000000  0.2500000000  0.2500000000  O
    0.2500000000  0.2500000000  0.7500000000  O
    0.2500000000  0.7500000000  0.2500000000  O
    0.2500000000  0.7500000000  0.7500000000  O
    0.7500000000  0.2500000000  0.2500000000  O
    0.7500000000  0.2500000000  0.7500000000  O
    0.7500000000  0.7500000000  0.2500000000  O
    0.7500000000  0.7500000000  0.7500000000  O
```

2.8 Modify KPOINTS file

```
Magnesium oxide
0
G
 3 3 3
 0 0 0
```

2.9 Phonons for MgO

Run VASP interactively (the run can be finished in less than one minute) or make a PBS batch job script and then submit your job. After your supercell calculation job is done, you need to type the following Linux commands sequentially:

```
vasp_fij
```

Note: if one does not have VASP.5, one can copy the superfij.out and dielecfij.out files from the YPHON package under the folder "Ydemo/MgO."

To calculate the phonon dispersions, you run

```
Yphon -Born dielecfij.out -pdis dfile.fcc -bvec -expt exp01.dat –plot <superfij.out
```

Note: dfile.fcc is a phonon dispersion file to instruct YPHON and exp01.dat contains the experimental neutron scattering data. You can get the dfile.fcc and exp01.dat files from the exercise folder "MgO." The data in the dfile.fcc are like

```
0 0 0 0 0 .5 Gamma X 1 $1 2
0 .5 .5 0 .5 .0 X X 3 $1 2
0 .5 .5 0 0 0 X Gamma 0 (1-$1) 2
0 0 0 .25 .25 .25 Gamma L 2 (2*$1) 2
.25 .25 .25 0 .5 .5 L X
```

(*cont.*)

The key "– expt" instructs YPHON to plot the experimental data contained in the file "exp01.dat" together with the calculated phonon dispersions. The key "–plot" instructs YPHON to plot the figure in the terminal using gnuplot for you to check the calculated results.

Exercise 3. Fe_2O_3

The folder "Fe2O3" comes together with the YPHON package under the subfolder "Ydemo." Go to the folder and you can play around by running

```
pos2s Symmetry.pos
Yphon -pdis dfile.rho -Born dielecfij.out -plot -Gfile symmetry.mode <superfij.out
```

You can see some outputs from the screen, where the lines after the line "Solving frequencies considering LO-TO splitting:" contain the vibrational mode analysis shown as

2 A1g Modes of raman_active

No	irrep	THz		(cm-1)		Z*(x)	Z*(y)	Zz(z)
0	A1g	14.7222	14.7222	(491.08	491.08)	0.0001	0.0001	-0.0000
1	A1g	6.7666	6.7666	(225.71	225.71)	0.0000	-0.0001	-0.0000

3 A2g Modes of silent_mode

No	irrep	THz	(cm-1)	Z*(x)	Z*(y)	Zz(z)
0	A2g	19.3966	(647.00	0.0000	-0.0000	0.0000
		19.3966	647.00)			
1	A2g	12.0187	(400.90	-0.0000	-0.0000	-0.0000
		12.0187	400.90)			
2	A2g	5.1938	(173.25	0.0000	0.0000	0.0000
		5.1938	173.25)			

5 Eg Modes of raman_active

No	irrep	THz	(cm-1)	Z*(x)	Z*(y)	Zz(z)
0	Eg	18.0561	(602.29	0.0001	-0.0000	-0.0000
		18.0561	602.29)			
1	Eg	12.3172	(410.86	0.0005	0.0000	0.0000
		12.3172	410.86)			
2	Eg	9.1456	(305.06	-0.0002	-0.0000	0.0000
		9.1456	305.06)			
3	Eg	8.5509	(285.23	0.0001	-0.0000	0.0000
		8.5509	285.23)			
4	Eg	7.3436	(244.96	0.0000	0.0000	-0.0000
		7.3436	244.96)			

2 A1u Modes of silent_mode

No	irrep	THz	(cm-1)	Z*(x)	Z*(y)	Zz(z)
0	A1u	16.9678	(565.99	0.0000	0.0000	0.0000
		16.9678	565.99)			
1	A1u	10.4414	(348.29	-0.0000	0.0000	-0.0000
		10.4414	348.29)			

(*cont.*)

3 A2u Modes of ir_active: P= 0.000, 0.000, 1.000 (0.577, 0.577, 0.577) with one translational mode

No	irrep	THz	(cm-1)	Z*(x)	Z*(y)	Zz(z)
0	A2u	15.3959	(513.55	0.0000	-0.0000	-1.6197
		19.1088	637.40)			
1	A2u	8.9912	(299.92	-0.0000	0.0000	0.6251
		11.5315	384.65)			
2	A2u	-0.0002	(-0.01	-0.0000	0.0000	-0.0000
		-0.0002	-0.01)			

5 Eu Modes of ir_active: P= 0.866, 0.500, 0.000 (0.000, 0.707,-0.707) with one translational mode

No	irrep	THz	(cm-1)	Z*(x)	Z*(y)	Zz(z)
0	Eu	15.4084	(513.97	-1.6366	-0.9449	0.0000
		19.0377	635.03)			
1	Eu	12.9891	(433.27	-0.3671	-0.2119	-0.0000
		14.7124	490.75)			
2	Eu	8.9470	(298.44	0.3687	0.2129	-0.0000
		10.9725	366.00)			
3	Eu	6.9920	(233.23	-0.0170	-0.0098	-0.0000
		7.0000	233.49)			
4	Eu	0.0000	(0.00	0.0000	-0.0000	0.0000
		0.0000	0.00)			

Note: for the LO-TO splitting analysis, do not use the output lines after the line "Frequencies in Gamma point without & with NA term)" that have been calculated by diagonalization of the force constant matrix in real space using a polarization direction which might not be along the polarization direction of all infrared modes. For example, the polarization direction of the Eu mode is different from that of the A2u mode for Fe_2O_3 as shown above.

Exercise 4. MnO

The salient feature of YPHON is best shown using the data contained in the subfolder MnO that we have published previously [14]. The idea is that, for many materials, measurements are usually made on the high symmetry structure, which may not be mechanically stable at low temperature. And if one employs the high symmetry structure in the calculation, one would get some imaginary phonon modes [15]. Other cases include the case when the magnetic ordering breaks the crystal symmetry. The YPHON solution is to restore the symmetry, or in other words, "unfold" the Brillouin zone.

First, one can run YPHON as follows and get a plot showing that there are in total 12 dispersion curves along each direction since the primitive unit cell of MnO in the antiferromagnetic structure contains four atoms:

```
Yphon -Born dielecfij.out -pdis dfile.scc -expt exp05.dat -plot -thr2 0.01 –bvec <superfij.out
```

Next, one can run YPHON as follows and get a total of six dispersion curves along each direction of the cubic MnO by averaging the force constants calculated

(*cont.*)

from the hexagonal structure through "restore translational symmetry," resulting in two atoms (six dispersion curves) in the primitive unit cell:

```
Yphon -Born dielecfij.fcc -pdis dfile.fcc -expt exp05.dat -plot -thr2 0.01 -bvec <superfij.out
```

At the same time, one can run YPHON in the following form and get the phonon dispersions of MnO in perfect fcc symmetry by further averaging the force constants calculated from the hexagonal structure by means of "restore rotational symmetry," recovering the degeneracy of the dispersion curves:

```
Yphon -Born dielecfij.fcc -pdis dfile.fcc -expt exp05.dat -plot -thr2 0.01 -bvec -Rfile Rotation.
sym <superfij.out
```

Exercise 5. Cu_3Au

This exercise is for the phonon dispersion calculation of disordered Cu_3Au using the SQS structure with the following YPHON command:

```
Yphon -pdis dfile.fcc -Born dielecfij.out -thr2 0.10 -expt exp01.dat -plot -bvec -nof -noNA
-sqs-mall -Mass mass.1 <superfij.out
```

For the detailed mechanism and formulations, see reference [7].

Exercise 6. Al_2O_3, $BiFeO_3$, GaAs, $MgAl_2O_4$, NaCl, ZrW_2O_8, BeO, and $ZrSiO_4$

These subfolders contain the force constants already calculated and the dielectric data. Some of them are not published yet. We hope that the users can find some useful settings from these subfolders on gaining experience in running YPHON.

6 CALPHAD modeling of thermodynamics

The CALPHAD modeling of thermodynamics was pioneered by Kaufman and Bernstein [47] and has been reviewed in detail by Saunders and Miodownik [48] and Lukas, Fries, and Sundman [49]. Information on features of software tools for CALPHAD modeling can be found in two series of publications in the CALPHAD journal [50, 51]. The key feature of the CALPHAD method is the modeling of the Gibbs energy of individual phases using both thermodynamic and phase equilibrium data. The main significance of the CALPHAD method is the following.

i. It enabled the development of the concept of lattice stability, i.e. the energy difference between the stable and non-stable crystal structures of a pure element.
ii. The Gibbs energy expression of each phase covers the full temperature, pressure, and composition spaces including both stable and non-stable regions of the phase. This enables the evaluation of the Gibbs energy of a system as a function of non-equilibrium state, i.e. with ξ as an independent variable.
iii. Thermodynamic data are usually obtained by measurements of heat such as the enthalpy of transition and heat capacity, as discussed in Section 4.2, which have large uncertainties typically in the range of kilojoules per mole-of-atoms. On the other hand, phase equilibrium data as discussed in Section 4.1, though more accurate, only contain information on compositions of phases at equilibria, i.e., the relative Gibbs energy of phases at equilibrium. The combination of these two sets of data is foundational in CALPHAD modeling and allows for the accurate modeling of thermodynamic properties of individual phases and reliable calculations of phase stability and driving forces.
iv. CALPHAD provides a framework to model thermodynamic properties of multi-component systems of industrial importance, enabling computational materials design. It has also been extended to model a range of properties of individual phases in multi-component systems such as diffusion coefficients, elastic coefficients, and thermal expansion, supplying input data for computational simulations of phase transformations during materials processing.

In this chapter, the basics of CALPHAD modeling of the Gibbs energy of individual phases are presented. For detailed implementations in various software packages and modeling procedures, readers are referred to the references given above.

6.1 Importance of lattice stability

For modeling of the Gibbs energy of individual phases, it is necessary to define the values of ${}^0G_i$ in Eq. 2.48. However, the independent component i may not be stable in the crystal structure of the phase under consideration, in which case its Gibbs energy cannot be obtained directly from experiments and must be estimated with respect to the Gibbs energy of the stable crystal structure. In the pioneering work by Kaufman and Bernstein [47], this Gibbs energy difference was termed the lattice stability and was obtained through extrapolations in either temperature–pressure or temperature–composition phase diagrams. It is evident from Eq. 2.48 that the values of ${}^0G_i$ and ${}^MG$ jointly contribute to the Gibbs energy of the solution, and Kaufman and Bernstein had to simplify the treatment of ${}^MG$ in order to show the importance of the concept of lattice stability. Using ideal or regular solution models, they were able to define the lattice stability for pure elements and remarkably reproduce many features of binary phase diagrams by introducing the interaction parameters afterwards.

Over the years, there have been various revisions of lattice stability values for common crystal structures [48], and every revision necessitates the re-evaluation of interaction parameters in the solution phase shown in Eq. 2.49. It was not until lattice stability values were established by the Scientific Group Thermodata Europe (SGTE) [52] that the development of binary thermodynamic models using the same thermodynamic models as for pure elements became possible, and those binary models as developed in different groups around the world can be combined to create thermodynamic models of ternary and multi-component systems. Clearly, any further modifications of the SGTE pure element database will require the re-modeling of all binary and ternary systems in which the models of pure elements are changed. This challenge is briefly addressed in the later part of this chapter.

An issue less addressed is the Gibbs energy of the end-members in non-stoichiometric compounds, i.e. Eq. 2.129, where each sublattice contains only one element. In the latter case, this is the lattice stability of the element in the structure of the compound. Since the stable composition ranges of non-stoichiometric compounds are typically small, the existing method cannot be used to reliably evaluate the Gibbs energy of end-members, and currently there is no commonly accepted lattice stability database for compounds. Most values used in the existing databases have been either roughly estimated or computed from first-principles calculations. Such a standard database is highly desirable in order to make the various models of compounds compatible.

In an effort to compare the lattice stability obtained from CALPHAD models and first-principles calculations, Wang *et al.* [53] systematically calculated the total energies of 78 pure elements at 0 K in the face-centered-cubic (fcc), body-centered-cubic (bcc), and hexagonal-close-packed (hcp) crystal structures using the projector augmented-wave (PAW) method within the generalized gradient approximation (GGA). The calculated values are compared with the values in the SGTE database in Table 6.1 and Table 6.2. For non-transition metal elements, the differences between the SGTE data and the PAW-GGA data are typically around 1 ~ 2 kJ/mole-of-atoms or less, while for

Table 6.1 Lattice stability $E^{bcc-fcc}$ (kJ/mol atom), from [53] with permission from Elsevier

Li	Be											B	C	N	O	F
0.25	2.19											34.73	– 19.71			
0.11	0.04												– 6.00			
0.11	0.50															
Na	**Mg**											**Al**	**Si**	**P**	**S**	**Cl**
0.10	1.37	←				VASP-PAW-GGA					→	9.21	– 1.89	– 16.04	– 17.65	
0.05	0.50	←				SGTE data					→	10.08	– 4.00	7.95		
0.05	0.50	←				Saunders *et al.*					→	10.08	– 4.00			
K	**Ca**	**Sc**	**Ti**	**V**	**Cr**	**Mn**	**Fe**	**Co**	**Ni**	**Cu**	**Zn**	**Ga**	**Ge**	**As**	**Se**	**Br**
0.08	1.62	5.80	4.79	– 23.95	– 36.76	7.41	– 8.45	8.31	9.15	2.87	5.94	1.48	0.71	– 10.71	– 14.68	– 2.85
– 0.05	1.41	– 3.02	0.48	– 7.50	– 6.13	0.78	– 7.97	1.71	7.99	4.02	– 0.08	0.70	– 1.90			
– 0.05	0.93			– 15.30	– 9.19	1.80		4.20	7.49	4.02	6.03	0.70	– 1.90			
Rb	**Sr**	**Y**	**Zr**	**Nb**	**Mo**	**Tc**	**Ru**	**Rh**	**Pd**	**Ag**	**Cd**	**In**	**Sn**	**Sb**	**Te**	**I**
0.07	0.43	10.02	3.61	– 31.20	– 38.74	19.04	48.93	32.39	3.74	2.27	4.92	1.02	0.99	– 8.96	– 11.19	– 1.26
– 0.20	1.33	1.19	– 0.29	– 13.50	– 15.20	8.00	9.00	19.00	10.50	3.40		0.64	– 1.11			
– 0.20	0.75			– 22.00	– 28.00	8.00	14.00	19.00	10.50	3.40		0.65	0.25			
Cs	**Ba**		**Hf**	**Ta**	**W**	**Re**	**Os**	**Ir**	**Pt**	**Au**	**Hg**	**Tl**	**Pb**	**Bi**	**Po**	**At**
0.12	– 1.62		10.14	– 23.75	– 45.02	24.87	70.92	59.39	7.85	1.90	– 1.43	– 1.41	4.20	– 4.70		
– 0.50	– 1.80		2.38	– 16.00	– 19.30	6.00	14.50	32.00	15.00	4.25		– 0.09	2.40	1.40		
– 0.50	– 1.80		– 4.14	– 26.50	– 33.00	18.20	30.50	32.00	15.00	4.25		0.07	2.40			
		La	**Ce**	**Pr**	**Nd**	**Pm**	**Sm**	**Eu**	**Gd**	**Tb**	**Dy**	**Ho**	**Er**	**Tm**	**Yb**	**Lu**
		12.22	22.40	11.56	12.00	12.53	12.89	– 1.61	13.08	12.97	12.73	12.36	11.86	11.29		9.90
Fr	**Ra**	**Ac**	**Th**	**Pa**	**U**	**Np**	**Pu**	**Am**	**Cm**	**Bk**	**Cf**	**Es**	**Fm**	**Md**	**No**	**Lr**
		12.56	13.95	17.09	– 10.36	– 23.17	11.73									

Table 6.2 Lattice stability $E^{hcp-fcc}$ (kJ/mol atom), from [53] with permission from Elsevier

Li	**Be**											**B**	**C**	**N**	**O**	**F**
– 0.05	– 7.91											– 67.84	– 6.18			
– 0.05	– 6.35												– 3.00			
– 0.05	– 6.35															
Na	**Mg**											**Al**	**Si**	**P**	**S**	**Cl**
0.03	– 1.22	←——— VASP-PAW-GGA ———→										2.85	– 3.26	– 3.77	– 43.63	
– 0.05	– 2.60	←——— SGTE data ———→										5.48	– 1.80			
– 0.05	– 2.60	←——— Saunders *et al.* ———→										5.48	– 1.80			
K	**Ca**	**Sc**	**Ti**	**V**	**Cr**	**Mn**	**Fe**	**Co**	**Ni**	**Cu**	**Zn**	**Ga**	**Ge**	**As**	**Se**	**Br**
0.26	0.29	– 4.48	– 5.51	0.53	– 0.91	– 3.01	– 7.76	– 1.99	2.13	0.52	– 2.88	0.69	– 0.27	– 4.83	– 35.88	3.00
0.00	0.50	– 5.00	– 6.00	– 3.50	– 2.85	– 1.00	– 2.24	– 0.43	2.89	0.60	– 2.97	0.70	– 1.00			
	0.50		– 6.00	– 4.80	– 1.82	– 1.00		– 0.43	1.50	0.60		0.70	– 1.00			
Rb	**Sr**	**Y**	**Zr**	**Nb**	**Mo**	**Tc**	**Ru**	**Rh**	**Pd**	**Ag**	**Cd**	**In**	**Sn**	**Sb**	**Te**	**I**
– 0.02	0.38	– 2.13	– 3.69	– 3.08	1.14	– 6.53	– 10.79	3.26	2.50	0.29	– 1.00	0.35	– 0.50	– 3.94	– 23.24	0.99
0.00	0.25	– 6.00	– 7.60	– 3.50	– 3.65	– 10.00	– 12.50	3.00	2.00	0.30	– 0.89	0.37	– 1.61			
	0.25		– 7.60	– 5.00	– 5.00	– 10.00	– 12.50	3.00	2.00	0.30		0.65	– 0.25			
Cs	**Ba**		**Hf**	**Ta**	**W**	**Re**	**Os**	**Ir**	**Pt**	**Au**	**Hg**	**Tl**	**Pb**	**Bi**	**Po**	**At**
– 0.07	– 0.40		– 6.82	3.06	– 1.78	– 6.26	– 13.26	6.55	5.02	0.08	– 1.92	– 1.81	1.80	– 4.20		
0.00	0.20		– 10.00	– 4.00	– 4.55	– 11.00	– 13.00	4.00	2.50	0.24	– 2.07	– 0.31	0.30			
	0.20		– 10.00	– 6.50	– 6.00	– 11.00	– 13.00	4.00	2.50	0.55		– 0.31	0.30			
		La	**Ce**	**Pr**	**Nd**	**Pm**	**Sm**	**Eu**	**Gd**	**Tb**	**Dy**	**Ho**	**Er**	**Tm**	**Yb**	**Lu**
		2.63	8.50	2.08	1.94	1.77	1.53	0.24	0.74	0.24	– 0.41	– 1.18	– 1.97	– 2.68		– 3.86
Fr	**Ra**	**Ac**	**Th**	**Pa**	**U**	**Np**	**Pu**	**Am**	**Cm**	**Bk**	**Cf**	**Es**	**Fm**	**Md**	**No**	**Lr**
		0.80	4.00	0.49	– 15.79	– 14.01	0.69									

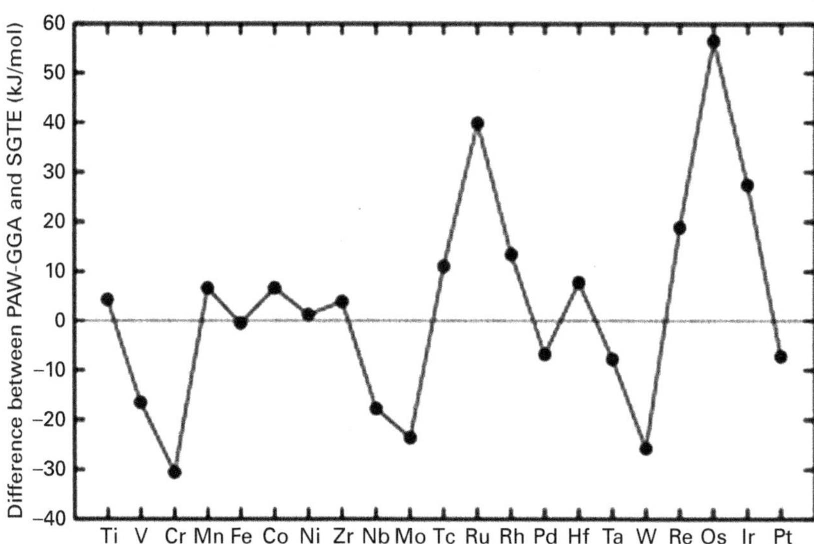

Figure 6.1 *Differences between the PAW-GGA and SGTE data for $E^{bcc-fcc}$, for selected elements, from [53] with permission from Elsevier.*

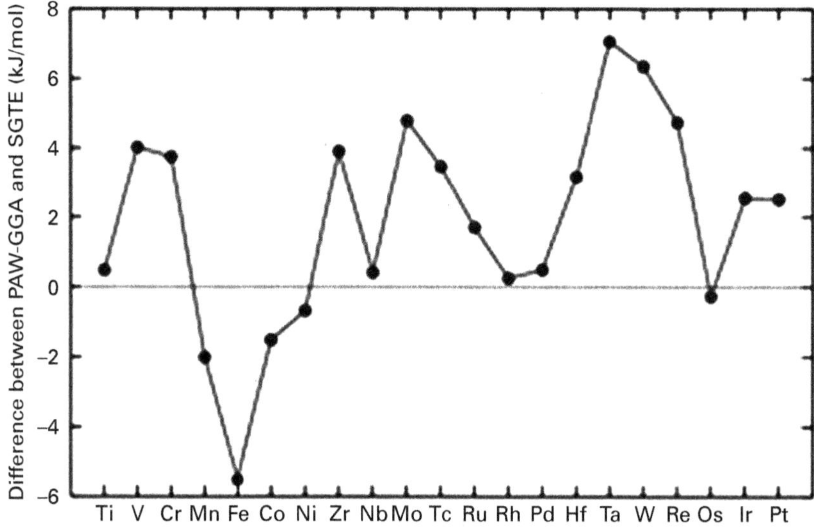

Figure 6.2 *Differences between the PAW-GGA and SGTE data for $E^{hcp-fcc}$, for selected elements, from [53] with permission from Elsevier.*

some transition metal elements the differences can be quite large, for example, as high as about 54 kJ/mole-of-atoms for $E_{Os}^{bcc-fcc}$ and about 40 kJ/mole-of-atoms for $E_{Ru}^{bcc-fcc}$. Figure 6.1 and Figure 6.2 present the differences between the PAW-GGA data and the SGTE data, for elements from the Ti group to the Ni group, respectively.

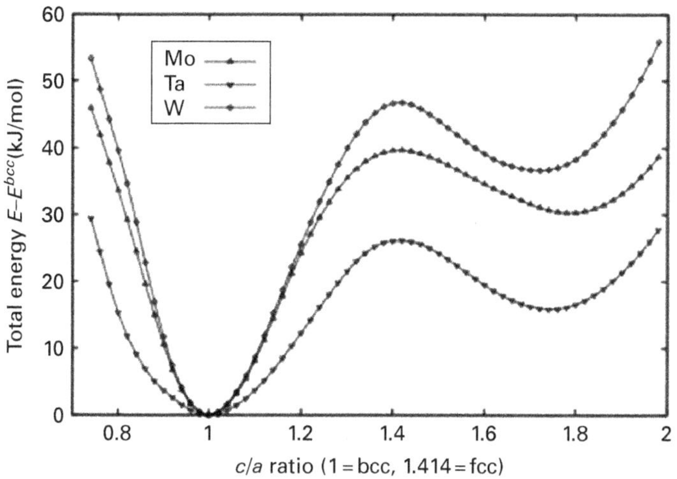

Figure 6.3 *Total energy, $E - E^{bcc}$, along the Bain deformation path between bcc and fcc structures for Mo, Ta, and W, from [53] with permission from Elsevier*

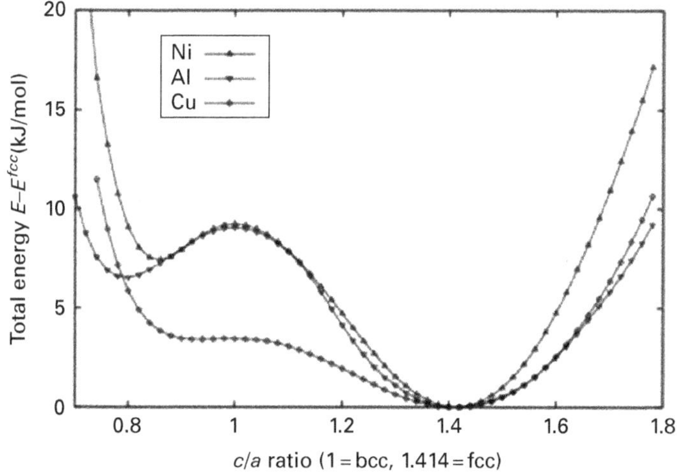

Figure 6.4 *Total energy, $E - E^{fcc}$, along the tetragonal transformation path between bcc and fcc structures for Ni, Al, and Cu, from [53] with permission from Elsevier*

The large differences between the first-principles calculations and the SGTE data could partly be attributed to the instability of the higher-energy phases, the entropies of which become abnormal at finite T. The lattice instabilities along the tetragonal transformation path between fcc and bcc structures with the continuous change of the c/a ratio defined in a bcc-based tetragonal lattice are demonstrated for bcc Mo, Ta, W in Figure 6.3 and for fcc Al, Cu, Ni in Figure 6.4. It is shown that the fcc structure of bcc Mo, Ta, and W is a local maximum with respect to the tetragonal transformation, and that the higher the maximum is, the larger the discrepancy between the SGTE data and the present PAW-GGA data. For fcc Al, Cu, Ni, the bcc structure is at a local maximum.

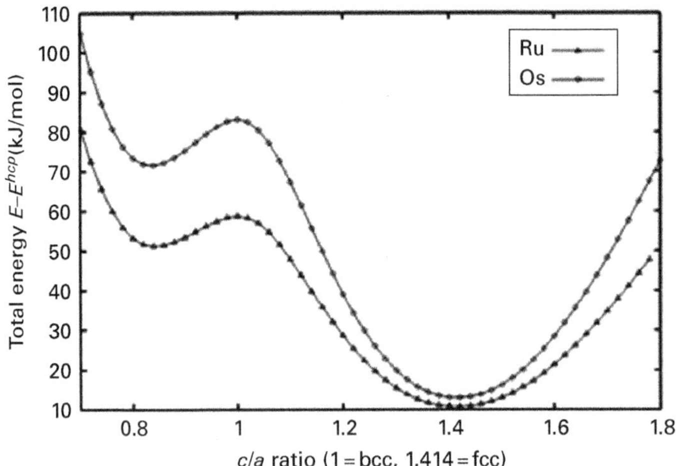

Figure 6.5 *Total energy, $E - E^{hcp}$, along the tetragonal transformation path for Ru and Os, from [53] with permission from Elsevier*

Similarly, the lattice instabilities along the tetragonal transformation path for the hcp metals Ru and Os are shown in Figure 6.5. The behavior of energy against the c/a ratio of these two hcp metals is very similar to that of the fcc elements.

It can be concluded that a fcc structure for elements with ground state bcc, or a bcc structure for elements with ground state fcc, is unstable with respect to the tetragonal transformation. For an unstable structure, the harmonic description of its vibrational entropy is thermodynamically incorrect since the potential surface seen by the lattice ion can no longer be approximated by a parabola. If an unstable structure of a pure element is stabilized at high temperatures, its entropy will be abnormal. The instability issue has been recently addressed by ab initio molecular dynamics simulations at high temperatures using W as an example [54]; this is beyond the scope of the book and thus not discussed here.

6.2 Modeling of pure elements

In modeling the Gibbs energy of pure elements in the SER structure in terms of Eq. 2.38, the coefficients in Eq. 2.35 are evaluated using the heat capacity data, b' in Eq. 2.36 is evaluated from the value of $S_{298.15}$, and a in Eq. 2.37 is evaluated from $H_{298.15}^{SER} = 0$. For the high temperature phase, the enthalpy of transformation from the low temperature phase to the high temperature phase, ΔH_{trans}, can be measured by the calorimetry methods discussed in Section 4.2, and the entropy of transformation, ΔS_{trans}, is then calculated using the equilibrium condition of equal Gibbs energy of the two phases, i.e.

$$\Delta S_{trans} = \frac{\Delta H_{trans}}{T_{trans}} \tag{6.1}$$

where T_{trans} is the transition temperature. The quantities ΔH_{trans} and ΔS_{trans} are then used to evaluate the integration constants b' and a, in place of $S_{298.15}$ and $H_{298.15}^{SER}$, for the structure in the SER state.

This works well for the stable temperature range of each phase. However, there is an issue in extrapolating above and below the melting temperature (T_m). It is known that the heat capacity of the solid phase, C_P^s, increases with temperature, while that of the liquid phase, C_P^l, is typically constant. The extrapolation of the Gibbs energy of a solid phase to temperatures above its melting temperature can result in the solid phase becoming more stable than the liquid phase at high temperatures. By the same token, extrapolation of the Gibbs energy of a liquid phase to temperatures below its melting temperature can result in the liquid phase becoming more stable than the solid phase at low temperatures. To address this problem, it was proposed by SGTE that the heat capacity of the solid phase approaches that of the liquid at high temperatures, and that of the liquid phase approaches that of the solid phase at low temperatures, using the following equations.

- For solid at $T > T_m$,

$$C_P^s = C_P^l + \left(C_P^s(T_m) - C_P^l(T_m)\right)\left(\frac{T}{T_m}\right)^{-10} \qquad 6.2$$

$$G^s = G^l + \left(C_P^s(T_m) - C_P^l(T_m)\right)\left\{-\frac{T - T_m}{10} + \left(1 - \left(\frac{T}{T_m}\right)^{-9}\right)\frac{T_m}{90}\right\} \qquad 6.3$$

- For liquid at $T < T_m$,

$$C_P^l = C_P^s + \left(C_P^l(T_m) - C_P^s(T_m)\right)\left(\frac{T}{T_m}\right)^{6} \qquad 6.4$$

$$G_m^l = G_m^s + \left(C_P^l(T_m) - C_P^s(T_m)\right)\left\{\frac{T - T_m}{6} + \left(1 - \left(\frac{T}{T_m}\right)^{7}\right)\frac{T_m}{42}\right\} \qquad 6.5$$

As an example, the heat capacities of solid fcc Al and liquid Al in the SGTE pure element database are plotted in Figure 6.6. It can be seen that the heat capacity of fcc Al approaches that of liquid Al at high temperatures, while the heat capacity of liquid Al approaches that of fcc Al at low temperatures. This ensures that the liquid Al is stable at high temperatures, and fcc Al is stable at low temperatures. However, this simple model for liquids is often not satisfactory, in comparison with the available experimental data for supercooled liquids, particularly those systems with glass transitions. New models are thus needed and are being developed in the CALPHAD community.

6.3 Modeling of stoichiometric phases

The Gibbs energy of a stoichiometric phase can be modeled in the same way as that of pure elements discussed above using the data for heat capacity, $S_{298.15}$, and enthalpy of formation at 298.15 K (Eq. 2.43). When these data are not available from experiments,

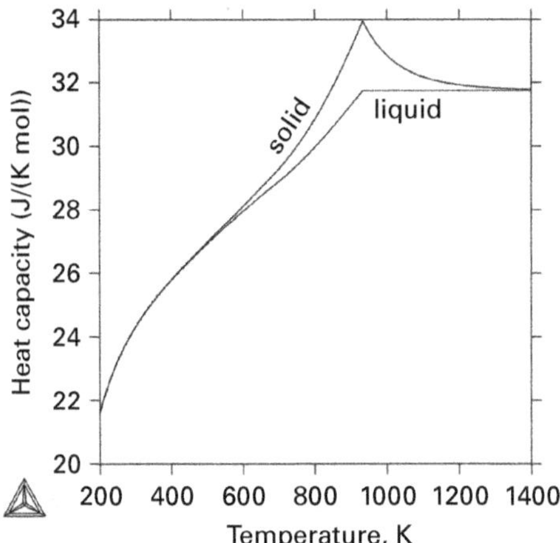

Figure 6.6 *Heat capacities of fcc Al solid and liquid as a function of temperature.*

they can be predicted by first-principles calculations as discussed in Chapter 5. It should be pointed out that constraints placed on the heat capacity of stoichiometric compounds above melting temperatures, i.e. Eq. 6.2 and Eq. 6.3, have not been rigorously implemented in the literature for such modeling, because the heat capacity of the corresponding liquid solution is not well established and the heat capacity of a compound is often not available.

When the data for heat capacity are not available, a simple approach, commonly referred to as the Neumann–Kopp rule, assuming that the heat capacity of formation of Eq. 2.45 is zero, can be used. The Gibbs energy of the compound is written as

$$G = \sum N_i {}^0G_i^{ref} + \Delta_f H - T\Delta_f S \qquad 6.6$$

with $\Delta_f H$ and $\Delta_f S$ modeled as constants. An drawback of Eq. 6.6 is that the melting temperature of the compound can be higher than those of the pure elements, and the heat capacity of the compound may thus be questionable at temperatures higher than the melting temperatures of pure elements due to the non-physical contributions from pure elements based on Eq. 6.2.

6.4 Modeling of random solution phases

Depending on the degree of short-range ordering in a solution phase, various Gibbs energy models are available as discussed in Section 2.3.1. When the short-range ordering is weak, it can be accounted for by the composition dependence of the excess Gibbs energy in a binary system. This is discribed in terms of the Redlich–Kister polynomial as follows:

$$^EG_m = x_i x_j \sum_{k=0} {}^kL_{i,j}(x_i - x_j)^k = x_i x_j \left[{}^0L_{i,j} + {}^1L_{i,j}(x_i - x_j) + {}^2L_{i,j}(x_i - x_j)^2 + \ldots\right] \qquad 6.7$$

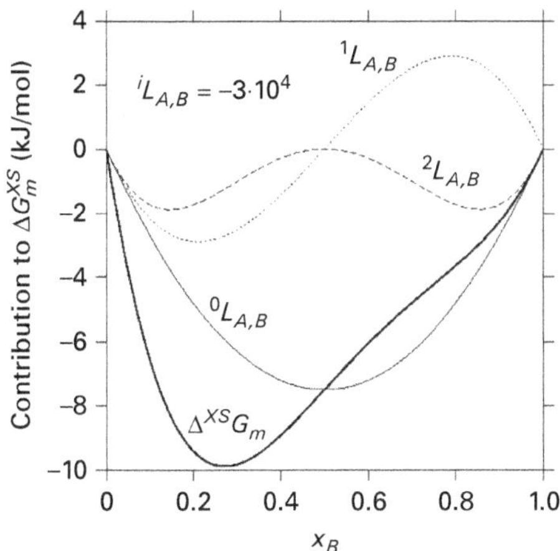

Figure 6.7 *Contributions of interactions parameters to the excess Gibbs energy.*

where the interaction parameters, ${}^kL_{i,j}$, can be temperature dependent or even have contributions from the heat capacity, as in Eq. 2.38, when data are available. Equation 6.7 shows that ${}^0L_{i,j}$ and ${}^2L_{i,j}$ are symmetrical with respect to composition, while ${}^1L_{i,j}$ is asymmetrical. Their individual contributions to the excess Gibbs energy are shown in Figure 6.7 with all interaction parameters taken as−30000 J/mole-of-atoms.

It can be seen in Figure 6.7 that even though all interaction parameters are negative, the asymmetrical shape of ${}^1L_{i,j}$ results in a change in curvature in the excess Gibbs energy as a function of composition. This indicates the tendency to form a miscibility gap at low temperatures. The interaction parameters are evaluated from the data of enthalpy, entropy, and heat capacity of mixing. The experimental data on the enthalpy of mixing are available for the liquid phase in some systems, but typically are very limited for solid solution phases. First-principles calculations can predict the enthalpy, entropy, and heat capacity of mixing in solid solution phases using the dilute solution approach with one solute atom in a supercell and the CPA/CE/SQS approach for concentrated solutions as discussed in Section 5.5. This demonstrates again that the interaction parameters and the lattice stability jointly determine the Gibbs energy of an individual phase. The change in lattice stability requires the re-evaluation of interaction parameters.

For individual phases with strong short-range ordering, quasi-chemical or associated models can be used. As discussed in Section 2.3.1, with fixed composition in the system, the amounts of various chemical bonds or associates are related through mass conservation in the system and are calculated through minimization of the Gibbs energy of the phase with given temperature, pressure, and amount of each independent component. The model parameters include the formation energy of bonds or associates and the interactions between various bonds or associates, noting that pure elements can be considered as the simplest associates. The interactions between pure elements can be predicted from first-principles calculations as mentioned above, but currently there are no efficient approaches to predict the interactions between associates from first-principles calculations.

6.5 Modeling of solution phases with long-range ordering

A commonly used Gibbs energy model is shown in Section 2.3.2, with the crystal lattice divided into sublattices; it is often referred to as the compound energy formalism [55]. The Gibbs energy of end-members represented by Eq. 2.129 plays the same important role for solution phases with sublattices as the lattice stability for random solution phases. The end-members are modeled in the same way as the stoichiometric phases discussed in Section 6.3. The enthalpy and entropy of mixing in each sublattice can be predicted by first-principles calculations using the dilute solution and SQS approaches discussed in Chapter 5 and modeled in the same way as the random solution discussed in Section 6.4.

It is important to realize that with a simple two-sublattice model $(A,B)_a(C,D)_b$, the miscibility gap can easily form even without any interaction parameters when the Gibbs energies of end-members differ from each other significantly. The contribution of end-members to the Gibbs energy of the phase, i.e. Eq. 2.130, is re-written as follows, and schematically shown in Figure 6.8,

$$^0G_{mf} = y_A^I y_C^{II}\,{}^0G_{A:C} + y_A^I y_D^{II}\,{}^0G_{A:D} + y_B^I y_C^{II}\,{}^0G_{B:C} + y_B^I y_D^{II}\,{}^0G_{B:D} \qquad 6.8$$

where the superscript denotes the sublattice, and colon and comma separate sublattices and interaction components, respectively. From Figure 6.8b, it is evident that there is a

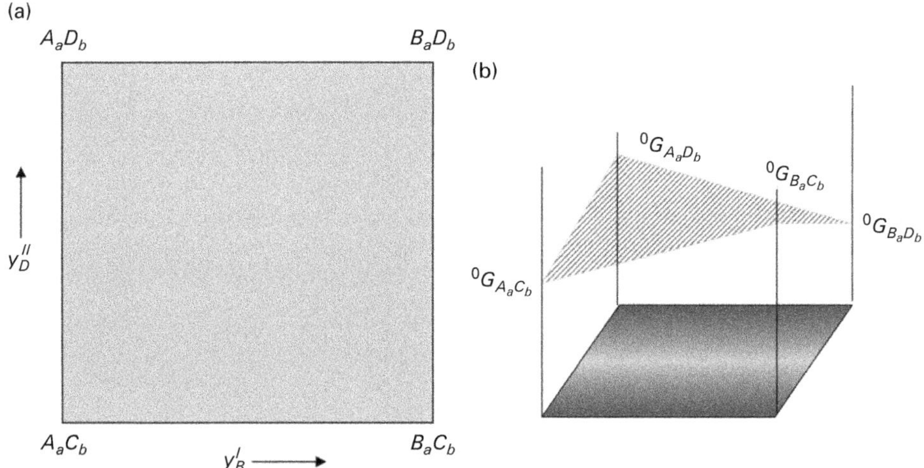

Figure 6.8 *Schematic diagrams depicting (a) the concentration square with the site fractions of B and D on the horizontal and vertical axes, respectively, and (b) Gibbs energy reference plane for $(A,B)_a(C,D)_b$, as represented by Eq. 6.8.*

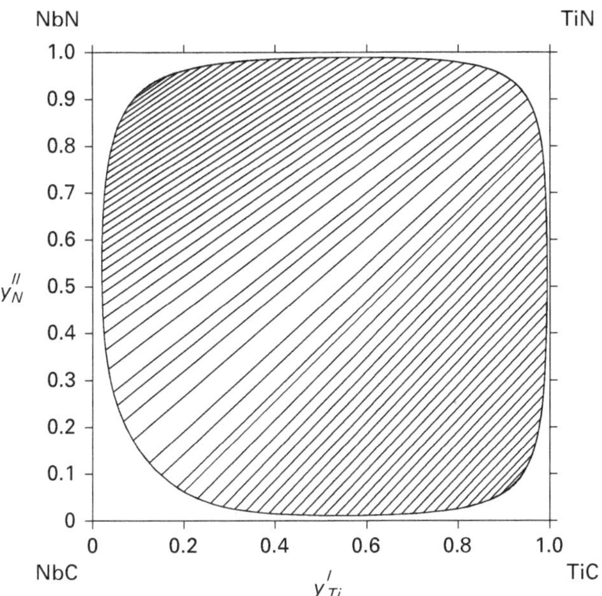

Figure 6.9 *Miscibility gap in (Ti,Nb)(C,N) at* 1673 K *. The straight lines in the middle of the plot are tie-lines.*

strong tendency to form a miscibility gap between the composition sets $(A)_a(C)_b$ and $(B)_a(D)_b$ because of their lower Gibbs energies compared with the other two end-members. Since it would be rare for all four end-members to have equal Gibbs energy values, a miscibility gap in this type of phase is practically inevitable at low temperatures. An example is shown in Figure 6.9 for the complex titanium niobium carbonitride (Ti,Nb)(C,N). The lines parallel to the direction from NbC to TiN are tie-lines. The Gibbs energy values of TiC, TiN, NbC, and NbN are −144495, −229236, −132324, and −179772 J/mole-of-atoms, respectively. The Gibbs energy value of TiN is significantly lower than the other values, resulting in the tie-lines originating from the TiN corner.

The order–disorder transitions can be similarly described, the simplest case being a two-sublattice model $(A, B)_a(A, B)_b$. When the site fractions of A or B in both sublattices are the same, it becomes a random solution model; when they are different, the phase is partially ordered; and when there is only one component in each sublattice, the phase is fully ordered as a stoichiometric compound. The Gibbs energy of this phase is obtained from Eq. 2.131 as follows:

$$\begin{aligned} G_{mf} = {} & y^I_A y^{II}_A {}^0G_{A:A} + y^I_A y^{II}_B {}^0G_{A:B} + y^I_B y^{II}_A {}^0G_{B:A} \\ & + y^I_B y^{II}_B {}^0G_{B:B} + aRT(y^I_A \ln y^I_A + y^I_B \ln y^I_B) + bRT \\ & + (y^{II}_A \ln y^{II}_A + y^{II}_B \ln y^{II}_B) + y^{II}_A y^I_A y^I_B L_{A,B:A} + y^{II}_B y^I_A y^I_B L_{A,B:B} + y^I_A y^{II}_A y^{II}_B L_{A:A,B} \\ & + y^I_B y^{II}_A y^{II}_B L_{B:A,B} + y^I_A y^I_B y^{II}_A y^{II}_B L_{A,B:A,B} \end{aligned} \qquad 6.9$$

The relationship between site fraction and overall atomic fractions in such a two-sublattice model can be represented by Eq. 2.138 and is schematically shown in

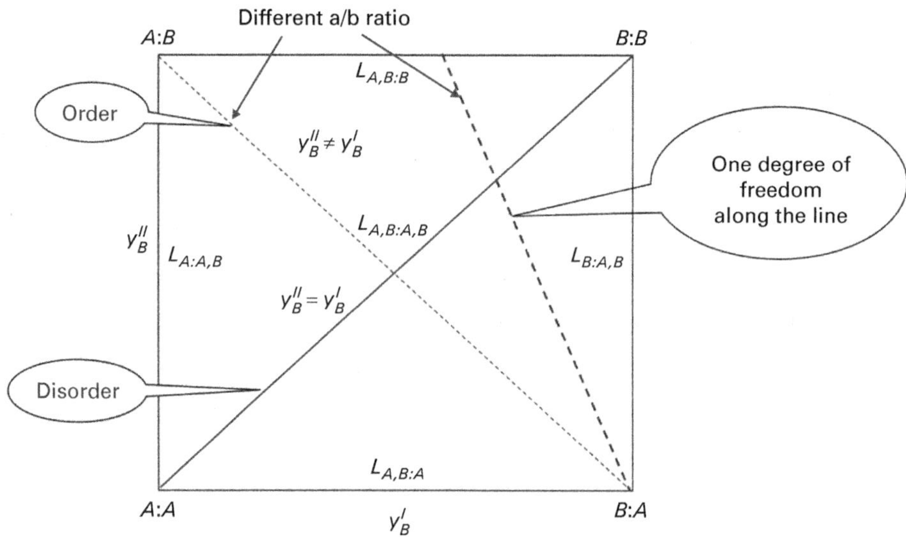

Figure 6.10 *Schematic composition square of* $(A,B)_a(A,B)_b$.

Figure 6.10. The two dashed lines represent the phase with $x_B = a/(a+b)$ but different a/b ratios. Along the dashed lines, the phase can adjust the site fraction to minimize its Gibbs energy, i.e. it has one internal degree of freedom, i.e. ξ, to be either disordered on the diagonal line between *A*:*A* and *B*:*B* or ordered anywhere else. The interplay of interaction parameters and site fractions is depicted: $L_{A,B:A}$, $L_{A,B:B}$, $L_{A:A,B}$, and $L_{B:A,B}$ affect the four sides, and $L_{A,B:A,B}$ influences the center part.

When fully disordered with $y_A^I = y_A^{II} = x_A$ and $y_B^I = y_B^{II} = x_B$, Eq. 6.9 becomes

$$\begin{aligned}
G_{mf} &= x_A(1-x_B)^0G_{A:A} + x_Ax_B\,{}^0G_{A:B} + x_Ax_B\,{}^0G_{B:A} + x_B(1-x_A)\,{}^0G_{B:B} \\
&\quad + (a+b)RT(x_A\ln x_A + x_B\ln x_B) + x_Ax_Bx_AL_{A,B:A} + x_Ax_Bx_B\,L_{A,B:B} \\
&\quad + x_Ax_Bx_A\,L_{A:A,B} + x_Ax_Bx_B\,L_{B:A,B} + x_Ax_Ax_Bx_BL_{A,B:A,B} \\
&= x_A{}^0G_{A:A} + x_B\,{}^0G_{B:B} + (a+b)RT(x_A\ln x_A + x_B\ln x_B) \\
&\quad + x_Ax_B\,[({}^0G_{A:B} + {}^0G_{B:A} - {}^0G_{A:A} - {}^0G_{B:B}) + x_A(L_{A,B:A} + L_{A:A,B}) \\
&\qquad + x_B\,(L_{A,B:B} + L_{B:A,B}) + x_Ax_BL_{A,B:A,B}] \\
&= (a+b)[x_A{}^0G_A + x_B\,{}^0G_B + RT(x_A\ln x_A + x_B\ln x_B) + x_Ax_BL_{A,B}]
\end{aligned} \tag{6.10}$$

with

$${}^0G_{A:A} = (a+b)^0G_A \tag{6.11}$$

$${}^0G_{B:B} = (a+b)^0G_B \tag{6.12}$$

$$\begin{aligned}
L_{A,B} = &\left[({}^0G_{A:B} + {}^0G_{B:A} - {}^0G_{A:A} - {}^0G_{B:B}\right) + x_A(L_{A,B:A} + L_{A:A,B}) \\
&+ x_B\,(L_{A,B:B} + L_{B:A,B}) + x_Ax_BL_{A,B:A,B}]/(a+b)
\end{aligned} \tag{6.13}$$

where ${}^0G_A$, ${}^0G_B$, and $L_{A,B}$ are the molar Gibbs energy of pure *A* and *B* and the molar interaction parameter in the disordered solid solution, respectively. It is evident that the

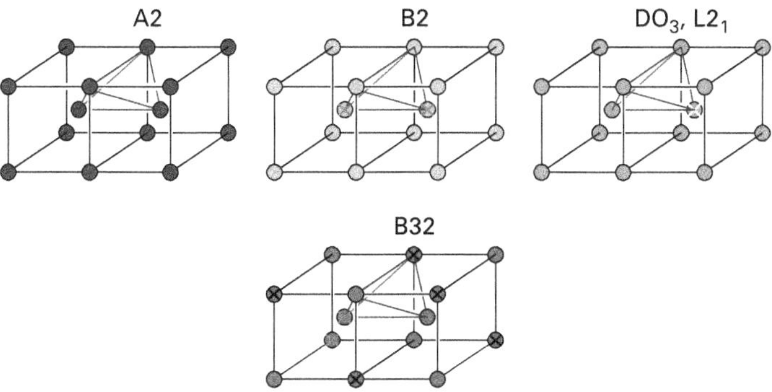

Figure 6.11 *Atomic structures and four sublattice tetrahedrons of bcc disordered and ordered phases.*

interaction parameter $L_{A,B}$ is fully determined by the parameters in the ordered phase, but the parameters in the ordered phase are not uniquely determined by the interaction parameters in the disordered phase.

Due to crystal symmetry, some of the parameters in Eq. 6.9 are related. For example, in the bcc *A2*/*B2* ordering with $a = b = 0.5$, the bcc corner or center lattice sites are favored by one type of atom but the two sublattices are equivalent crystallographically, resulting in the following relations:

$$^{0}G_{A:B} = {}^{0}G_{B:A} \tag{6.14}$$

$$L_{A,B:A} = L_{A:A,B} \tag{6.15}$$

$$L_{A,B:B} = L_{B:A,B} \tag{6.16}$$

For more complex orderings of a bcc lattice such as B32, D0$_3$, and L2$_1$ shown in Figure 6.11, with ideal compositions *AB*, A_3B, and A_2BC, respectively, more sublattices are needed, noting that the L2$_1$ Heusler structure exists in ternary systems only. To use one model to describe all orderings in the bcc lattice, the minimum cluster is an irregular tetrahedron with four sublattices, depicted in Figure 6.11, as discussed in the modeling of the Al–Fe system [56]. In such a four-sublattice model $(A,B)^{I}_{0.25}(A,B)^{II}_{0.25}(A,B)^{III}_{0.25}(A,B)^{IV}_{0.25}$, the site fractions of A2, B2, B32, and D0$_3$ are represented by $y^{I}_{i} = y^{II}_{i} = y^{III}_{i} = y^{IV}_{i}$, $y^{I}_{i} = y^{II}_{i} \neq y^{III}_{i} = y^{IV}_{i}$, $y^{I}_{i} = y^{III}_{i} \neq y^{II}_{i} = y^{IV}_{i}$, and $y^{I}_{i} = y^{II}_{i} \neq y^{III}_{i} \neq y^{IV}_{i}$, respectively. The site fractions of L2$_1$ are the same as those of D0$_3$ except that they have at least three components.

Another common ordering phenomenon occurs for the fcc lattice including the disordered A1 structure and ordered L1$_0$ and L1$_2$ structures as shown in Figure 6.12. In the L1$_0$ structure, the neighboring (001) planes are favored by different atoms, respectively, resulting in an ideal composition *AB*. In the L1$_2$ structure, however, the corners and faces are favored by different atoms, respectively, resulting in an ideal composition A_3B. In a four-sublattice model $(A,B)^{I}_{0.25}(A,B)^{II}_{0.25}(A,B)^{III}_{0.25}(A,B)^{IV}_{0.25}$, the

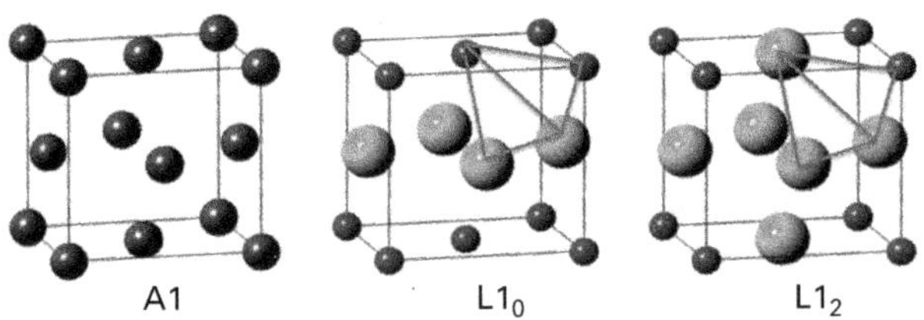

Figure 6.12 *Atomic structures and four-sublattice tetrahedrons of fcc disordered and ordered phases.*

site fractions of A1, L1$_0$, and L1$_2$ are represented by $y_i^I = y_i^{II} = y_i^{III} = y_i^{IV}$, $y_i^I = y_i^{II} \neq y_i^{III} = y_i^{IV}$, and $y_i^I = y_i^{II} = y_i^{III} \neq y_i^{IV}$, respectively [57].

As mentioned above, the interaction parameters in each sublattice can be predicted by first-principles calculations using the dilute solution and SQS approaches when there is only one component in each of the remaining sublattices. For interactions involving two components in two or more sublattices, i.e. $L_{A,B:A,B}$ in Eq. 6.9, applicable to four-sublattice models [58], the energetics from the cluster expansion (CE) approach discussed in Section 5.5 can be used to evaluate the interaction parameters.

6.6 Modeling of magnetic and electric polarizations

The elastic, magnetic, and electric energy contributions to the Gibbs energy discussed in Chapter 2.5 originate from the changes of strain, magnetization, and polarization in the system due to the external elastic, magnetic, and electric fields. The CALPHAD modeling of the elastic compliance coefficients, s_{ijkl}, permeability, μ_{ij}, and permittivity, k_{ij}, as functions of temperature, stress, magnetic and electric fields, and composition has not been reported in the literature, except for recent work on the modeling of elastic stiffness coefficients [59], which are the inverse of elastic compliance coefficients.

In addition to the contributions from the external magnetic and electric fields, some phases are spontaneously polarized due to unpaired electron spins, such as in ferromagnetic bcc Fe, or internal electric dipoles, such as in ferroelectric $PbTiO_3$, or both, such as in multi-ferroric $BiFeO_3$. The CALPHAD modeling of the spontaneous magnetization contribution to the Gibbs energy is based on the Inden–Hillert–Jarl model with an empirical constraint [60], recently modified by Xiong *et al.* [61]. This contribution is important in Fe-based alloys, as it is the reason for the return of the bcc structure at low temperatures. The CALPHAD modeling of the spontaneous electric polarization contribution to the Gibbs energy has not been reported in the literature. The existing thermodynamic modeling of the spontaneous electric polarization is based on the Landau–Ginsburg–Devonshire formalism in a power series of the polarization [62]. Our recent approach in modeling critical phenomena in general is discussed in Section 5.2.5 and Chapter 9.

7 Applications to chemical reactions

A chemical reaction can be viewed as a framework dividing a system into two closed subsystems: reactants and products. The phases and species of reactants and products are selected from the possible phases and species that may form from the independent components of the system. A chemical reaction can thus be considered as an internal process to transfer heat and work between the two subsystems of reactants and products. It is evident that this subset of phases and species only represents partial equilibrium information for the system under given external conditions, and more stable equilibrium states may exist if more phases and species are included in the global equilibrium, depicted by the phase diagrams discussed in previous chapters where all known phases and species are included.

7.1 Internal process and differential and integrated driving forces

The driving force for an internal process can be defined as follows, using U, H, F, or G as discussed in Sections 1.2 and 1.4 depending on which system variables are kept constant:

$$-D = \left(\frac{\partial U}{\partial \xi}\right)_{S,V,N_i} = \left(\frac{\partial H}{\partial \xi}\right)_{S,P,N_i} = \left(\frac{\partial F}{\partial \xi}\right)_{T,V,N_i} = \left(\frac{\partial G}{\partial \xi}\right)_{T,P,N_i} \qquad 7.1$$

The quantity D can be termed the differential driving force as it relates the derivative of an energy with respect to an internal process. For a system under constant T and P, let us consider an internal process for forming a new phase α with the composition x_i^α and Gibbs energy $G_m^\alpha(x_i^\alpha)$. The differential driving force for such an internal process can thus be defined as

$$-D = G_m^\alpha(x_i^\alpha) - \sum_i x_i^\alpha \mu_i = \sum_i x_i^\alpha(\mu_i^\alpha - \mu_i) \qquad 7.2$$

where μ_i is the chemical potential of component i in the system. It may also be called the nucleation driving force if the α phase is nucleating in the system. As discussed in Section 2.2.1, chemical potentials are the intercepts on the Gibbs energy axis of the tangent plane of the Gibbs energy. Equation 7.2 thus represents the distance between the tangent planes of the original system and the α phase, at the composition of the α phase. Evidently, this distance is at its maximum when the two tangent planes are parallel to each other; then it is commonly called the parallel tangent construction when evaluating the nucleation driving force.

The situation is different for chemical reactions where the amount of each component in the reactants is the same as that in the products, i.e. there is a mass balance between reactants and products. The driving force for a chemical reaction is defined by the Gibbs energy difference between the reactants and products, if all the reactants are transferred to the products. This driving force may thus be called an integrated driving force as it describes the energy difference of a system under two different states, i.e.

$$-\int D d\xi = \Delta G = \sum_p n_p G_p - \sum_r n_r G_r \qquad 7.3$$

where the subscripts p and r denote products and reactants, n_p and n_r are the corresponding numbers of moles, and G_p and G_r are the Gibbs energies per mole of formula of their respective stoichiometries as written in the chemical reaction. Conventionally, the products and reactants are represented by species or stoichiometric compounds rather than individual phases. This is particularly evident when various gaseous species are considered in a chemical reaction. Consequently, G_p and G_r in Eq. 7.3 represent the chemical potentials of the product and reactant species. For a species with a fixed composition, its chemical potential is the same as its Gibbs energy as shown by Eq. 2.21. For a species in a solution phase, its chemical potential is related to its activity as shown in Eq. 2.66. Equation 7.3 can thus be further written as

$$\Delta G = \sum_p n_p{}^0G_p - \sum_r n_r{}^0G_r + RT\ln\frac{\prod_p (a_p)^{n_p}}{\prod_r (a_r)^{n_r}} = \Delta^0 G + RT\ln\frac{\prod_p (a_p)^{n_p}}{\prod_r (a_r)^{n_r}} \qquad 7.4$$

It is evident from Eq. 2.21 that for stoichiometric phases in a chemical reaction, their activities are equal to one. At equilibrium, the integrated driving force becomes zero, i.e. $\Delta G = 0$, and Eq. 7.4 can be re-arranged to become

$$-RT\ln\frac{\prod_p (a_p)^{n_p}}{\prod_r (a_r)^{n_r}} = -RT\ln K_e = \Delta^0 G = \Delta^0 H - T\Delta^0 S \qquad 7.5$$

where K_e is often called the reaction constant relating the activities of products and reactants at equilibrium. In a system with $\prod_p (a_p)^{n_p} / \prod_r (a_r)^{n_r} > K_e$ the chemical reaction goes to the left, and the products are decomposed, while if $\prod_p (a_p)^{n_p} / \prod_r (a_r)^{n_r} < K_e$ the chemical reaction goes to the right, and the products are formed. Equation 7.5 is often recast into the following form by dividing by $-RT$ on both sides of the equation:

$$\ln K_e = -\frac{\Delta^0 H}{RT} + \frac{\Delta^0 S}{R} \qquad 7.6$$

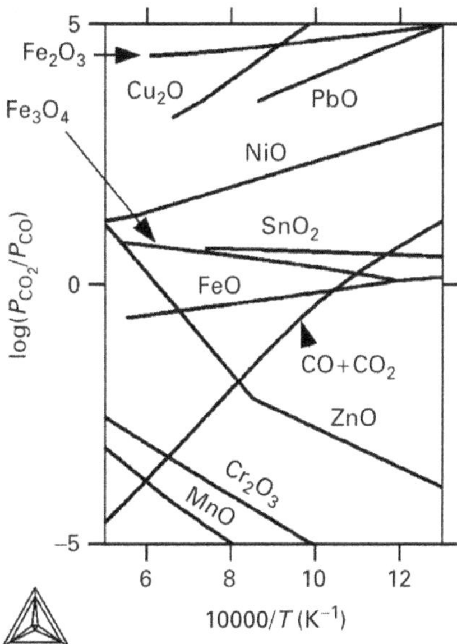

Figure 7.1 *The logarithm of K_ε plotted with respect to $1/T$ for several M–O systems at* 1 bar K_e *represented by the partial pressure ratio of CO_2 and CO, from [1] with permission from Cambridge University Press.*

With $\ln K_\varepsilon$ plotted with respect to $1/T$, Eq. 7.6 indicates that the slope is $-\Delta^0 H/R$ and the intercept on the y axis is $\Delta^0 S/R$, as shown in Figure 7.1.

7.2 Ellingham diagram and buffered systems

One type of chemical reaction is that between a pure element in liquid or solid state and its oxides involving one mole of oxygen gas, i.e.

$$\frac{4}{v_M} M + O_2(\text{gas}) = M_{4/v_M} O_2 \qquad 7.7$$

with v_M the valence of the element M in the oxide. Taking the pure M and gaseous O_2 at the reaction temperature and one atmospheric pressure as their respective reference states, the activities of both the pure M solid or liquid phase and its oxide are unity, and the activity of O_2 equals its partial pressure in an ideal gas. Equation 7.5 becomes

$$RT \ln P_{O_2} = \Delta^0 G = \Delta^0 H - T\Delta^0 S \qquad 7.8$$

Based on Eq. 7.8, one can plot $\Delta^0 G$ as a function of temperature for various oxidation reactions. This is called an Ellingham diagram. The intercept at $T = 0\text{K}$ is given by $\Delta^0 H$, and the slope is represented by $-\Delta^0 S$ and given by the following equation:

$$\Delta^0 S = S_{M_{4/v_M} O_2} - S_{O_2} - \frac{4}{v_M} S_M \qquad 7.9$$

Since the entropy of one mole of O_2 gas is significantly larger than those of the pure element and its oxide when they are in solid or liquid states and than the entropy

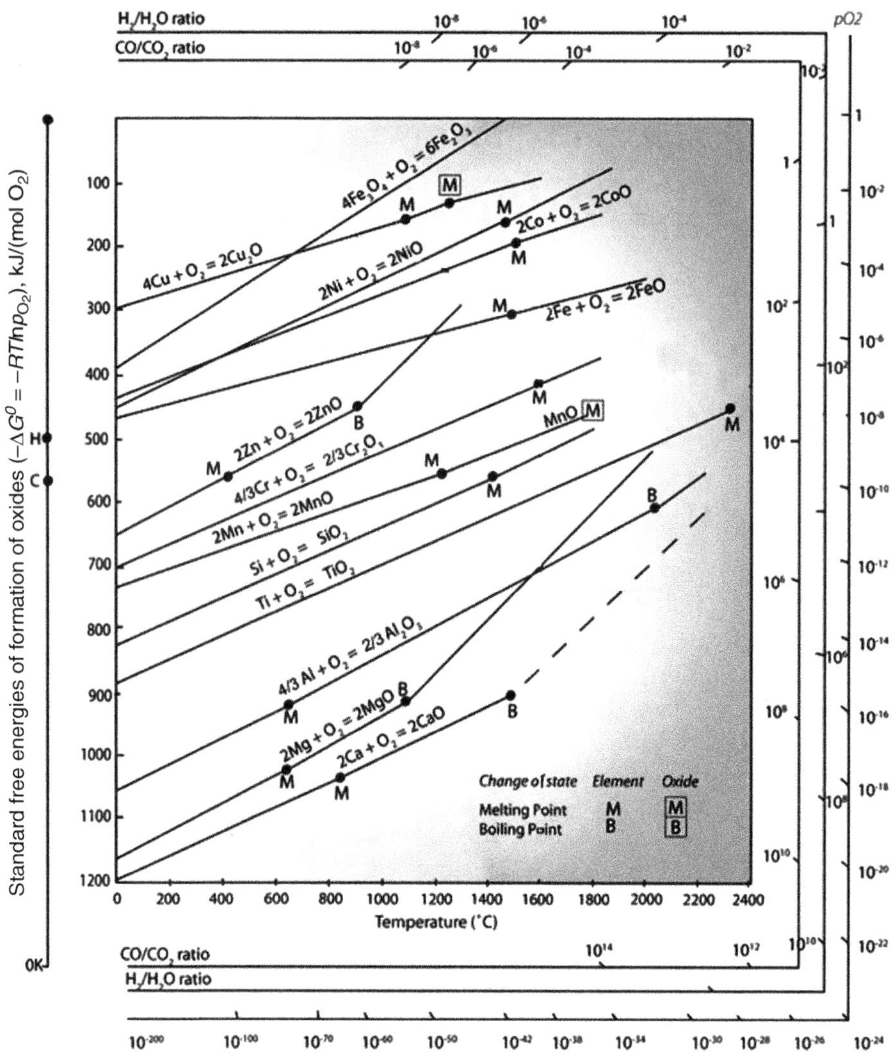

Figure 7.2 *Ellingham diagram for a number of metal–oxide systems, from http://www.doitpoms.ac.uk/tlplib/ellingham_diagrams/ellingham.php, with permission from D.ITP.MS of Cambridge University Press.*

difference between the pure element and its oxide, the entropy of reaction is dominated by the reduction of the entropy corresponding to one mole of O_2 gas. Consequently, the entropies of reaction are approximately the same for most reactions when the pure elements and their oxides are solid, as seen in Figure 7.2 where most lines have similar slopes.

For a chemical reaction on the Ellingham diagram at a given temperature, $\Delta^0 G$ can be read from the y axis of the diagram, and the equilibrium partial pressure of O_2 gas can be calculated using Eq. 7.8. Alternatively, one can plot the left part of Eq. 7.8 for a given P_{O_2} as a function of temperature on the Ellingham diagram, i.e.

$$\Delta^0 G = RT\ln P_{O_2} \tag{7.10}$$

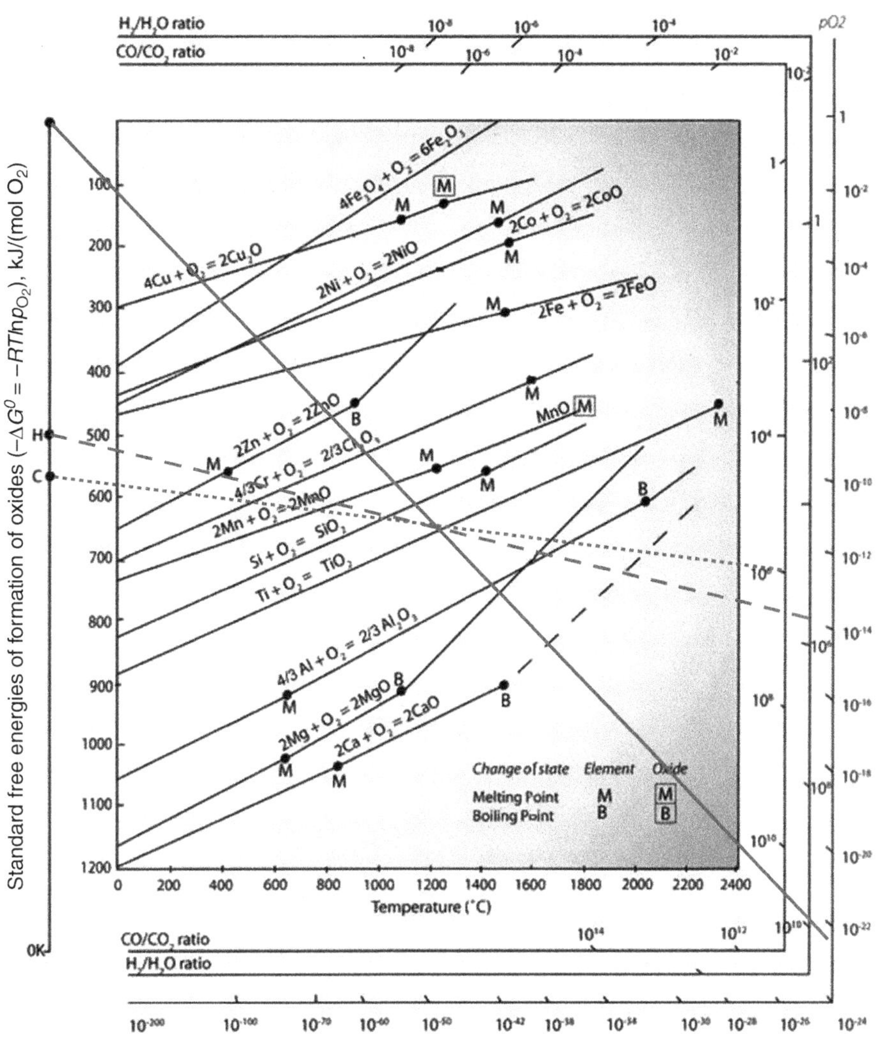

Figure 7.3 *Intersection of iso-partial-pressure lines of O_2 and equilibrium lines in the Ellingham diagram, http://www.doitpoms.ac.uk/tlplib/ellingham_diagrams/ellingham.php, with permission from D.ITP.MS of Cambridge University Press.*

This results in straight lines, representing iso-partial-pressure lines of O_2, with intercepts zero at $T = 0K$ and slopes $R\ln P_{O_2}$, which are negative for P_{O_2} lower than one atmospheric pressure (1 atm). The values of P_{O_2} are marked on the secondary axis on the right side of the Ellingham diagram. The intersection of the isoactivitiy line and the equilibrium line of the chemical reaction thus gives the relation between the equilibrium temperature and equilibrium partial pressure of O_2. This is demonstrated in Figure 7.3.

For each chemical reaction in the Ellingham diagram, the three phases are in equilibrium on the line represented by Eq. 7.8, i.e. metal, metal oxides, and O_2 gas. For conditions above the line, the value of P_{O_2} is larger than its equilibrium value, and the metal will be oxidized. For conditions below the line, the value of P_{O_2} is lower than its equilibrium value, and the metal oxide will be reduced. Therefore, the metal oxides in the upper part of the Ellingham diagram can be reduced by the metals in the lower part of the diagram, and, vice versa, metals in the lower part of the diagram can be oxidized by the metal oxides in the upper part of the diagram. For example, in Figure 7.2, Ca can reduce all oxides, and Cu_2O is the least stable oxide.

From the Ellingham diagram, it may be noted that the equilibrium partial pressures of O_2 are very low for most chemical reactions, with many of them lower than 10^{-12} atm. One approach to obtaining such a low pressure is to use auxiliary reactions containing O_2 that are easy to control and are independent of the equilibrium system of interest except for a shared oxygen partial pressure. Two common auxiliary reactions are those of H_2/H_2O and CO/CO_2 mixtures. For an H_2/H_2O mixture, the chemical reaction is

$$2H_2(\text{gas}) + O_2(\text{gas}) = 2H_2O(\text{gas}) \quad 7.11$$

The equilibrium oxygen partial pressure is obtained as

$$-RT\ln\left\{\frac{1}{P_{O_2}}\left(\frac{P_{H_2O}}{P_{H_2}}\right)^2\right\} = \Delta^0G = \Delta^0H - T\Delta^0S = -498488 + 112.972T \text{ (J)} \quad 7.12$$

where the values of Δ^0H and Δ^0S are taken from the substance thermodynamic database (SSUB4) compiled by Scientific Group Thermodata Europe (SGTE) [63], and are slightly dependent on temperature, and the values in Eq. 7.12 are evaluated at 1273 K using Thermo-Calc [64]. At any given temperature, one has the following relation:

$$RT\ln P_{O_2} = \Delta^0H - T\Delta^0S - 2RT\ln\frac{P_{H_2}}{P_{H_2O}} = -498488 + \left(112.972 - 2R\ln\frac{P_{H_2}}{P_{H_2O}}\right)T \quad 7.13$$

Its intercept at $T = 0$ K is given by $\Delta^0H = -498488$ (J), and the slope by $-\Delta^0S - 2R\ln(P_{H_2}/P_{H_2O}) = 112.972 - 2R\ln(P_{H_2}/P_{H_2O})$. The values of the P_{H_2}/P_{H_2O} ratio are marked on a secondary axis on the right side of the Ellingham diagram. The intersection of the iso-partial-pressure ratio line and the equilibrium line of the chemical reaction gives the relation between equilibrium temperature and equilibrium partial pressure ratio P_{H_2}/P_{H_2O} for the desired P_{O_2} of the chemical equilibrium as shown in Figure 7.3 by the dashed line marked by H at T = 0 K.

For an CO/CO_2 mixture, the chemical reaction is

$$2CO(\text{gas}) + O_2(\text{gas}) = 2CO_2(\text{gas}) \quad 7.14$$

Similarly, from the SSUB database, the following equation is obtained:

$$-RT\ln\left\{\frac{1}{P_{O_2}}\left(\frac{P_{CO_2}}{P_{CO}}\right)^2\right\} = \Delta^0G = \Delta^0H - T\Delta^0S = -562,927 - 172.020T \text{ (J)} \quad 7.15$$

with Δ^0H and Δ^0S calculated at 1273 K; they are also slightly temperature dependent as in the case of the H_2/H_2O mixture. The iso-partial-pressure ratio line is written as

$$RT\ln P_{O_2} = \Delta^0H - T\Delta^0S - 2RT\ln\frac{P_{CO}}{P_{CO_2}} = -562,927 + \left(172.020 - 2R\ln\frac{P_{CO}}{P_{CO_2}}\right)T \quad 7.16$$

Its intercept at $T = 0$ K is given by $\Delta^0 H = -562927$ (J), and its slope by $172.020 - 2R\ln(P_{CO}/P_{CO_2})$. The values of the P_{CO}/P_{CO_2} ratio are marked on the third secondary axis on the right of the Ellingham diagram. The intersection of the iso-partial-pressure ratio line and the equilibrium line of the chemical reaction gives the relation between equilibrium temperature and equilibrium partial pressure ratio P_{CO}/P_{CO_2} for the desired P_{O_2} of the chemical equilibrium as shown in Figure 7.3 by the dotted line marked by C at T = 0 K.

Therefore, one can use the H_2/H_2O system or the CO/CO_2 system to obtain the desired low P_{O_2} values using Eq. 7.13 and Eq. 7.16 or the calculation from the SSUB database. For example, for $P_{O_2} = 10^{-15}$atm at 1273 K, the calculated values from the SSUB database are $P_{H_2}/P_{H_2O} \approx 1.67$ and $P_{CO}/P_{CO_2} \approx 2.78$, respectively, in which the temperature dependences of $\Delta^0 H$ and $\Delta^0 S$ and the many more gaseous species in the gas phase are taken into account. On the other hand, reading from the Ellingham diagram gives $P_{H_2}/P_{H_2O} \approx 2.0$ and $P_{CO}/P_{CO_2} \approx 2.4$, and the agreement with the more accurate calculations above is remarkable keeping in mind the uncertainties in graphically drawing the lines and reading both values in the logarithmic scales from the diagram, indicating the robustness of the Ellingham diagram.

7.3 Trends of entropies of reactions

The reaction entropy, $\Delta^0 S$ in Eq. 7.5, plays an important role in determining the equilibria of high temperature reactions. The most important single factor that determines the entropy of a reaction is the net change in the number of moles of gas, as briefly mentioned in the discussion of the Ellingham diagram above. The reason this is true can be explained as follows.

The entropy of a substance can be thought of as being the sum of four parts: (i) translational, (ii) rotational, (iii) vibrational, and (iv) electronic. The translational entropy of a gas is the largest entropy term under most conditions. To the extent that the other contributions cancel between reactants and products, the entropy of reaction is determined by the change in the number of moles of gaseous molecules. Based on the literature data or calculations from the SSUB database, the net change in the number of moles of gas in a reaction results in an entropy of reaction of about 175±45 J/K per mole of gas at 298 K for many halides and oxides. The chemical reactions of Eq. 7.11 and Eq. 7.14 discussed above both reduce the gas by one mole, and their entropies of reaction are −113 and −172 J/K at 1273 K, and −89 and −173 J/K at 298 K, respectively, indicating that the chemical reaction of Eq. 7.11 is an exception to the empirical rule. For the chemical reactions shown in the Ellingham diagram, the entropies of reaction follow this empirical rule pretty well with some of them, shown in Table 7.1, calculated from the SGTE database.

Since the entropy of a reaction is primarily determined by the net change in the number of moles of gas, the entropies for reactions involving only condensed phases must be small. The entropies of fusion of monatomic solids are usually in the range 8–15 J/K per mole of atoms, as shown for some elements in Table 7.2. Most metals and many ionic salts have values that lie in this range when given per mole of atoms. There

Table 7.1 Entropies of reactions with gas at 298.15 K

Reaction	Entropy (J/K)
$Si+O_2=SiO_2$	−182
$Ti+O_2=TiO_2$	−185
$2Mg+O_2=2MgO$	−217
$2Ca+O_2=2CaO$	−212
$2Mn+O_2=2MnO$	−150

Table 7.2 Entropies of reactions of condensed phases at 298.15K

Reaction	Entropy (J/K)
Si(s)=Si(l)	29.762
Ti(s2)=Ti(l)	7.288
Mg(s2)=Mg(l)	9.184
Ca(s2)=Ca(l)	7.659
Mn(s2)=Mn(l)	11.443
W(s)=W(l)	14.158
B(s)=B(l)	21.380
$3Fe+C=CFe_3$	17.060
S+Mn=MnS	13.909
$NiO+Fe_2O_3=Fe_2NiO_4$	0.464

are a few exceptions such as silicon and boron, as shown in the table. For solid state reactions, the average values can be approximated as 0±8 J/K per mole of atoms, as also shown in the table.

7.4 Maximum reaction rate and chemical transport reactions

Equilibrium thermodynamics can be used to calculate the maximum rates of reaction that are possible in dynamically reacting systems, such as when a corrosive gas passes over a heated sample. Other examples of such reactions include the reduction of a metal oxide in flowing hydrogen, the evaporation of a material in a vacuum, and the deposition of a thin film by a chemical vapor deposition process. The basic assumption used in calculating these maximum reaction rates is that local equilibrium exists at the location of the considered reaction.

This section examines several examples in which maximum reaction rates are calculated. The system can typically be divided into three regions: (i) the input region, which is usually near room temperature, (ii) the high temperature region in which the primary reaction of interest is occurring, and (iii) the exit region. Such a system is almost always at constant pressure throughout the system. Since the three regions have different temperatures, the key is to use the mass conservation of the carrier gas in all three regions. With the input gas at $T = 298$ K and $P = 1$ atm, the volume of

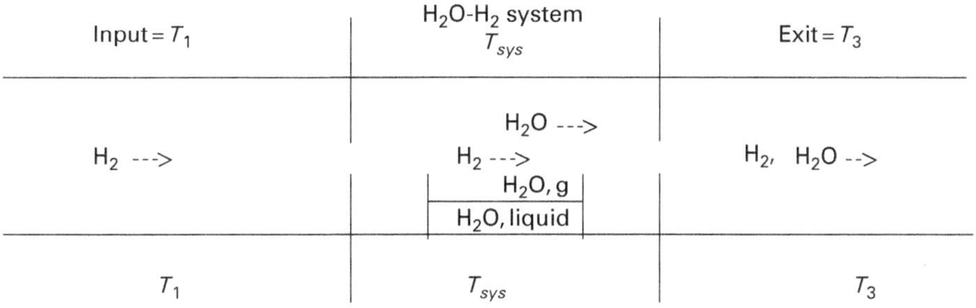

Input = T_1	H_2O-H_2 system T_{sys}	Exit = T_3
H_2 --->	H_2O ---> H_2 ---> H_2O, g H_2O, liquid	H_2, H_2O -->
T_1	T_{sys}	T_3

Figure 7.4 *Schematic diagram of the vaporization of water in a flowing stream of dry hydrogen.*

one mole of ideal gas is 0.0244 m^3 = 24.4 liters. The number of moles of input gas flowing through the system can be written as

$$n^0_{gas} = \frac{Pf_{gas}}{RT} = \frac{f_{gas}}{24.4} = 0.0409 f_{gas} \qquad 7.17$$

where f_{gas} is the input gas flow rate in liters per unit time at $T = 298$ K and $P = 1$ atm, and R is the gas constant.

The first example is the evaporation of water in a flowing stream of dry hydrogen. A schematic diagram of this system is given in Figure 7.4. The goal of the calculation is to determine the maximum rate of H_2O(liquid) loss, n_{H_2O}, through vaporization in a flowing stream of H_2, for example for the purpose of generating a given H_2O/H_2 ratio in order to produce a given O_2 pressure in a high temperature system, as related to the Ellingham diagram discussed in Section 7.2. The maximum rate is determined by saturating the H_2 with H_2O(liquid) at equilibrium vapor pressure, the number of moles of H_2 being $n^0_{H_2} = 0.0409 f_{H_2}$ and f_{H_2} being the input flow rate of H_2 in liters per unit time.

In the first input region, the total number of moles of gas is $N = n^0_{H_2}$. In the second, high temperature, region, with temperature and pressure T_{sys} and P_{sys}, the total number of moles of gas is $N = n^0_{H_2} + n_{H_2O}$, which is the same for the third, exit, region with T_{exit}. To avoid condensation, one needs to maintain $T_{exit} > T_{sys}$. The vapor pressure of H_2O at T_{sys}, P_{H_2O}, can be obtained from equilibrium thermodynamic calculations. Since the H_2 and H_2O are occupying the same volume at the same temperature, the maximum number of moles of H_2O can be calculated from the following relation, based on the ideal gas law:

$$n_{H_2O} = \frac{P_{H_2O}}{P_{H_2}} n^0_{H_2} = \frac{P_{H_2O}}{P_{sys} - P_{H_2O}} n^0_{H_2} = \frac{P_{H_2O}}{P_{sys} - P_{H_2O}} \frac{f_{H_2}}{24.4} \qquad 7.18$$

For $P_{sys} = 101325$ Pa, one can calculate P_{H_2O} and plot n_{H_2O}/f_{H_2} as a function of temperature from the SSUB database, as shown in Figure 7.5. The corresponding partial pressure ratio, P_{H_2O}/P_{H_2}, is plotted in Figure 7.6, which can be used to calculate P_{O_2} at any given temperature. An example is shown in Figure 7.7 with $T_{sys} = 348.15$ K and $P_{H_2O}/P_{H_2} = 0.607$.

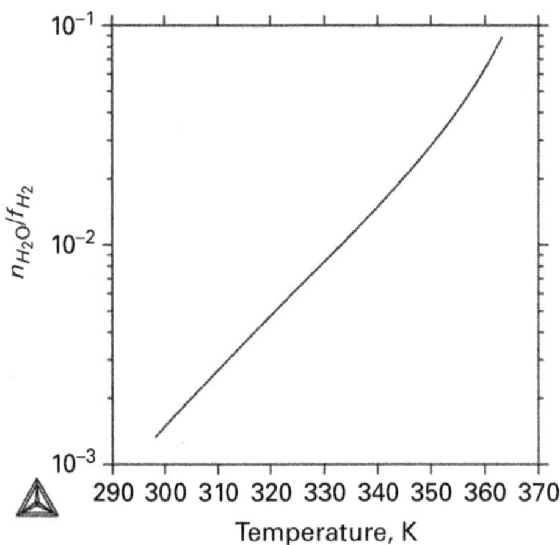

Figure 7.5 *Ratio of the maximum number of moles of H_2O with respect to the hydrogen flow rate, n_{H_2O}/f_{H_2}, plotted as a function of temperature.*

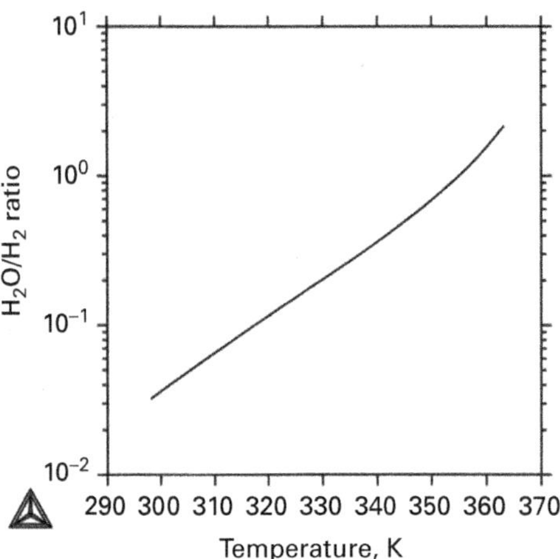

Figure 7.6 *Partial pressure ratio, P_{H_2O}/P_{H_2}, corresponding to Figure 7.5.*

The second example is the corrosion of SiO_2 (s) by flowing H_2 gas at high temperatures. Taking $T_{sys} = 1700$ K, $P_{sys} = 101325$ Pa, and $f_{H_2} = 1$ liter/min, the system is thus defined with 0.0409 moles of H_2 and an equilibrium between the gas phase and the tridymite SiO_2. The equilibrium calculation gives 0.04092 moles of gas with constitution H_2:H_2O:SiO = $0.998887:4.833\times10^{-4}:4.832\times10^{-4}$. The corrosion rate of SiO_2(s) is thus $0.04092\times4.832\times10^{-4} = 1.98\times10^{-5}$ mol/min = 1.19 gram/min.

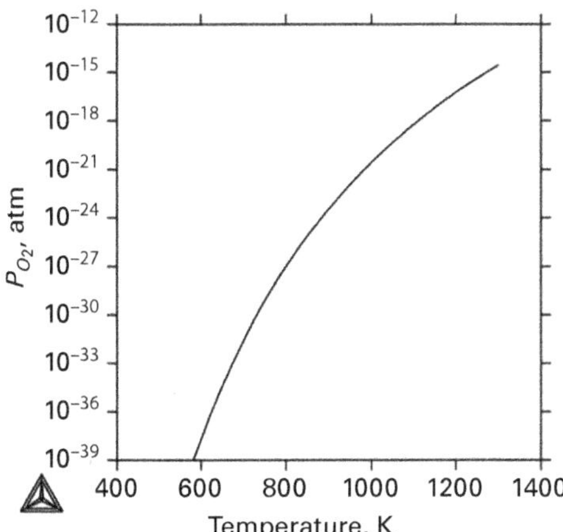

Figure 7.7 *Partial pressure of oxygen, P_{O_2}, as a function of temperature with $T_{sys} = 348.15\text{K}$ and $P_{H_2O}/P_{H_2} = 0.607$.*

Another example is that of CO gas at 298 K and 1 atm (101325 Pa) flowing at a rate of 1 liter/min through and equilibrating with single phase C(s) at 1500 K. The equilibrium system is defined by $T = 1500$ K, $P = 1$ atm, and 0.0409 moles of CO, the equilibrium being between the gas phase and C(s). The equilibrium calculation gives 0.040872 moles of gas phase with mole fraction of CO equal to 0.999327, resulting in the loss of CO or the deposition of C(s) at a rate of $0.0409 - 0.040872 \times 0.999327 = 5.55 \times 10^{-5}$ mol/min.

A chemical transport reaction is a reaction in which a condensed phase reacts with a gas phase to form vapor-phase products, which in turn undergo the reverse reaction to the condensed phase. Two well-known examples of such reactions are

$$\text{M(s)} + \frac{n}{2}\text{I}_2\text{(g)} = \text{MI}_n\text{(g)} \qquad 7.19$$

$$\text{Ni(s)} + 4\text{CO(g)} = \text{Ni(CO)}_4\text{(g)} \qquad 7.20$$

which are used in the purification of metals by the iodide process and in the purification of nickel by the Mond–Langer process. In both processes the forward reaction is favored by lower temperatures and the reverse reaction by higher temperatures, resulting in the deposition of the metal. The most common technique for causing chemical transport of a condensed substance makes use of the temperature dependence of the equilibrium constant. As was discussed previously, the enthalpy of reaction, $\Delta^0 H$, determines the manner in which K_e changes with temperature (see Eq. 7.6). The value of K_e increases with increasing T for $\Delta^0 H > 0$, decreases with increasing T for $\Delta^0 H < 0$, and is independent of T for $\Delta^0 H = 0$. The $\Delta^0 H$ and $\Delta^0 S$ values for chemical transport reactions may be either positive or negative.

In a typical experiment the starting solid is located at the point in a temperature gradient that corresponds to the largest K_e value for the experimental conditions. As the gaseous species migrate to other locations in the system with temperatures corresponding to lower K_e values, the reverse reaction occurs to satisfy the new equilibrium requirements, and the solid phase is deposited. The dependence of K_e on $\Delta^0 H$ results in material transport from hot to cold for $\Delta^0 H > 0$ (the same as for vaporization–condensation reactions), from cold to hot for $\Delta^0 H < 0$, and in no transport for $\Delta^0 H = 0$.

The success of a particular reaction in causing appreciable transport of a condensed phase depends mainly upon the partial pressure gradients or concentration gradients of the gaseous species in the system. A reaction whose equilibrium is extreme toward either the reactant side or the product side will not give an appreciable transport of material. The concentration gradients are too small in such a system. Reactions with equilibrium constants near unity at the experimental temperatures usually give the largest transport since small changes in K_e cause large changes in concentration. The general condition required to obtain a K_e value near unity at a reasonable temperature is that $\Delta^0 H$ and $\Delta^0 S$ both have the same sign; this results from the equalities of Eq. 7.5.

Exercises

1. Estimate the entropy of formation at 298 K ($\Delta_f S$ at 298 K) for the following species based on the trends of the entropy of reactions. Some of them can be found in the SSUB database; calculate them and compare them with your estimations.

   $BaSO_4(s)$ $LaN(s)$ $UC_2(s)$ $ZrBr_4(g)$ $LiF(g)$ $Mg_2SiO_4(s)$

   $TiO_2(g)$ $A_{12}Cl_6(g)$ $SiC(s)$ $U_2N_3(s)$ $UC_2(g)$

2. A defect mechanism that may occur simultaneously with oxygen vacancies in $LaCoO_{3-\delta}$ is charge disproportionation, where two Co^{3+} ions transform into a Co^{2+} and Co^{4+} pair. Perform defect analysis on $LaCoO_{3-\delta}$ to predict the slope of a plot of $\log(P_{O_2})$ versus $\log(y_{Va})$, assuming the valence of Co^{3+} changes when oxygen vacancies form. Do this with and without charge disproportionation present. Compare your two results to the experimental plot shown below (from [65] with permission from Elsevier). Which result best describes the experiments?

(*cont.*)

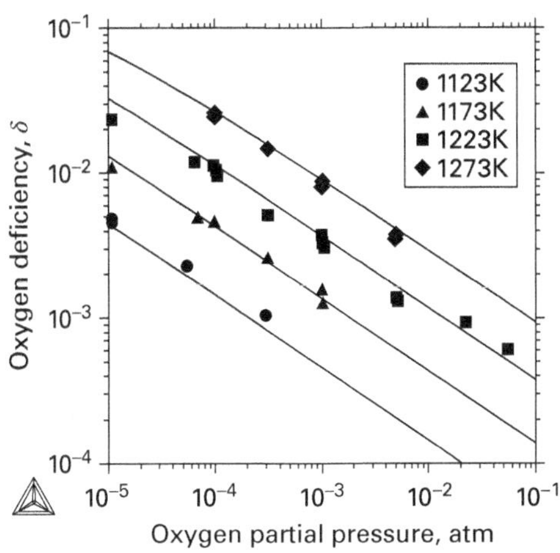

3. Use the oxygen potential diagram given to answer the following questions.
   a. Will an Nb/NbO mixture maintained at 2000 K in a furnace containing ~1 bar $H_2(g)$ oxidize Cr metal that is located in another part of the same furnace chamber at 1000 K?
   b. Can a Cr_2O_3 crucible be used to contain Mo metal at 1500 K? Why or why not?

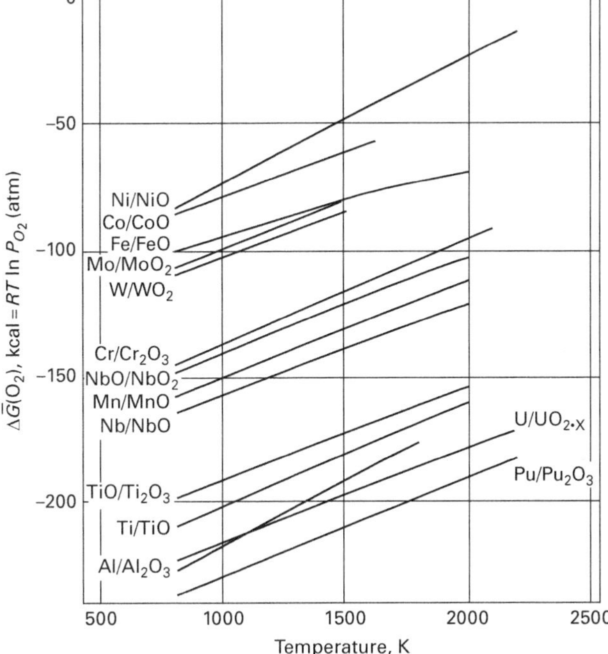

(*cont.*)

4. Ar(g) flowing at a rate of 2 liter/min (T = 273 K and P = 1 atm) contains 1% H_2(g) impurity. It is passed through a furnace containing ZnO(s) at 1000 K. Assume the 1 atm gas equilibrates with the ZnO(s) before passing out of the system. The equilibrium calculation results from Thermo-Calc using the SSUB4 database are given below. The given constants may also be useful.
   - At T = 273 K and P = 1 atm, the volume of 1 mol gas is equal to 22.4 liters.
   - Molar weights (g/mol): H 1.008, O 15.9994, Zn 65.38, ZnO 81.3794, and Ar 39.948.

   a. What is the primary chemical reaction causing the erosion of ZnO(s)?

   b. What is the maximum rate of mass loss of ZnO(s) in units of mg/min?

```
          Conditions:
T=1000, P=100000, N(H)=2E-2, N(AR)=9.9E-1, N(ZN)=2, N(O)-N
(ZN)=0
DEGREES OF FREEDOM 0

Temperature 1000.00, Pressure 1.000000E+05
Number of moles of components 5.01000E+00, Mass 2.02327E+02
Total Gibbs energy -9.99490E+05, Enthalpy -6.18436E+05,
Volume 8.32438E-02

Component Moles      W-Fraction Activity   Potential   Ref
                                                        .stat

AR          9.9000E-01 1.9547E-01 2.2663E-09 -1.6550E+05 SER
H           2.0000E-02 9.9631E-05 1.4841E-05 -9.2442E+04 SER
O           2.0000E+00 1.5815E-01 1.9969E-17 -3.1971E+05 SER
ZN          2.0000E+00 6.4628E-01 8.3890E-06 -9.7185E+04 SER

GAS#1       Status ENTERED Driving force 0.0000E+00
Number of moles 1.0124E+00, Mass 3.9665E+01 Mass fractions:
AR 9.97057E-01 ZN 1.95593E-03 H 5.08203E-04 O 4.78644E-04
Constitution:
AR   9.88827E-01  H1O1ZN1  1.25762E-11  O      1.84294E-21
H2   8.80291E-03  H2O2ZN1  1.26148E-12  O2     1.37349E-22
ZN   1.18523E-03  H1ZN1    7.35625E-13  H2O2   3.76372E-23
H2O1 1.18523E-03  H1O1     6.32162E-14  H1O2   1.86169E-26
H    2.13129E-10  O1ZN1    5.41021E-16  O3     1.00000E-30
```

(cont.)

```
O1ZN1_S#1    Status ENTERED    Driving force       0.0000E+00
Number of moles 3.9976E+00, Mass 1.6266E+02 Mass fractions:
ZN 8.03397E-01 O 1.96603E-01 AR 0.00000E+00 H 0.00000E+00
```

5. Assume a small, closed isothermal chamber at 1000 K with the following solid materials:

0.6 moles NiO	0.6 moles Mn	0.6 moles Sc	0.2 moles Cr_2O_3
0.5 moles Cu_2O	0.4 moles Fe_2O_3	0.5 moles Zn	0.3 moles Co

Each material is in its own container, so it cannot react with the other materials to form complex phases or solutions. By using a mixture of H_2O and H_2 as a means of efficiently moving oxygen around in the system, equilibrium is readily obtained. Use the figure to help in answering the following questions (K. E. Spear, private communication).

a. What is the equilibrium form of each material? Show the logic/calculations behind your answer. A simple list of metals and metal oxides is not sufficient.

b. What is the equilibrium O_2 partial pressure for this system?

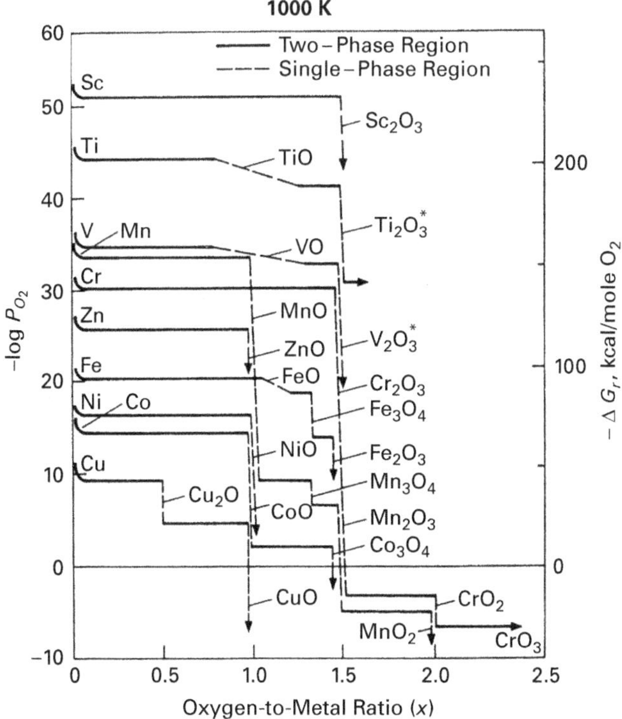

(cont.)

6. The W–C binary phase diagram is shown in the figure below. At T = 1250 °C, there is a three-phase equilibrium among WC, W, and W_2C, all the phases being solid. This three-phase equilibrium can be represented by the chemical reaction WC + W = W_2C. As the phase diagram shows, at temperatures above 1250 °C, the reaction goes to the right, and at temperatures below 1250 °C, the reaction goes to the left. Use the information given below along with the required estimates to calculate the enthalpy and entropy of formation for W_2C. Explain any estimates you make. $\Delta_f^0 H_m^{WC} = -21$ kJ/(mole atom), and $\Delta_f^0 S_m^{WC} = -3.14$ J/K/(mole atom).

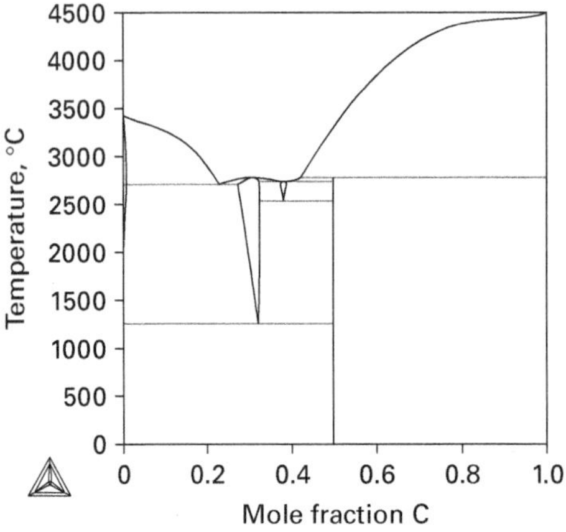

7. A waste material is found to contain 1 wt% of aluminum with the rest assumed to be aluminum oxide (Al_2O_3). The aluminum and the Al_2O_3 are thermally connected. If the waste material is stored at 298 K, what is the maximum temperature to which it may rise if all the metallic aluminum is oxidized by air? The entire mass may be assumed to rise to the same temperature. Specify any assumptions you make. You may need to use the following data. If you use Thermo-Calc for your work, include appropriate screen shots.
   a. Atomic weights: Al 27 and O 16 g/mol.
   b. Heat capacity: solid aluminum, 26 J/(k mol); solid Al_2O_3, 104 J/(k mol Al_2O_3).
   c. Enthalpy of formation for Al_2O_3: −1676000 J/(mol Al_2O_3).
8. Use the Ellingham diagrams to answer the following questions.
   a. What is the partial pressure of oxygen above pure Ti and TiO_2 at 1100 °C? What ratios of CO/CO_2 or H_2/H_2O are needed to obtain this partial pressure of oxygen?
   b. If the system in part (a) is in equilibrium, and then carbon is added to the system, would the oxygen pressure increase, decrease, or remain the same? Explain your answer. What are the equilibrium phases?

(cont.)

c. Will a Ti/TiO_2 mixture maintained at 1600 °C in a furnace containing ~1 bar H_2(g) oxidize Cr metal that is located in another part of the same furnace chamber at 1000 °C? Explain.

d. Can a Cr_2O_3 crucible be used to contain Mo metal at 1500 K? Why or why not?

8 Applications to electrochemical systems

The two basic types of chemical processes involving ions as reactant and/or product species are electrolyte reactions and electrochemical reactions. Electrolyte reactions are accompanied by the atomic scale movement of ionic species and possibly electrons. Chemical changes that produce changes in valence and electron and ion transport over finite distances constitute an area of science termed electrochemistry. The latter chemical changes occur in an electrochemical cell comprising two electrodes, an anode and a cathode, which are coupled by an electrolyte and an external electron conductor. Most thermodynamic concepts and analyses described in the previous chapters remain unchanged when applied to electrochemistry, but the analysis of electrochemical systems does require some new terminology, new definitions, and new conventions. The primary focus of this chapter is on applications of thermodynamics to electrochemical reactions that involve either aqueous electrolyte solutions or solid state electrolytes. Since all electrochemical systems include ionized species as reactant and/or product species, electrolyte reactions will also be discussed.

8.1 Electrolyte reactions and electrochemical reactions

Electrolytes that dissolve in a polar solvent such as water to produce ionic species do not necessarily exhibit changes in valence. A simple example is the strong electrolyte NaCl(s) dissolving in water to produce solvated ions:

$$\mathrm{NaCl(s)} = \mathrm{Na^+(aq)} + \mathrm{Cl^-(aq)} \qquad 8.1$$

where (aq) indicates that the ionic species is in an aqueous solution. In this system, the ion concentrations must become quite large before the solution is saturated and can exist in equilibrium with NaCl(s). Its reaction constant, defined by Eq. 7.5, is shown as $K_e = a_{Na^+}a_{Cl^-}$. If the product of the ion activities is less than K_e, the solution is not saturated, and more NaCl(s) can be dissolved.

The precipitation of AgCl(s), a weak electrolyte, occurs quite readily when Cl^- ions are added to an aqueous solution containing Ag^+(aq):

$$\mathrm{Ag^+(aq)} + \mathrm{Cl^-(aq)} = \mathrm{AgCl(s)} \qquad 8.2$$

The equilibrium constant for this reaction, $K_e = 1/\left(a_{Ag^+}a_{Cl^-}\right)$ is quite large, so the equilibrium product of the ion activities, proportional to their concentrations, is quite small. In the laboratory, the above reaction could occur as a result of adding hydrochloric acid to a silver nitrate solution. The accompanying $H^+(aq)$ (or $H_3O^+(aq)$) and $NO_3^-(aq)$ ions in solution are not directly involved in the silver chloride precipitation reaction so are not shown in the reaction represented by Eq. 8.2.

The above ionic equilibrium in the $AgCl(s)–H_2O$ system is not only important for understanding this electrolyte system, but is also critical in electrochemical systems in which Ag(s) undergoes a valence change at one electrode and reacts with a $Cl^-(aq)$ ion to produce AgCl(s) and an electron that is externally transported finite distances to another electrode. The oxidation reaction occurs at the Ag/AgCl electrode (an *anode* half-cell reaction where electrons are added into the system):

$$\mathrm{Ag(s) + Cl^-(aq) = AgCl(s) + e^-} \qquad 8.3$$

A reduction reaction occurs at the other electrode (a *cathode* half-cell reaction where electrons are consumed by the reaction):

$$\frac{1}{2}\mathrm{Cl_2(g) + e^- = Cl^-(aq)} \qquad 8.4$$

The net cell reaction results in the formation of AgCl(s) from its elements:

$$\mathrm{Ag(s)} + \frac{1}{2}\mathrm{Cl_2(g) = AgCl(s)} \qquad 8.5$$

Without knowledge of the physical system under which the reaction is occurring, it would not be possible to know whether the reaction of Eq. 8.5 is a result of chlorine gas reacting directly with Ag(s), or whether the reaction is part of an electrochemical cell with a transport of electrons and ions over finite distances. The addition of the two half-cell reactions gives the net cell reaction, which does not show electrons as either reactant or product species and may or may not include ionic species. A schematic diagram of an electrochemical cell for the above system is shown in Figure 8.1.

Oxidation and reduction can occur in electrolyte reactions without creating an electrochemical cell. This is the case when chlorine gas reacts directly with silver on an Ag(s) surface. The reaction of Eq. 8.5 above is the net reaction for this process, but the electrons produced from the oxidation of Ag(s) are not transported over finite

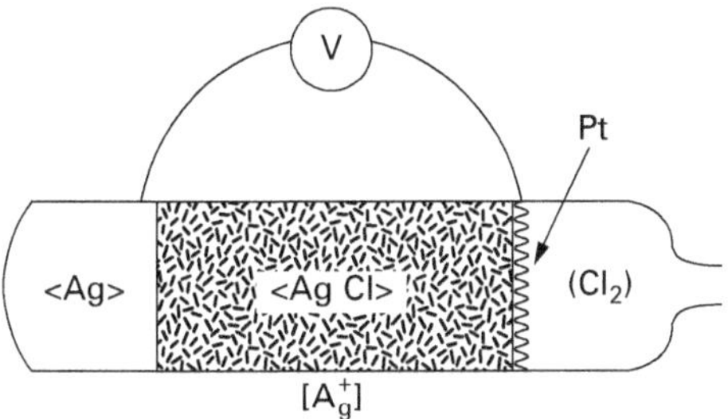

Figure 8.1 *Schematic diagram of an electrochemical cell consisting of a chlorine electrode and a silver–silver-chloride electrode.*

distances before combining with $Cl_2(g)$ in its reduction to $Cl^-(aq)$. No anode or cathode half-cell reactions exist in this system. The electrons and ions involved in the reaction move only over atomic scale distances.

8.2 Concentrations, activities, and reference states of electrolyte species

Thermodynamic descriptions of ionic species in solutions are different from those of neutral species, which leads to a need to define concentration units, standard states, activities, and activity coefficients of ionic solutions. In most studies of electrochemical corrosion and electrodeposition, and in the applied work of electrochemical engineers, ionic species concentrations are given in units of molarity, i.e., the number of moles of a species in a liter of solution (mol/l), symbolically represented in equations by either c_i or $[M^{Z+}]$. The other common concentration used for ionic species is molality, which is defined as the number of moles of a species in 1000 g of solvent. For dilute aqueous solutions, molarity and molality values are very similar.

As discussed in Section 2.2.1, a practical definition of the activity of a species i is the thermodynamic reactivity, or tendency to react, of species i in the system of interest as compared to i in its reference state form. The reference state of a species is typically chosen as a specific chemical/physical state of the species at 1 atm external pressure and the temperature of interest. Similarly, a typical reference state for ionic species in aqueous solutions is the 1 molar ideal solution at 1 bar external pressure and the temperature of interest. If an electrolyte solution behaves ideally, then the activity of species i in solution is

$$a_i = \frac{c_i(\text{mol/l})}{c_i^0(\text{mol/l})} = \frac{c_i(\text{mol/l})}{1(\text{mol/l})} = c_i(\text{dimensionless}) \quad 8.6$$

where c_i is the molar concentration of i in the solution divided by c_i^0, the 1 molar reference state ideal solution concentration. Thus, in ideal solutions, the activity of an electrolyte species is numerically equal to its molar concentration. The above treatment of ionic species is equivalent to the common practice of depicting the activity of a gas by the value of its ideal gas partial pressure in units of bars.

The activity coefficient corrects for the non-ideality of the species in solution as defined in Eq. 2.51. If the solution is ideal, $\gamma_i = 1$ for all concentrations of the species in solution. For all solutions, one expects $\gamma_i \to 1$ as $c_i \to 1$. It is not possible to measure γ_{i^+} or γ_{i^-} for individual charged ions, but only a geometric mean of the positive and negative ion values. Consider the following ionic solution,

$$A_u^{v+} B_v^{u-} = uA^{v+} v + B^{u-} \quad 8.7$$

Its chemical potential can be written as

$$\begin{aligned}\mu_{A_u^{v+} B_v^{u-}} &= u\mu_{A^{v+}} + v\mu_{B^{u-}} = u\mu^0_{A^{v+}} + v\mu^0_{B^{u-}} + uRT\ln a_{A^{v+}} + vRT\ln a_{B^{u-}} \\ &= \mu^0_{A_u^{v+} B_v^{u+}} + RT\ln a_{A_u^{v+} B_v^{u-}}\end{aligned} \quad 8.8$$

Its geometric average or mean activity and activity coefficient are defined as

$$a_{A_u^{v+}B_v^{u-}} = (a_{A^{v+}})^u (a_{B^{u-}})^v = u^u v^v c^{u+v} (\gamma_{A^{v+}})^u (\gamma_{B^{u-}})^v = u^u v^v (\gamma_\pm c)^{u+v} \quad 8.9$$

$$\gamma_\pm = (\gamma_{A^{v+}}^u \gamma_{B^{u-}}^v)^{1/(u+v)} \quad 8.10$$

For example, one can define $\gamma_\pm = (\gamma_{Na^+}\gamma_{Cl^-})^{1/2}$ and $\gamma_\pm = \left(\gamma_{Al^{3+}}^2 \gamma_{SO_4{}^{2-}}^3\right)^{1/5}$ for NaCl and $Al_2(SO_4)_3$, respectively. For ideal, weak, electrolytes, $\gamma_\pm = 1$, and for non-ideal, strong, electrolytes, $\gamma_\pm \neq 1$.

8.3 Electrochemical cells and half-cell potentials

An electrochemical system must fulfill certain requirements, including the following, in order to apply equilibrium thermodynamic descriptions of the system.

- The cell must be reversible when slight changes in conditions (potentials, concentrations, pressures, temperature) cause electrochemical reactions and an external flow of electrons to occur in the direction needed to re-establish equilibrium.
- All non-electrochemical reactions in the system must be prevented as such reactions would cause a shift in equilibrium without causing a shift in cell potential and thus a driving force for external electron flow.
- Chemical reactions must occur only when an external current flows. The finite distances for external electron transport can be as short as grain size dimensions in many corrosion reactions, or this transport may be through an external electrical conductor connecting the anode and cathode half-cell, as in batteries.
- Charge balance as well as mass balance is required of all reactions.

8.3.1 Electrochemical cells

A potential difference, i.e. voltage difference, can be generated between the electrodes in a cell from differences in the potentials of the half-cell reactions. This potential can originate from potential differences of two chemically different half-cells (a galvanic cell), or concentration differences in two otherwise identical half-cells (a concentration cell). Each type of cell is illustrated below.

The reaction between copper ions and zinc illustrated below represents the net cell reaction of a galvanic cell in which the oxidation of Zn(s) occurs at the anode electrode, and the reduction of Cu^{2+} occurs at the cathode electrode (see Figure 8.2):

$$Cu^{2+}(aq) + Zn(s) = Cu(s) + Zn^{2+}(aq) \quad 8.11$$

The reaction at each electrode, the half-cell reaction, includes ions and electrons as reactant and/or product species. The anode oxidation, reaction is represented by

$$Zn(s) = Zn^{2+}(aq) + 2e^- \quad 8.12$$

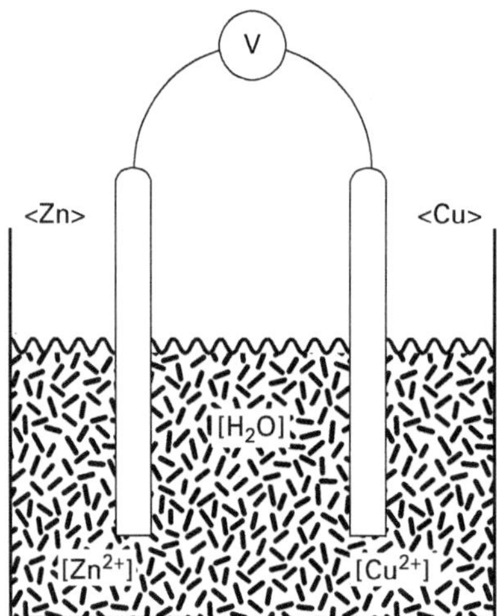

Figure 8.2 *Schematic diagram of a galvanic electrochemical cell consisting of a zinc electrode and a copper electrode.*

Electrons are products of anode reactions and flow externally from anode to cathode. By convention, the activities of the electrons in an equilibrium cell are taken as unity. The cathode, reduction, reaction is written as

$$Cu^{2+}(aq) + 2e^- = Cu(s) \tag{8.13}$$

Electrons are reactants of cathode reactions and are supplied by an external flow from the anode. In addition to consuming electrons at the cathode at the same rate as they are produced at the anode, charge balance is maintained in the electrolyte by the generation of Zn^{2+} ions at the same rate that Cu^{2+} ions are consumed. A schematic diagram in Figure 8.2 illustrates the simple physical relationships in such an electrochemical cell.

A concentration cell in which an electrochemical potential is developed because of concentration differences between otherwise equivalent anode and cathode reactions is illustrated below. Such a cell can be produced by the oxidation and reduction of copper at two separate electrodes as is depicted in the following reactions:

$$Cu(s) = Cu^{2+}(aq, c_a) + 2e^- \quad \text{(anode)} \tag{8.14}$$

$$Cu^{2+}(aq, c_b) + 2e^- = Cu(s) \quad \text{(cathode)} \tag{8.15}$$

where c_a and c_b are the respective concentrations of Cu^{2+} in the aqueous solutions at the anode and cathode, and $c_a < c_b$. The net cell reaction is

$$Cu^{2+}(aq, c_b) = Cu^{2+}(aq, c_a) \tag{8.16}$$

where the reaction occurs spontaneously to decrease c_b and to increase c_a until the two concentrations become the same, $c_b = c_a$. A schematic diagram of such a cell is shown in Figure 8.3.

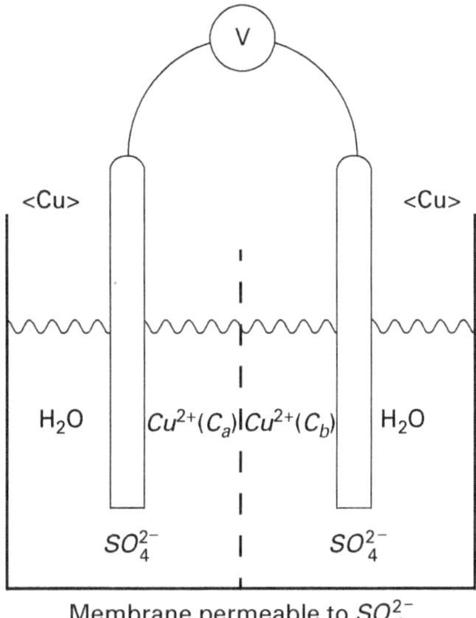

Figure 8.3 *Schematic diagram of a concentration electrochemical cell consisting of two copper electrodes.*

Anode Cathode

$| Cu(s) | Cu^{2+}(c_a) \,||\, Cu^{2+}(c_b) | Cu(s) |$

Figure 8.4 *Standard notation for an electrochemical cell.*

A semi-impermeable membrane, or a salt bridge, must exist in such a cell to maintain the charge balance. As Cu^{2+} ions are produced at the anode and consumed at the cathode, the negatively charged ions in the solution, for example SO_4^{2-}, must be transferred from the cathode region to the anode region of the cell to maintain an electrically neutral solution.

The above concentration cell provides a good example for illustrating standard notation for an electrochemical cell. This cell can be represented by Figure 8.4. The anode where oxidation occurs is always shown on the left, and the cathode where reduction occurs is on the right. A single line separating phases denotes an interface between two phases. The above anode electrode and reaction of Eq. 8.14 are symbolically represented by

$$|\mathrm{Cu(s)}|\mathrm{Cu}^{2+}(c_a) \qquad 8.17$$

The interface between the external conductor and Cu(s) is depicted by the single line to the left of Cu(s), while the single line between Cu(s) and $Cu^{2+}(c_a)$ depicts the interface between the anode electrode and the electrolyte solution. Similarly, the cathode electrode and reaction of Eq. 8.15 are symbolically represented by

$$Cu^{2+}(c_b)|Cu(s)| \quad 8.18$$

A double line between the two copper ions in the notation denotes a physical separation of two solution phases, the anode and cathode electrolyte regions, which exhibit different concentrations of copper ions:

$$Cu^{2+}(c_a)|\,|Cu^{2+}(c_b) \quad 8.19$$

These solution phases are physically connected by a semi-impermeable membrane or salt bridge that allows a common negative ion, for example SO_4^{2-}, of the solution phases to diffuse from one region to the other in order to maintain charge balance as the cell reaction occurs. The Cu^{2+} ions cannot be transported from one region to the other.

8.3.2 Half-cell potentials

When an electron current flows between electrodes, reactions are occurring at the electrodes, and concentration gradients causing polarization develop around the electrodes. These gradients result in extraneous potentials occurring at the electrodes. In such cases cell equilibrium is not established and the measured cell potentials are not those for true partial equilibrium. If a cell is short-circuited, with the electrodes connected by a conductor, current will flow until the external potential becomes zero, i.e. $\varepsilon_{ext} = 0$, and equilibrium is established with the same conditions as for non-electrochemical systems. If an external potential, ε_{ext}, is applied to the cell, chemical reactions occur until the cell potential balances to ε_{ext}, and no current flows. This potential is called the open-circuit voltage (OCV) in the literature. It is important to realize that the OCV includes all reactions that occur on the electrode surface when the electrode is in contact with the electrolyte, such as passivation, discussed in Section 8.5.1. Partial equilibrium in a cell is achieved when the cell potential is balanced by an applied external potential. In such partial equilibrium cases, equilibrium thermodynamic analyses can be used even though the cell potential is not zero, i.e. $\varepsilon_{cell} \neq 0$. This differentiates electrochemical systems from other equilibrium systems discussed previously.

The number of electrons involved in a net cell reaction is important in relating the cell potential and the Gibbs energy change for the cell reaction. As will be illustrated later in this section, this number is equal to the number of electrons involved in the half-cell reactions that were added to yield the net cell reaction. The electrical work achieved by the transport of an electrical charge through a cell potential can be written as

$$w = zf\,\varepsilon \quad 8.20$$

where z represents the number of moles of electrons in the cell reaction, f is the Faraday constant, equal to 96,485 J/V per mole of electrons, and ε is the potential

Table 8.1 Thermodynamic equations for electrochemical cells

$$\Delta G = -zf\varepsilon$$
$$\Delta S = -\left(\frac{\partial \Delta G}{\partial T}\right)_P = +zf\left(\frac{\partial \varepsilon}{\partial T}\right)_P$$
$$\Delta H = \left[\partial\left(\frac{\Delta G}{T}\right)/\partial\left(\frac{1}{T}\right)\right]_P = -zf\left[\partial\left(\frac{\varepsilon}{T}\right)/\partial\left(\frac{1}{T}\right)\right]_P = zf\left[T\left(\frac{\partial \varepsilon}{\partial T}\right)_P - \varepsilon\right]$$
$$\Delta C_P = \left(\frac{\partial \Delta H}{\partial T}\right)_P = Tzf\left(\frac{\partial^2 \varepsilon}{\partial T^2}\right)_P$$

difference, often referred to as the electromotive force (emf) in the literature. For a system at constant temperature, pressure, and composition, this work is the same as the Gibbs energy difference between the two electrodes, i.e. the Nernst equation

$$-\Delta G = w = zf\varepsilon \tag{8.21}$$

where the negative sign is needed because the system does work on the surroundings when the Gibbs energy of the system is decreased. When the applied external potential is larger than the cell potential, the surroundings does work on the system, and a common example is the charging of a battery. Thermodynamic relations discussed in previous chapters can thus be directly applied to electrochemical systems; some common equations are shown in Table 8.1.

A half-cell reaction potential cannot be measured directly, but only its potential relative to another half-cell reaction. By convention, a standard half-cell potential is measured relative to the standard hydrogen half-cell reduction reaction at 25 °C (298 K) and 1 bar, which has a defined standard potential of zero volts,

$$\mathrm{H^+(aq}, a = 1) + \mathrm{e^-} = \frac{1}{2}\mathrm{H_2(g,\ 1\ bar)} \tag{8.22}$$

with $\varepsilon^0(\mathrm{H^+/H_2,g}) = 0.00$ V. The standard half-cell reduction reactions of metals at 25 °C are for the general reaction

$$\mathrm{M^{z+}(aq}, a = 1) + ze^- = \mathrm{M(s)} \tag{8.23}$$

with $\varepsilon^0(\mathrm{M^{z+}/M})$ volts. Half-cell reactions with the most positive standard electrode potentials have a tendency to spontaneously proceed toward reduction (cathode reactions). Half-cell reactions with the most negative standard electrode potentials have a tendency to spontaneously proceed toward oxidation (anode reactions).

Consider, for example, a cell made up of a standard hydrogen electrode and a standard zinc electrode with $\varepsilon^0(\mathrm{H^+/H_2,g}) = 0.00$ V and $\varepsilon^0\ (\mathrm{Zn^{2+}/Zn}) = -0.762$ V. Thus, the H^+ would tend to be reduced, and the zinc metal would tend to be oxidized, and the spontaneous reaction if all species had unit activities would be

$$2\mathrm{H^+(aq}, a = 1) + \mathrm{Zn} = \mathrm{H_2(1\ bar)} + \mathrm{Zn^{2+}(aq}, a = 1) \tag{8.24}$$

with $\varepsilon^0_{cell} = 0.762$ V and $\Delta^0 G = -2 \times 96485 \times \varepsilon^0_{cell}$. The cathode half-cell reaction would be the same as Eq. 8.22, while the anode half-cell reaction would be

$$\mathrm{Zn} = \mathrm{Zn^{2+}(aq}, a = 1) + 2e^- \tag{8.25}$$

Table 8.2 Standard reduction potentials of some common metals at 25 °C

Electrode	E* Volts	Reaction
Li^+/Li	–3.00	$Li^+ + e^- \rightarrow Li$
Rb^+/Rb	–2.92	$Rb^+ + e^- \rightarrow Rb$
K^+/K	–2.92	$K^+ + e^- \rightarrow K$
Sr^+/Sr	–2.92	$Sr^+ + 2e^- \rightarrow Sr$
Ca^{2+}/Ca	–2.87	$Ca^{2+} + 2e^- \rightarrow Ca$
Na^+/Na	–2.71	$Na^+ + e^- \rightarrow Na$
Mg^{2+}/Mg	–2.39	$Mg^{2+} + 2e^- \rightarrow Mg$
Al^{3+}/Al	–1.67	$Al^{3+} + 3e^- \rightarrow Al$
Zn^{2+}/Zn	–0.76	$Zn^{2+} + 2e^- \rightarrow Zn$
Cr^{3+}/Cr	–0.60	$Cr^{3+} + 3e^- \rightarrow Cr$
Fe^{2+}/Fe	–0.44	$Fe^{2+} + 2e^- \rightarrow Fe$
Ni^{2+}/Ni	–0.24	$Ni^{2+} + 2e^- \rightarrow Ni$
Sn^{2+}/Sn	–0.14	$Sn^{2+} + 2e^- \rightarrow Pb$
H^+/H_2	0.00	$H^+ + e^- \rightarrow ½H_2$
Cu^{2+}/Cu	+0.34	$Cu^{2+} + 2e^- \rightarrow Cu$
Ag^+/Ag	+0.80	$Ag^+ + e^- \rightarrow Ag$

When the ion concentrations and H_2 gas do not all have unit activities, the Gibbs energy and cell potential of the cell reaction, Eq. 8.24, become

$$\Delta G = \Delta^0 G + RT\ln\frac{a_{Zn^{2+}}P_{H_2}}{(a_{H^+})^2} \qquad 8.26$$

$$\varepsilon_{cell} = \varepsilon^0_{cell} - \frac{RT}{zf}\ln\frac{a_{Zn^{2+}}P_{H_2}}{(a_{H^+})^2} \qquad 8.27$$

The standard reduction potentials of some common metals at 25 °C are given in Table 8.2.

A cell reaction can be established by different half-cell reactions. For example, for the following reaction, ε^0_1 can be derived from two different cells:

$$3\,Fe^{2+} = 2\,Fe^{3+} + Fe(s) \qquad 8.28$$

cell A

$$3Fe^{2+} + 6e^- = 3Fe(s) \quad \varepsilon^0_1 = -0.440\ V \qquad 8.29$$

$$2\,Fe(s) = 2\,Fe^{3+} + 6e^- \quad \varepsilon^0_2 = +0.036V \qquad 8.30$$

cell B

$$2Fe^{2+} = 2Fe^{3+} + 2e^- \quad \varepsilon^0_4 = -0.772\ V \qquad 8.31$$

$$Fe^{2+} + 2e^- = Fe(s) \quad \varepsilon^0_5 = -0.440\ V \qquad 8.32$$

Both cells give the same net reaction shown by Eq. 8.28, but with six and two electrons and standard cell potentials $\varepsilon^0_{cell\ A} = -0.404$ V and $\varepsilon^0_{cell\ B} = -1.212$ V, respectively. However, the standard Gibbs energies of both cells are the same, i.e.

$$\Delta^0 G_{cell\ A} = -6f(-0.404) = +2.424f \quad 8.33$$

$$\Delta^0 G_{cell\ B} = -2f(-1.212) = +2.424f \quad 8.34$$

This shows that $\Delta^0 G$ values are independent of half-cell reactions and depend only on the net reaction because the net reaction is neutral in electrons and balanced in mass.

8.4 Aqueous solution and Pourbaix diagram

The importance of aqueous solutions in all aspects of life is well known and needs not be discussed further. Since many electrochemical processes involve electrolyte solutions in an aqueous solvent, electrochemical processes including water, hydrogen, and/or oxygen are discussed in more detail. The hydrogen–oxygen cell can be described for both acidic electrolytes, and alkaline electrolytes. With acidic electrolytes, H^+ is in much higher concentrations than OH^-, and thus half-cell reactions with H^+ as an ionic transport species are more important than those involving OH^-. The reverse is true for alkaline electrolytes which contain high OH^- concentrations. Other than for nearly neutral acid–base systems, either H^+ or OH^- dominates the other by several orders of magnitude as can be seen from the value of the 298 K dissociation constant for H_2O:

$$H_2O(l) = H^+(aq) + OH^-(aq) \quad 8.35$$

with reaction constant $K_e = [H^+][OH^-] = 10^{-14}$ and $\Delta^0 G = -RT\ln K_e = +79{,}908$ J. By convention, one defines pH = $-\log\,[H^+]$ and pOH = $-\log\,[OH^-]$, and then pH + pOH = 14.

Under acidic electrolyte conditions of low pH (high $[H^+]$ concentrations) the anode reaction in a hydrogen–oxygen cell is:

$$\frac{1}{2}H_2(g) = H^+(aq) + e^- \quad 8.36$$

with $\varepsilon^0_1 = 0.0$ V and $\Delta^0 G_1 = 0$ J. The corresponding cathode (reduction) reaction is:

$$2H^+(aq) + \frac{1}{2}O_2(g) + 2e^- = H_2O(l) \quad 8.37$$

with $\varepsilon^0_2 = 1.229$ V and $\Delta^0 G_2 = -2\times 1.229\times 96485$ J $= -237160$ J. The net cell reaction for acidic electrolytes is:

$$H_2(g) + \frac{1}{2}O_2(g) = H_2O(l) \quad 8.38$$

with $\varepsilon^0_{cell} = 1.229$ V and $\Delta^0 G_{cell} = -2\times 1.229\times 96485$ J $= -237160$ J.

Under alkaline electrolyte conditions of high pH (high $[OH^-]$ concentrations) the anode reaction in a hydrogen–oxygen cell is:

$$2OH^-(aq) + H_2(g) = 2H_2O(l) + 2e^- \quad 8.39$$

with ε_3^0 = 0.828 V and $\Delta^0 G_3$= −2×0.828×96485 J = −159779 J. The corresponding cathode (reduction) reaction is:

$$H_2O(l) + \frac{1}{2}O_2(g) + 2e^- = 2OH^-(aq) \qquad 8.40$$

with ε_4^0 = 0.401 V and $\Delta^0 G_4$ = −2×0.401×96485 J = −77381 J. The net cell reaction for alkaline electrolytes is:

$$H_2(g) + \frac{1}{2}O_2(g) = H_2O(l) \qquad 8.41$$

with ε_{cell}^0 = 1.229 V and $\Delta^0 G_{cell}$ = −2×1.229×96,485 J = −237160 J.

Plots of ε versus pH for a given chemical system have been typically used to exhibit the stability relationships of ionic species and solid phases in aqueous-based electrochemical systems. These graphs are often called Pourbaix diagrams after their inventor and are constant temperature and constant pressure diagrams for a constant concentration, usually for one metallic element. By convention, the variable ε in a Pourbaix diagram corresponds to the potential for the cathode reduction reactions in the electrochemical half-cell with electrons as reactants. Pourbaix diagrams can be extended to multi-component materials when the thermodynamic properties of the components are available for both the materials and the aqueous solution.

An example of an ε versus pH diagram is shown in Figure 8.5 for the Ni–H_2O system at a 298 K, 1 bar, and $c_{Ni^{2+}} = 0.001$ molality. Three stability regions for Ni species are shown: Ni(s), NiO(s), and [Ni^{2+}]. The two lines on the upper and lower bounds of this diagram correspond to hydrogen reduction (Eq. 8.36) and oxygen reduction (Eq. 8.37) reactions, respectively.

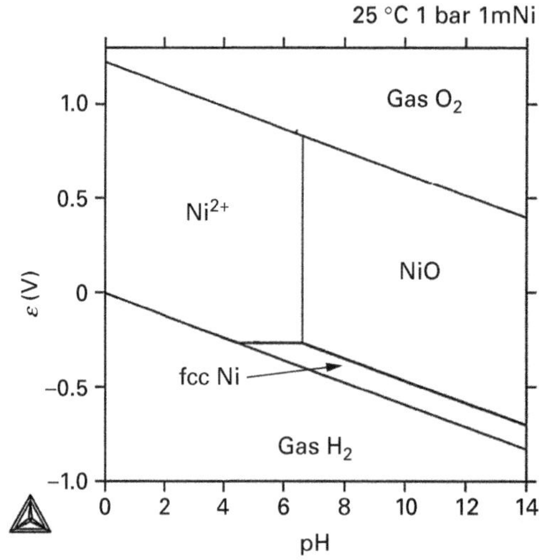

Figure 8.5 *An ε versus pH, Pourbaix diagram for Ni–H_2O at* 298 K, 1 bar, *and* $c_{Ni^{2+}} = 0.001$ *molality.*

For the ε and pH conditions within the boundaries of the Ni(s) region, no solid phase other than Ni(s) is stable, no ionic species with a concentration of 10^{-3} molarity is stable, and no gas species with a pressure of 1 bar is stable. Similar statements could be made about the NiO(s) and $[Ni^{2+}]$ regions on the diagram. In the $[Ni^{2+}]$ area, the introduction of Ni(s) or NiO(s) into the system would result in the dissolution of these solid phases since they are not stable with respect to the $[Ni^{2+}]$ aqueous solution. The corresponding chemical reactions proceed spontaneously to the right as follows until the solid phases are consumed:

$$\mathrm{Ni(s)} \rightarrow \mathrm{Ni^{2+}}(10^{-3}\ \mathrm{molarity}) + 2\mathrm{e}^- \tag{8.42}$$

$$\mathrm{NiO(s)} + 2\mathrm{H^+(aq)} \rightarrow \mathrm{Ni^{2+}}(10^{-3}\ \mathrm{molarity}) + \mathrm{H_2O(l)} \tag{8.43}$$

No H^+(aq) is involved in the first reaction, Eq. 8.42, so the boundary line separating Ni(s) and Ni^{2+} is independent of pH. No oxidation or reduction occurs in the second reaction, Eq. 8.43, i.e. no electrons are reactants or products in the reaction, so the boundary line separating NiO(s) and Ni^{2+} is independent of ε.

Note the convention that ε is the potential for a cathode reduction reaction, and also note that boundary lines between two stability regions depict conditions under which partial equilibrium of the two species occurs for the ε and pH values at any point on these lines. For the boundary line separating Ni(s) and Ni^{2+} in an ideal aqueous solution, i.e. the reverse of Eq. 8.42, the following equation is obtained.

$$\varepsilon = \varepsilon^0 = -0.268\ \mathrm{V} \tag{8.44}$$

For the NiO(s)–Ni^{2+} boundary line of an ideal solution, the reaction Eq. 8.43 is a complete equilibrium, and thus the relationship is

$$0 = \Delta^0 G + RT\ln\frac{1}{(c_{H^+})^2} = \Delta^0 G + 2 \times 2.303RT\mathrm{pH} \tag{8.45}$$

$$pH = -\frac{\Delta^0 G}{2 \times 2.303RT} \tag{8.46}$$

where $\Delta^0 G$ is obtained as follows and can be calculated from the SSUB database and the standard potential of Ni:

$$\begin{aligned}\Delta^\circ G &= {}^\circ G^{H_2o} + {}^\circ G^{Ni^{2+}} - {}^\circ G^{NiO} - 2{}^\circ G^{H^+} \\ &= \left({}^\circ G^{H_2O} - {}^\circ G^{H_2} - \frac{1}{2}{}^\circ G^{O_2}\right) - \left({}^\circ G^{NiO} - G^{Ni} - \frac{1}{2}{}^\circ G^{O_2}\right) \\ &\quad -\left[\left({}^\circ G^{Ni} - 2{}^\circ G^{e^-} - {}^\circ G^{Ni^{2+}}\right) - \left({}^\circ G^{H_2} - 2{}^\circ G^{e^-} - 2{}^\circ G^{H^+}\right)\right] \\ &= \Delta_f^\circ G^{H_2O} - \Delta_f^\circ G^{NiO} - \Delta^\circ G^{Ni^{2+}/Ni}\end{aligned} \tag{8.47}$$

At a specified temperature, only one standard free energy and only one equilibrium constant exists for this chemical reaction, and thus only one specific value of pH = 6.631 exists for the reaction represented by Eq. 8.43 in this Pourbaix diagram.

The diagonal line between Ni(s) and NiO(s) in Figure 8.5 represents the equilibrium between the two phases and holds for a partial equilibrium reaction that is the sum reactions Eq. 8.42 and Eq. 8.43:

$$NiO(s) + 2H^+(aq) + 2e^- = Ni(s) + H_2O(l) \qquad 8.48$$

The reduction of Ni from a divalent state in NiO to metallic Ni(s) occurs, but the reaction also depends on the H^+ concentration, the pH value. The corresponding Gibbs energy and Nernst equations are

$$\Delta G = \Delta^0 G + RT\ln\frac{1}{(c_{H^+})^2} = -23939 + 2 \times 2.303RTpH \qquad 8.49$$

$$\varepsilon = \varepsilon^0 - \frac{RT}{2f}\ln\frac{1}{(c_{H^+})^2} = 0.124 - \frac{2.303RT}{f}pH \qquad 8.50$$

where $\Delta^0 G$ can be calculated as follows:

$$\begin{aligned}\Delta^\circ G &= {}^\circ G^{H_2O} + {}^\circ G^{Ni} - {}^\circ G^{NiO} - 2{}^\circ G^{H+} - 2{}^\circ G^{e^-} \\ &= \left({}^\circ G^{H_2O} - {}^\circ G^{H_2} - \frac{1}{2}{}^\circ G^{O_2}\right) - \left({}^\circ G^{NiO} - {}^\circ G^{Ni} - \frac{1}{2}{}^\circ G^{O_2}\right) \\ &\quad + \left({}^\circ G^{H_2} - 2{}^\circ G^{e^-} - 2{}^\circ G^{H+}\right) \\ &= \Delta_f^\circ G^{H_2O} - \Delta_f^\circ G^{NiO}\end{aligned} \qquad 8.51$$

The bottom and top lines in Figure 8.5 correspond to the reduction reactions related to H_2 and O_2 gases, i.e. the stability of H_2O. The lower one is for the reverse of Eq. 8.36 under $\varepsilon^0 = 0$ and $P_{H_2} = 1$ with the Nernst equation being

$$\varepsilon = \varepsilon^0 - \frac{RT}{f}\ln\frac{(P_{H_2})^{1/2}}{c_{H^+}} = -\frac{2.303RT}{f}\text{pH} \qquad 8.52$$

As the pH increases from 0, ε becomes more negative, as depicted. The top line corresponds to the oxygen reduction reaction represented by Eq. 8.37 with $\varepsilon^0 = 1.225$ and calculated from the aqueous solution database in Thermo-Calc [64] and $P_{O_2} = 1$ with the Nernst equation being

$$\varepsilon = \varepsilon^0 - \frac{RT}{2f}\ln\frac{(P_{O_2})^{1/2}}{(c_{H^+})^2} = 1.225 - \frac{2.303RT}{f}\text{pH} \qquad 8.53$$

The dependence of ε on pH is identical for the reduction reactions Eq. 8.37 and Eq. 8.48, and their intercepts at pH $= 0$ differ by the difference in their ε^0 values.

In this simple Pourbaix diagram for Ni in an ideal aqueous solution, all boundary lines are straight because there is only one ionic species of Ni in the aqueous solution, i.e. Ni^{2+}. When there is more than one ionic species in the aqueous solution, the boundary lines may no longer be straight due to the competition between species. One example is Cu with two main ionic species of Cu^{2+} and $CuOH^+$, and the reduction reaction between the metallic Cu and the aqueous solution involves both species, i.e.

$$xCu^{2+} + (1-x)CuOH^+ + (1-x)H^+ + 2\,e^- - = Cu(s) + (1-x)H_2O \qquad 8.54$$

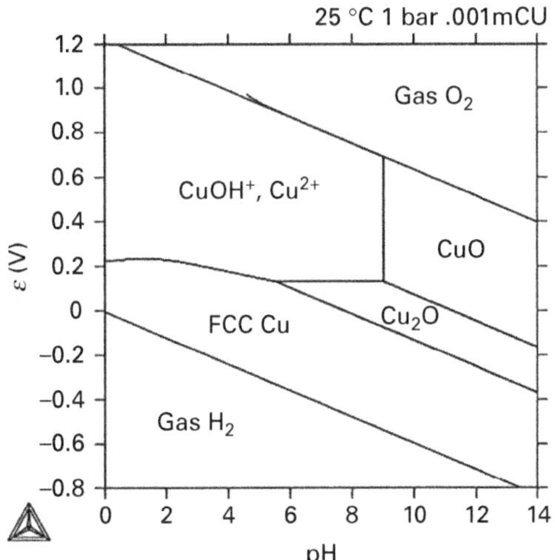

Figure 8.6 *An* ε *versus pH Pourbaix diagram for the Cu–H*$_2$*O system at* 298 K, 1 bar, *and* $c_{Cu} = 0.001$ *molality.*

with

$$\begin{aligned}\Delta G &= \Delta^0 G + RT\ln\frac{1}{\left(c_{Cu^{2+}}\right)^x \left(c_{CuOH^+} c_{H^+}\right)^{1-x}} \\ &= \Delta^0 G + RT\ln\frac{1}{\left(c_{Cu^{2+}}\right)^x \left(c_{CuOH^+}\right)^{1-x}} + 2.303(1-x)RT\text{pH}\end{aligned} \qquad 8.55$$

$$\varepsilon = \varepsilon^0 - \frac{RT}{2f}\ln\frac{1}{\left(c_{Cu^{2+}}\right)^x \left(c_{CuOH^+}\right)^{1-x}} - \frac{2.303(1-x)\times RT}{2f}\text{pH} \qquad 8.56$$

It is evident that both the slope and the intercept at pH = 0 are a function of the concentration of $CuOH^+$, which is a function of the pH value. Consequently, the boundary between metallic Cu and the aqueous solution is no longer a straight line, as shown in Figure 8.6. The concentrations of various species in the aqueous solution, i.e. commonly called the speciation, are plotted in Figure 8.7, showing the change of dominant species as a function of pH value.

In Pourbaix diagrams for alloys with two or more elements, the activities of individual elements are used in calculating the potentials of reduction reactions. Considering a Fe–Ni alloy with Fe^{2+} and Ni^{2+} in the aqueous solution, the reduction reactions for Fe and Ni can be written separately as

$$Ni^{2+}(c_{Ni}) + 2e^- \rightarrow Ni\ (a_{Ni}\text{ in alloy}) \qquad 8.57$$

$$Fe^{2+}(c_{Fe}) + 2e^- \rightarrow Fe\ (a_{Fe}\text{ in alloy}) \qquad 8.58$$

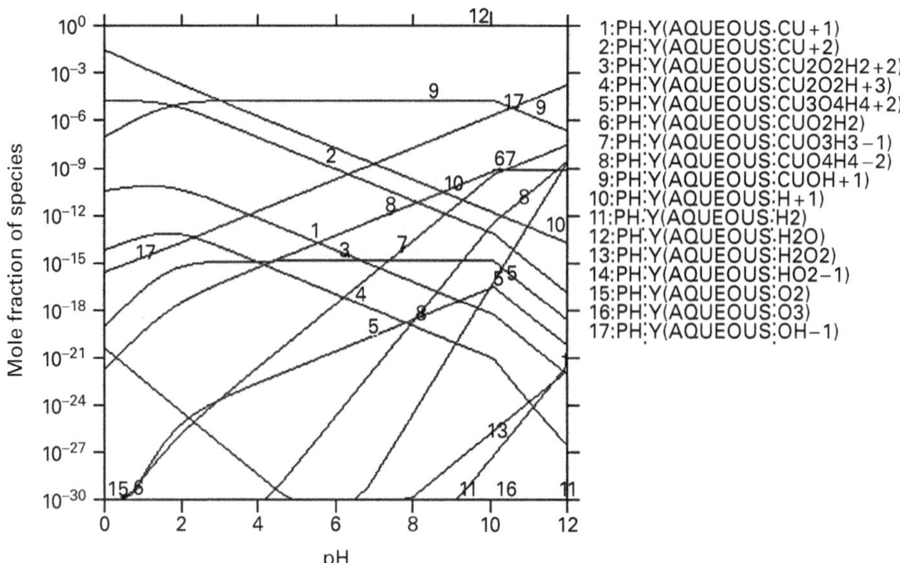

Figure 8.7 *Concentrations Y of ionic species in the aqueous solution at $\varepsilon = 0.3V$ from Figure 8.6.*

with potentials

$$\varepsilon_{Ni} = \varepsilon_{Ni}^0 - \frac{2.303RT}{2f}\ln\frac{a_{Ni}}{c_{Ni}} = -0.268 - \frac{2.303RT}{2f}\ln\frac{a_{Ni}}{c_{Ni}} \qquad 8.59$$

$$\varepsilon_{Fe} = \varepsilon_{Fe}^0 - \frac{2.303RT}{2f}\ln\frac{a_{Fe}}{c_{Fe}} = -0.441 - \frac{2.303RT}{2f}\ln\frac{a_{Fe}}{c_{Fe}} \qquad 8.60$$

In principle, there are two scenarios for a given set of a_{Ni} and a_{Fe} for the alloy. The first scenario is at the limit of a dilute aqueous solution, i.e. $c_{Ni} = c_{Fe} = 0.001$ molarity, ε_{Ni} and ε_{Fe} can be calculated, and the element with the lower potential has the tendency to dissolve first, which can result in the so-called dialloying effect. The second scenario is for equal potentials, i.e. $\varepsilon_{Ni} = \varepsilon_{Fe}$ due to an externally imposed potential, and the equilibrium concentrations of Fe^{2+} and Ni^{2+} can be calculated from Eq. 8.59 and Eq. 8.60.

8.5 Application examples

Among many applications of electrochemistry, several are briefly discussed in this section.

8.5.1 Metastability and passivation

Our modern industrial society is built on various metals such as Fe, Ni, Al, Ti, and Zr alloys which are reactive, but exhibit extraordinary kinetic stability in

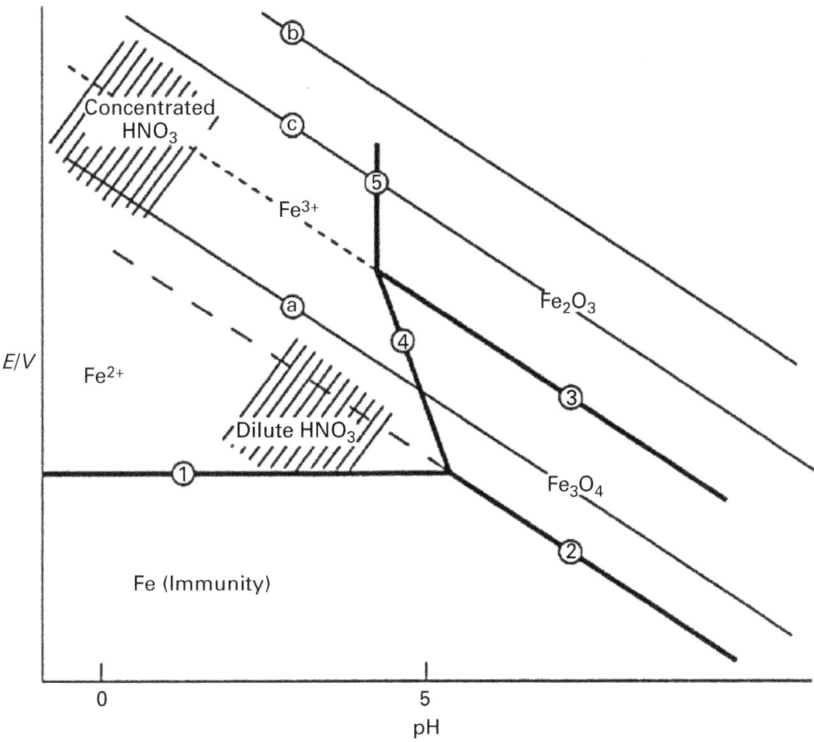

Figure 8.8 *Schematic Pourbaix diagram for iron illustrating the resolution of the Faraday paradox in the corrosion of iron in nitric acid [66]. Lines (a) (b), and (c) correspond to the following equilibria: (a)* $H^+ + e^- = \frac{1}{2}H_2$; *(b)* $O_2 + 4H^+ + 4e^- = 2H_2O$; *(c)* $NO_3^- + 3H^+ + 2e^- = HNO_2 + H_2O$, *respectively, with permission from IUPAC.*

oxidizing environments due to the existence of a thin reaction product film on the surface. This film effectively isolates the metal from the corrosive environment, a phenomenon called passivation. One interesting experiment was reported by Faraday in 1836 who found that iron corrodes freely in dilute nitric acid, while in concentrated nitric acid no reaction apparently occurred. To understand this phenomenon, let us examine a simple, schematic Pourbaix diagram for an iron–water system, shown in Figure 8.8.

For iron in deaerated acid solution, the partial anodic and cathodic reactions are given by line 1 (Fe/Fe^{2+}) and line (a), respectively, resulting in a corrosion potential that lies between lines 1 and (a). In oxygenated (aerated) solutions, the corrosion potential may lie between lines 1 and (b), because the reduction of oxygen is a possible (likely) cathodic reaction. Since dilute HNO_3 is only a weak oxidizing agent, the principal cathodic reaction is most likely hydrogen evolution, and hence the corrosion potential is expected to lie between lines 1 and (a) at relatively high pH, as shown. Since the Fe/Fe^{2+} reaction is relatively fast compared with that for

H^+/H_2 on iron, if the corrosion potential is situated below the extension of line 2 (Fe/Fe_3O_4) into the Fe^{2+} stability region then Fe_3O_4 cannot form on the surface, even as a metastable phase.

On the other hand, concentrated HNO_3 is a strong oxidizing agent due to the reaction of $NO_3^- + 3H^+ + 2e^- = HNO_2 + H_2O$, so that the corrosion potential can lie anywhere between lines 1 and (c) at low pH. Since reaction (2) is likely to be fast, the corrosion potential will be high and certainly will be more positive than the extension of line 2 into the stability region for Fe^{2+} at low pH. Therefore, Fe_3O_4 becomes metastable and can form between the aqueous solution and iron. The thickness of this Fe_3O_4 film depends on its dissolution rate into the aqueous solution and its growth rate at the interface with iron, which in turn depends on the diffusion of ionic species across the film. Its existence results in passivity and the observed kinetic inactivity of iron in this medium. When the potential becomes even more positive above the extension of line 3, Fe_2O_3 may form on Fe_3O_4 as an additional metastable phase, resulting in the commonly observed bilayer structure.

8.5.2 Galvanic protection

A galvanic reaction takes place between two different materials at the two respective electrodes each with a different tendency to hold on to electrons. Consider the following electrochemical cell used to protect Cu tanks against oxidation by using a "sacrificial" Fe electrode:

anode	solution	cathode
$Fe(s) \mid Fe^{2+}\mid$	$SO_4^{=}\mid$	$Cu^{2+}\mid Cu(s) \mid$

Cathode reduction:	$Cu^{2+} + 2e^- = Cu(s)$	ε^0 (volts) = 0.34
Anode oxidation:	$Fe(s) = Fe^{2+} + 2e^-$	ε^0 (volts) = 0.44
Net reaction:	$Cu^{2+} + Fe \rightarrow Cu + Fe^{2+}$	ε^0 (volts) = 0.78

If the cell has a direct connection between the electrodes, i.e. it has a short circuit, then $\Delta G \rightarrow 0$ and thus $\varepsilon_{cell} \rightarrow 0$. Since we have the value $\varepsilon^0_{cell} > 0$ for the net cell reaction, the equilibrium constant $K_e > 1$, which means $[Fe^{2+}]/[Cu^{2+}] > 1$. By assuming an ideal electrolyte solution, the activities in K_e can be represented by concentrations (in molar concentration units), further assuming that solid Fe and Cu are present at unit activities. If the electrodes of Cu and Fe are short circuited while in contact with the same "electrolyte solution," the final equilibrium concentrations can be calculated by the standard equation

$$K_e = [Fe^{2+}]/[Cu^{2+}] = \exp(-\Delta^0 G/RT) \qquad 8.61$$

or, using the Nernst equations,

$$\varepsilon = \varepsilon^0 - (RT/zf) \ln ([Fe^{2+}]/[Cu^{2+}]) = 0 \qquad 8.62$$

Thus the above standard cell potential $[Fe^{2+}]/[Cu^{2+}] = 2.4 \times 10^{26}$.

With this large ratio it can be seen that the tendency to produce Cu^{2+} ions, i.e. the tendency to corrode the Cu(s), is extremely small if a sacrificial Fe electrode is configured in an electrochemical cell with the Cu tank.

8.5.3 Fuel cells

Fuel cells are devices to convert chemical energy to electricity and heat through electrochemical reactions, with the fuel and oxygen supplied to the anode and cathode, respectively. Typical ions migrating through the electrolyte are H^+, OH^-, CO_3^{2-}, and O^{2-}. In fuel cells with H^+ as migrating ions, H_2 molecules are dissociated into H^+ on the anode, which combine with O_2 on the cathode to form H_2O and release heat, on the basis of the half-cell and the net cell reactions in their simplest form as shown by Eq. 8.36 for the anode, Eq. 8.37 for the cathode, and Eq. 8.38 for the net cell, respectively. Commonly used electrolytes are polymer and phosphoric acid, and both anode and cathode reactions are facilitated by a catalyst, typically platinum. The thermodynamic limit for the power which can be generated by the fuel cell is represented by

$$w = -\Delta G = -\Delta^0 G_{cell} + RT\ln\left(P_{H_2} P_{O_2}^{1/2}\right) \qquad 8.63$$

For fuel cells with anions as migrating ions, the anions are generated on the cathode, with H_2O formed and heat generated on the anode. Their representative cathode reactions are

$$\frac{1}{2}O_2 + H_2O + 2e^- = 2OH^- \qquad 8.64$$

$$\frac{1}{2}O_2 + CO_2 + 2e^- = CO_3^{2-} \qquad 8.65$$

$$\frac{1}{2}O_2 + 2e^- = O^{2-} \qquad 8.66$$

The anode reaction Eq. 8.64 is the reaction represented by Eq. 8.39, operating at low temperatures and using a catalyst for both electrodes. The anode reactions Eq. 8.65 and Eq. 8.66 are

$$CO_3^{2-} + H_2 = H_2O + CO_2 + 2e^- \qquad 8.67$$

$$O^{2-} + H_2 = H_2O + 2e^- \qquad 8.68$$

respectively. To enable the diffusion of CO_3^{2-} and O^{2-} through the cathode and the electrolyte, both fuel cells are operated at relatively high temperatures, with the former typically in molten carbonate solutions and the latter through solid oxides. Due to the high operating temperatures, fuels are converted to hydrogen within the fuel cell itself by a process called internal reforming, removing the need for a precious-metal catalyst and enabling the use of a variety of fuels. The net cell reaction for all three fuel cells is the same as in the case of H^+, represented by Eq. 8.38.

8.5.4 Ion transport membranes

Ion transport membranes (ITMs) are ceramic membranes conducting both electrons and oxygen ions, but no other species. The chemical potential difference of oxygen between the two sides of a membrane provides the driving force for oxygen to diffuse through the membrane. Commonly used ITM oxides include perovskite and fluorite, with chemical formulas ABO_3 and AO_2, respectively, typically with more than one element in the A-site and/or the B-site to modify the electron and ionic conductivities. Key thermodynamic properties of ITM oxides are their stability in service environments, their vacancy concentrations in the oxygen and cation sites, and their cation valences their. On the high oxygen partial pressure side, the reaction is the following:

$$\frac{1}{2}O_2 + 2e^- = O^{2-} \tag{8.69}$$

At the same time, the number of oxygen vacancies is reduced, resulting in a lower concentration of oxygen vacancies and higher oxygen activity in the oxide on the high oxygen partial pressure side. On the low oxygen partial pressure side, the reaction is reversed to produce oxygen molecules, i.e.

$$O^{2-} = \frac{1}{2}O_2 + 2e^- \tag{8.70}$$

This reaction results in a higher oxygen vacancy concentration and lower oxygen activity. On both sides, the charge neutrality is compensated for by the valence changes of the cations, resulting in electron flow in the opposite direction of oxygen diffusion. The ionic conductivity is dictated by the oxygen transportation across the membrane by the driving force of the following net reaction:

$$\frac{1}{2}O_2\left(P_{high}\right) = \frac{1}{2}O_2\left(P_{low}\right) \tag{8.71}$$

with change in Gibbs energy

$$\Delta G = 0.5RT\ln\left(\frac{P_{low}}{P_{high}}\right) \tag{8.72}$$

The oxygen transportation is closely related to the concentration of oxygen vacancies in the membrane, which is obtained by minimizing the Gibbs energy of the phase under given temperature and oxygen partial pressure conditions. High vacancy concentrations can be obtained by cation dopants with lower valences or small energy differences between various valence states. However, at the same time, high vacancy concentrations reduce the thermodynamic stability of the membrane, which may result in its decomposition into more stable phase and can be tailored by alloying in A and B sites.

8.5.5 Electrical batteries

Batteries utilize electrochemical reactions to generate electricity for various devices. The theoretical voltage of a battery can be calculated from Eq. 8.20 and Eq. 8.21 as

$$\varepsilon = -\frac{\Delta G}{zf} = \varepsilon^0 - \frac{RT\ln Q}{zf} \tag{8.73}$$

with ΔG the driving force of the net cell reaction and Q the reaction activity quotient. The actual voltage of a battery is lower than the theoretical value due to kinetic limitations of cell reactions and resistance to ion diffusion through the electrolyte. Based on whether the cell reactions are reversible or not, batteries are typically categorized as either primary disposable or secondary rechargeable batteries. The net cell reactions in primary disposable batteries are not easily reversible, and electrode materials may not return to their original forms on application of a higher external potential of opposite sign. Consequently, primary batteries cannot be reliably recharged. On the other hand, the net cell reactions in secondary batteries are easily reversible. Furthermore, two half-cells in batteries may use different electrolytes with each half-cell enclosed in a container and a separator permeable to conducting ions but not to the bulk of the electrolytes.

One common primary battery is the zinc–carbon battery, with a zinc anode cylinder and a carbon cathode central rod. The electrolytes are ammonium or zinc chloride next to the zinc anode and a mixture of ammonium chloride and manganese dioxide next to the carbon cathode. The half-cell and net reactions with ammonium chloride are as follows:

$$Zn + 2NH_3 \rightarrow Zn(NH_3)_2^{\ 2+} + 2e^- \quad 8.74$$

$$2NH_4Cl + 2MnO_2 + 2e^- \rightarrow 2NH_3 + Mn_2O_3 + H_2O + 2Cl^- \quad 8.75$$

$$Zn + 2MnO_2 + 2NH_4Cl \rightarrow Mn_2O_3 + Zn(NH_3)_2Cl_2 + H_2O \quad 8.76$$

The electric potential of the reaction is, treating all compounds as stoichiometric,

$$\begin{aligned} \varepsilon &= -\frac{\Delta G}{2f} = -\frac{\Delta^0 G}{2f} \\ &= \frac{1}{2f}\left({}^0G^{Zn} + 2\,{}^0G^{MnO_2} + 2\,{}^0G^{NH_4Cl} - {}^0G^{H_2O} - {}^0G^{Zn(NH_3)_2Cl_2} - {}^0G^{Mn_2O_3}\right) \end{aligned} \quad 8.77$$

The Gibbs energy of $Zn(NH_3)_2Cl_2$ is not available in current databases but has been estimated to be -505375 J per mole of formula [67]. The value of Eq. 8.77 at 298.15 K is thus obtained as 1.67 V, which is pretty close to the actual operating voltage of the battery, around 1.5 V.

With zinc chloride, the cell reactions and electric potential may be written as

$$Zn + ZnCl_2 + 2OH^- \rightarrow 2ZnOHCl + 2e^- \quad 8.78$$

$$MnO_2 + H_2O + e^- \rightarrow MnOOH + OH^- \quad 8.79$$

$$Zn + 2MnO_2 + ZnCl_2 + 2H_2O \rightarrow 2MnOOH + 2ZnOHCl \quad 8.80$$

$$\varepsilon = \frac{1}{2f}\left({}^0G^{Zn} + 2\,{}^0G^{MnO_2} + {}^0G^{ZnCl_2} + 2\,{}^0G^{H_2O} - 2\,{}^0G^{MnOOH} - 2\,{}^0G^{ZnOHCl}\right) \quad 8.81$$

Secondary batteries can be recharged by applying an external electrical potential, which reverses the net cell reaction that occurs during discharging. The oldest form of rechargeable battery is the lead–acid batteries used in automotives, and the latest development is the lithium-ion (Li-ion) batteries. A lead–acid battery typically uses Pb and PbO_2 as the cathode and anode electrodes and a 35% sulfuric acid and 65% water solution as the electrolyte. Its anode and cathode reactions can be simplified as follows

$$Pb + SO_4^{2-} = PbSO_4 + 2e^- \tag{8.82}$$

$$PbO_2 + 4H^+ + SO_4^{2-} + 2e^- = PbSO_4 + 2H_2O \tag{8.83}$$

The net cell reaction is

$$Pb + PbO_2 + 2H_2SO_4 = 2PbSO_4 + 2H_2O \tag{8.84}$$

Its electric potential is represented by the following equation:

$$\varepsilon = -\frac{1}{2f}\left(2\,{}^0G^{H_2O} + 2\,{}^0G^{PbSO_4} - {}^0G^{Pb} - {}^0G^{PbO_2} - 2\,{}^0G^{H_2SO_4}\right) \tag{8.85}$$

the value being 2.651 V at 298.15 K as calculated from Thermo-Calc [64]. During discharge, the reaction Eq. 8.84 goes to the right, and $PbSO_4$ is formed on both the anode and cathode. During charging, the reaction Eq. 8.84 goes to the left, and Pb and PbO_2 are restored. In practical applications, other ionic species such as H_3O^+ and HSO_4^- may form in the electrolyte, complicating the reactions and affecting its potential.

In lithium-ion batteries, during charging and discharging lithium ions migrate in electrolytes between electrodes made of intercalated lithium compounds; $LiCoO_2$ and $LiFePO_4$ are two of the several cathode materials used in lithium-ion batteries, and the anode is typically made of carbon or metallic lithium. The anode and cathode reactions for $LiCoO_2$ batteries can be written in simple form as follows:

$$Li_xC_6 = xLi^+ + xe^- + 6C \tag{8.86}$$

$$xLi^+ + xe^- + Li_{1-x}CoO_2 = LiCoO_2 \tag{8.87}$$

with the net reaction and electric potential being

$$Li_xC_6 + Li_{1-x}CoO_2 = LiCoO_2 + 6C \tag{8.88}$$

$$\begin{aligned}\varepsilon &= -\frac{1}{xf}\left\{6\,{}^0G^{C} + {}^0G^{LiCoO_2} - G^{Li_xC_6} - G^{Li_{1-x}CoO_2}\right\}\\ &= -\frac{1}{f}\left\{\left(\mu_{Li}^{Li_{1-x}CoO_2} - \mu_{Li}^{Li_xC}\right) - \frac{1}{x}\left(\mu_{LiCoO_2}^{Li_{1-x}CoO_2} - {}^0G^{LiCoO_2}\right)\right\}\end{aligned} \tag{8.89}$$

The electric potential is a function of x. The value in the first parentheses in the second line of the above equation denotes the chemical potential difference of Li between two electrodes, and the value in the second parentheses represents the chemical potential difference of $LiCoO_2$ between the states in the solution phase of $Li_{1-x}CoO_2$ and by itself. The Gibbs energies of Li_xC_6 and $Li_{1-x}CoO_2$ need to be obtained as a function of x in order to calculate the electric potential of the battery.

The compound $LiFePO_4$ uses metallic lithium as the anode with the following half-cell and net cell reactions:

$$x\text{Li} = x\text{Li}^+ + xe^- \quad 8.90$$

$$x\text{Li}^+ + xe^- + \text{Li}_{1-x}\text{FePO}_4 = \text{LiFePO}_4 \quad 8.91$$

$$x\text{Li} + \text{Li}_{1-x}\text{FePO}_4 = \text{LiFePO}_4 \quad 8.92$$

Its electric potential is also a function of x, i.e.

$$\begin{aligned}\varepsilon &= -\frac{1}{xf}\left\{{}^0G^{LiFePO_4} - x\,{}^0G^{Li} - G^{Li_{1-x}FePO_4}\right\}\\ &= -\frac{1}{f}\left\{\left(\mu_{Li}^{Li_{1-x}FePO_4} - {}^0\mu_{Li}\right) - \frac{1}{x}\left(\mu_{LiFePO_4}^{Li_{1-x}FePO_4} - {}^0G^{LiFePO_4}\right)\right\}\end{aligned} \quad 8.93$$

The value in the first parentheses in the second line of the above equation denotes the chemical potential difference of Li between the two electrodes, and the value in the second parentheses represents the chemical potential difference of $LiFePO_4$ between the states in the solution phase of $Li_{1-x}FePO_4$ and by itself. Consequently, the Gibbs energy of $Li_{1-x}FePO_4$ needs to be obtained as a function of x in order to calculate the electric potential of the battery. It is known that there are several miscibility gaps in the $FePO_4$ and $LiFePO_4$ pseudo-binary system, in which the chemical potentials are constants, and therefore so is the electric potential, resulting in a stable battery output.

Exercises

1. The following solid state electrochemical cell, which is based on a calcia stabilized zirconia electrolyte, is operated at 800 K and produces a cell potential $\varepsilon = 0.10$ V.

   $$<\text{Pt}><\text{Co,CoO}>|<\text{ZrO}_2 - 12\%\text{CaO}>|<\text{Cu}-\text{Co,CoO}><\text{Pt}>$$

   a. Which species transports charge through the electrolyte?
      What is the half-cell reaction at the anode?
      What is the half-cell reaction at the cathode?
      What is the net cell reaction?
   b. For the net cell reaction, calculate ΔG^0 and ΔG.
      What is the activity of Co in the Cu–Co alloy?

2. The emf of a cell having the following reaction,

   $$\text{Zn(s)} + 2\text{AgCl(s)} = \text{ZnCl}_2\text{(s)} + 2\text{Ag(s)}$$

   is 1.010 V at 298 K, and 1.021 V at 273 K.

(cont.)

a. Assume the temperature coefficient is constant, and calculate the 298 K values of $\Delta^0 G$, $\Delta^0 H$, $\Delta^0 S$ for the reaction.

b. What is the anode reaction? The cathode reaction?

3. Create your own lithium ion battery by selecting anode and cathode materials. Write down the half-cell reactions and the net cell reaction. Find their thermodynamic properties and estimate the voltage of your battery.

4. The Pourbaix diagram (ε versus pH) for the Cu–H_2O system is given below. This diagram is for a temperature of 298 K and ion concentrations of 10^{-3} molality.

a. Write down the chemical reaction depicting the equilibrium between Cu^{2+} ions and CuO(s).

b. Calculate the change in the position of the vertical line between Cu^{2+} and CuO(s) on this diagram when the Cu^{2+} concentration for the diagram is changed to 1 molality.

c. Calculate the standard reduction potential of Cu^{2+}/Cu using the ε value at pH = 0.

d. Describe what the possible forms of Cu in the aqueous solution are on the boundary between the aqueous solution and Cu_2O. Give your reasons.

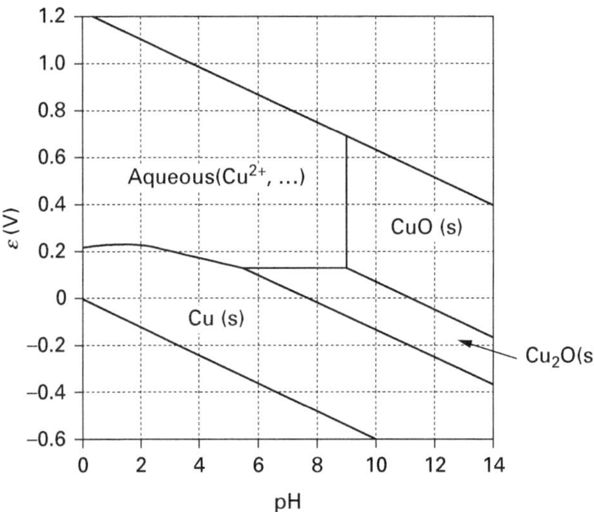

5. The Pourbaix diagram (ε versus pH) for the Ni–H_2O system is given here. This diagram is for a temperature of 298 K and ion concentrations of 10^{-3} molality.

a. Calculate the change in the position of the vertical line between Ni^{2+} and NiO(s) on this diagram when the Ni^{2+} concentration for the diagram is changed to 0.1 molality.

b. Calculate the standard reduction potential of Ni^{2+}/Ni.

(*cont.*)

c. Write down the chemical reaction depicting the equilibrium between Ni^{2+} ions and NiO(s).

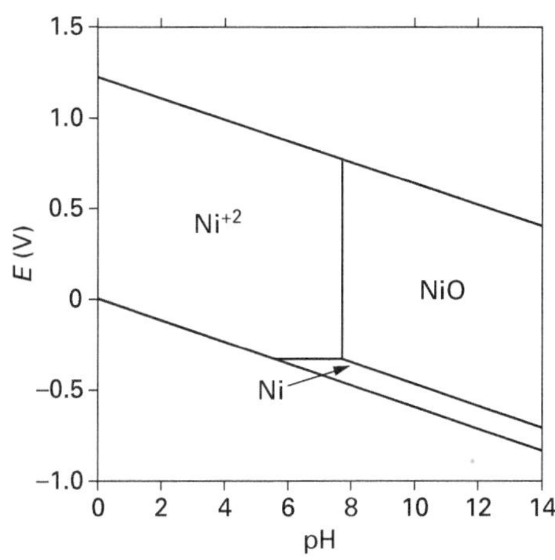

9 Critical phenomena, thermal expansion, and Materials Genome®

In Section 1.3, it was shown that all molar quantities of a homogeneous system diverge at the critical point, i.e. the limit of stability, including the additional quantities shown in Eq. 2.148. As illustrated by Eq. 1.43, even though each molar quantity changes in the same direction as its conjugate potential, i.e. with the same sign, its dependence with respect to a non-conjugate potential can be of either the same sign or opposite sign. It is often considered to be normal when the quantities change in the same direction, while abnormal when they change in different directions.

In this chapter, the thermal expansion defined by Eq. 2.8 is used as an example for a detailed discussion, based on the MMS model presented in Section 5.2.5, of those extraordinary phenomena in the context of a critical point. The MMS model is first discussed, in terms of thermal expansion and is then applied to elemental cerium (Ce), with colossal positive thermal expansion (CPTE), and Fe_3Pt with negative thermal expansion (NTE).

9.1 MMS model applied to thermal expansion

As shown in Eq. 1.44, the thermal expansion of a system can be positive, zero, or negative depending on the pressure dependence of the entropy of the system. Let us carry out a virtual experiment by analyzing a system starting with one microstate only, α, when a metastable microstate β has a higher entropy than the microstate α, i.e. $S^\beta > S^\alpha$, and the relative stability of the β microstate thus increases with temperature. Cases with $S^\beta < S^\alpha$, starting with a mixture of the α and β microstate, will be discussed after this.

When a metastable microstate, β, is introduced by changing the pressure under constant temperature, based on Eq. 5.51 and Eq. 5.52 the entropy change of the system can be written as

$$\Delta S = S - S^\alpha = p^\beta\left(S^\beta - S^\alpha\right) - k_B\left[\left(1 - p^\beta\right)\ln\left(1 - p^\beta\right) + p^\beta \ln p^\beta\right] \qquad 9.1$$

where p^β represents the statistical probability of the microstate β in the system. With $S^\beta > S^\alpha$, this would result in a positive entropy change of Eq. 9.1, i.e. $\Delta S > 0$, since $0 < p^\beta < 1$. If this entropy increase is due to the decrease of pressure, i.e. $V^\beta > V^\alpha$ because volume and its conjugate potential (negative pressure) change in the same direction, the volume thermal expansion of the system is positive due to the increase of the population of the β microstate with a larger volume. In this case, $\Delta V^{\alpha\beta}/\Delta S^{\alpha\beta} > 0$, and the volume and entropy of the two microstates change in the same direction.

On the other hand, if this entropy increase is realized by increasing the pressure, i.e. $V^{\beta} < V^{\alpha}$, the volume thermal expansion of the system is negative due to the increase of the population of the β microstate with a smaller volume. In this case, $\Delta V^{\alpha\beta}/\Delta S^{\alpha\beta} < 0$, and the volume and entropy of the two microstates change in opposite directions.

Therefore, the sign of $\Delta V^{\alpha\beta}/\Delta S^{\alpha\beta}$ for the two microstates can be used as a criterion to determine whether a system has NTE, since a positive value of $\Delta V^{\alpha\beta}/\Delta S^{\alpha\beta}$ means positive thermal expansion and a negative value of $\Delta V^{\alpha\beta}/\Delta S^{\alpha\beta}$ means NTE. At a critical point, the entropy change with respect to temperature is infinite, resulting in either infinite positive or infinite negative thermal expansion correspondingly. When the system moves away from the critical point into the macroscopically homogeneous single-phase region, the thermal expansion becomes less positive or negative. A number of systems with $\Delta V^{\alpha\beta}/\Delta S^{\alpha\beta} < 0$, thus potentially NTE, are listed in the supplementary information of reference [68].

Now let us consider the case when the metastable β microstate has lower entropy than the α microstate, i.e. $S^{\beta} < S^{\alpha}$, and the β microstate is thus more stable at low temperatures. The system at higher temperatures thus contains only the α microstate and has positive thermal expansion. When the metastable β microstate is introduced, the sign of the entropy change in Eq. 9.1 can be either positive or negative because the first term is negative and the second term is positive, and its sign thus depends on the value of the entropy difference between two microstates and the probability of the metastable β microstate. The virtual experiment should thus be carried out in a system with the highest microstate configurational entropy in Eq. 9.1, i.e. $p^{\beta} = 0.5$ when the two microstates have the same free energy and are in equilibrium with each other. From Eq. 5.51 and Eq. 5.52, the system entropy can be written as

$$S = S^{\alpha} + 0.5\left(S^{\beta} - S^{\alpha}\right) + k_B \ln 2 \tag{9.2}$$

With a change of pressure, p^{β} will either increase or decrease, and the entropy of the system becomes

$$S = S^{\alpha} + p^{\beta}\left(S^{\beta} - S^{\alpha}\right) - k_B\left[\left(1 - p^{\beta}\right)\ln\left(1 - p^{\beta}\right) + p^{\beta}\ln p^{\beta}\right] \tag{9.3}$$

The difference of Eq. 9.3 and Eq. 9.2 is obtained as

$$\Delta S = \left(p^{\beta} - 0.5\right)\left(S^{\beta} - S^{\alpha}\right) - k_B\left[\left(1 - p^{\beta}\right)\ln\left(1 - p^{\beta}\right) + p^{\beta}\ln p^{\beta} + \ln 2\right] \tag{9.4}$$

The second term in Eq. 9.4 is always negative, and the first term is also negative if $p^{\beta} > 0.5$ because $S^{\beta} < S^{\alpha}$. It is thus evident that if p^{β} is increased by decreasing the pressure, the entropy of the system decreases, and the system would possess negative thermal expansion because Eq. 9.4 is negative, and the entropy is reduced by the decrease of pressure. At the same time, $V^{\beta} > V^{\alpha}$ and $\Delta V^{\alpha\beta}/\Delta S^{\alpha\beta} < 0$, the latter being the same condition for a negative thermal expansion as in the first virtual experiment with $S^{\beta} > S^{\alpha}$. One can thus conclude that for a two-phase equilibrium line with $dT/dP = \Delta V^{\alpha\beta}/\Delta S^{\alpha\beta} < 0$, both phases can display negative thermal expansion. On the other hand, if p^β is increased the by increasing pressure, the system would possess positive thermal expansion because the entropy is reduced by the increase of pressure.

Furthermore, the thermal expansion of a system can be approximated as follows, using the rule for the mixture of volumes:

$$V = \left(1 - p^\beta\right) V^\alpha + p^\beta V^\beta \tag{9.5}$$

$$a_v = \left\{ a_v^\alpha V^\alpha + p^\beta \left(a_v^\beta V^\beta - a_v^\alpha V^\alpha\right) + \left(V^\beta - V^\alpha\right) \frac{\partial p^\beta}{\partial T} \right\} \Big/ V \tag{9.6}$$

where a_V, a_V^α, a_V^β, V, V^α, and V^β are the thermal expansion coefficients and volumes of the system and the α and β microstates, respectively. For simplification, let us assume both microstates have similar positive thermal expansion, i.e. $a_V^\alpha \approx a_V^\beta$, and Eq. 9.6 becomes

$$a_V = \left\{ a_v^\alpha V^\alpha - \left(V^\alpha - V^\beta\right) \left(p^\beta a_V^\beta + \frac{\partial p^\beta}{\partial T} \right) \right\} \Big/ V \tag{9.7}$$

Equation 9.7 shows that it is the combination of volume difference and $\partial p^\beta / \partial T$ value that determines the macroscopic thermal expansion. By setting $a_V = 0$, one obtains

$$\frac{\partial p^\beta}{\partial T} = a_V^\alpha \left(\frac{1}{1 - V^\beta / V^\alpha} - p^\beta \right) \tag{9.8}$$

For $V^\beta > V^\alpha$, $\partial p^\beta / \partial T < 0$ for $p^\beta \geq 0$, and for $V^\beta < V^\alpha$, $\partial p^\beta / \partial T > 0$ at $p^\beta \to 0$. Readers are reminded that the sign of $\partial p^\beta / \partial T$ is the same as the sign of $\left(S^\beta - S^\alpha\right)$.

9.2 Application to cerium

Cerium (Ce) displays intriguing physical and chemical properties of which the most fascinating is its first-order isostructural phase transition. This involves a magnetic, high temperature/high volume "γ-phase" and a non-magnetic, low temperature/low volume "α-phase," both in the same face centered cubic (fcc) lattice structure. At 298 K and 0.7 GPa, the $\gamma \rightarrow \alpha$ transition is accompanied by a 14%–17% volume collapse. The Ce phase transition has been studied extensively; this includes our own works on systems with with two or three microstates [69, 70].

The simplest model for the system has two microstates: ferromagnetic and non-magnetic. The first-principles calculations of the free energy of two Ce microstates are problematic in the absence of strong correlation of the *f*-electrons in the DFT Hamiltonian. The stability of the non-magnetic ("delocalized") Ce 4*f* state relative to that of the magnetic ("localized") Ce 4*f* state is greatly overestimated in the generalized gradient approximation [31, 32] with spin polarization. A usual approach to surmount this is the Dudarev DFT + U method [71] with on-site Coulomb and exchange interactions described by a Hartree–Fock approximation added to the DFT Hamiltonian. This method offers the advantage that only the difference between the Hubbard U (due to the energy

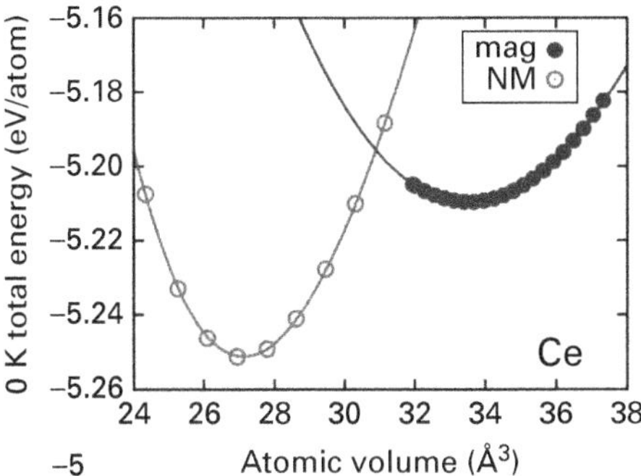

Figure 9.1 *Variation of cell energy* (eV) *with atomic volume* ($Å^3$) *for Ce computed with strong correlation using Dudarev's method with* $U - J = 1.6$ eV.

increase from addition of an electron to a specific site) and J (due to the screened exchange energy) need to be specified a priori.

Evaluation of numerous $U - J$ values over a 1.0–6.0 eV range revealed that 1.6 eV gives the most consistent prediction of the non-magnetic Ce and magnetic Ce energetics over a range of atomic volumes that includes both microstates at 0 K. The energy–volume curve thus obtained is plotted in Figure 9.1, showing that the non-magnetic microstate is the ground state, and the equilibrium between the two microstates at 0 K is at the negative pressure of −0.87GPa. Since $S^{\alpha} < S^{\gamma}$ and $V^{\alpha} < V^{\gamma}$, NTE does not exist in the system.

To take into account the possible magnetic disordering in the ferromagnetic microstate at finite temperatures, the following contribution is added to the free energy of the ferromagnetic microstate:

$$F_{mag}(V,T) = -k_B T \ln\left[1 + M_S(2l - M_S)\right] \qquad 9.9$$

where M_S is the spin moment, and $l = 3$ the orbital angular momentum of an f-electron. Equation 9.9 is a generalization of Hund's rule, with total angular momentum $J = M_S(2l - M_S)/2$. The Helmholtz energies thus obtained for both microstates and the system are shown in Figure 9.2 at several temperatures, with the tie-lines included. In the figures, the dot-dashed curves are for the non-magnetic microstate, the solid curves are for the ferromagnetic microstate, the shading shows the entropy of mixing between two microstates, and the circle in (e) is the critical point. The numbers below the black dashed lines, representing the common tangent curves, mark the transition pressures. The 0 K static energies of the non-magnetic microstate and the magnetic microstate are also plotted in (a) using solid circles and dotted lines.

The temperature versus volume phase diagram is plotted and compared with available experimental data (triangles) in Figure 9.3 [72]. In this figure, the volume (V) is normalized to its equilibrium volume (V_N) at atmospheric pressure and room temperature. In the pressure range 2.25–3.5 GPa, the system is within the single-phase region at all temperatures considered, as shown by the five continuous isobaric volumes as functions of temperature. In this pressure range, normal thermal expansion is observed at both low and high temperatures on each isobaric curve, where the

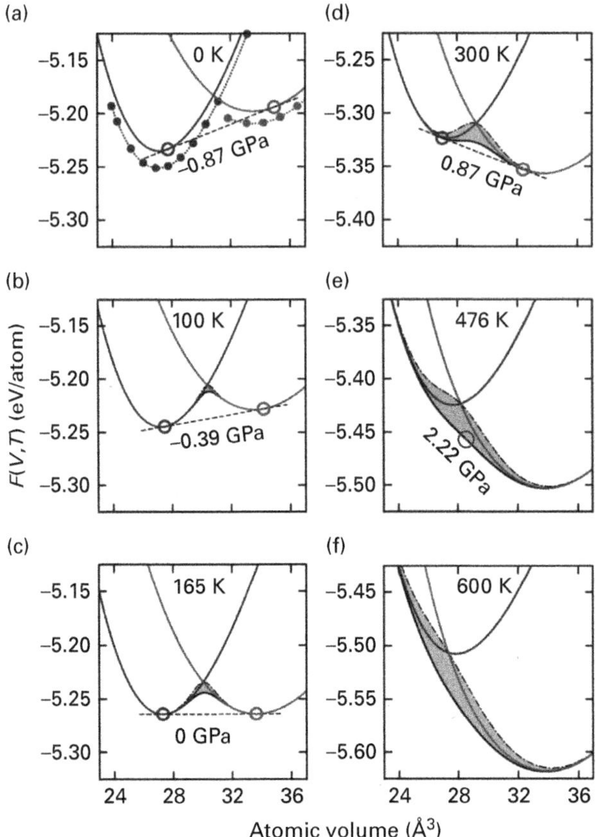

Figure 9.2 *Helmholtz energies at (a)* 0 K, *(b)* 100 K, *(c)* 165 K, *(d)* 300 K, *(e)* 476 K, *and (f)* 600 K, *from [69] with permission from the American Physical Society.*

probability of each microstate does not change significantly with temperature. However, in the middle temperature range on each isobaric curve, colossal positive thermal expansion (CPTE), highlighted by the open diamond symbols, exists due to the fast increase of the probability of the metastable ferromagnetic microstate with temperature, i.e. $S^{\alpha} < S^{\gamma}$, $V^{\alpha} < V^{\gamma}$, and $\partial p^{\gamma}/\partial T > 0$. This CPTE is much higher than the individual positive thermal expansions of the stable and metastable microstates, respectively.

With decreasing pressure, the system reaches a critical point (circle) where the homogeneous single phase becomes unstable, represented by $(\partial S/\partial T)_P = \infty$ and $(\partial V/\partial P)_T = \infty$, and both the entropy and volume change infinitely. At even lower pressure a miscibility gap forms, and the single phase separates into two phases with the same fcc crystal structure, but different magnetic spin structures. Inside the miscibility gap, the volume changes discontinuously with respect to temperature by a so-called first-order transition, as shown by the isobaric curve at zero pressure. This compares well with the experimental volume data (solid squares) under ambient pressure.

The fraction of the ferromagnetic microstate, x^{mag}, in α-Ce (dark line) and γ-Ce (light line) calculated using $x^{\sigma} = Z^{\sigma}/Z$ is plotted in Figure 9.4 as a function of pressure along

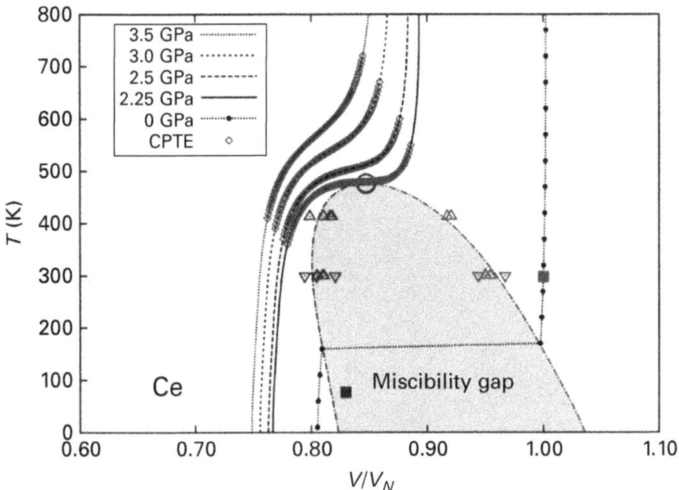

Figure 9.3 *Calculated temperature–volume phase diagram of Ce, from [72] with permission from the Nature Publishing Group.*

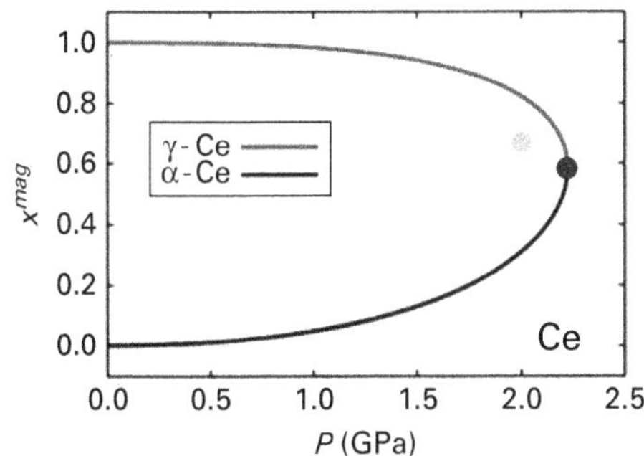

Figure 9.4 *Fraction of the ferromagnetic microstate in α-Ce (dark line) and γ-Ce (light line) along the γ–α phase boundary, from [69] with permission from the American Physical Society.*

the miscibility gap phase boundary. It can be seen that the fraction of the ferromagnetic microstate in α-Ce increases with increasing pressure while the fraction of the ferromagnetic microstate in γ-Ce decreases. At the critical point, the fraction of ferromagnetic microstate is calculated to be 0.58. This is in qualitative agreement with the 0.67 value (solid circle) estimated experimentally at the critical point.

The relative volume, V/V_N, as a function of pressure, is plotted as the thin black solid lines in Figure 9.5 from 200 to 600 K at 50 K increments. The dark and light thicker solid lines correspond to α-Ce and γ-Ce, respectively. The symbols denote experimental data in the literature, except for the open circle which is the calculated critical point, in good agreement with the computed isotherms. In the two-phase miscibility gap region, the $\gamma \rightarrow \alpha$ volume collapse is again noted, with the magnitude of the collapse increasing with decreasing T. This is shown explicitly by the dashed vertical lines at T = 200, 250, 300, 350, 400, and 450 K. For $T > 476$ K, the

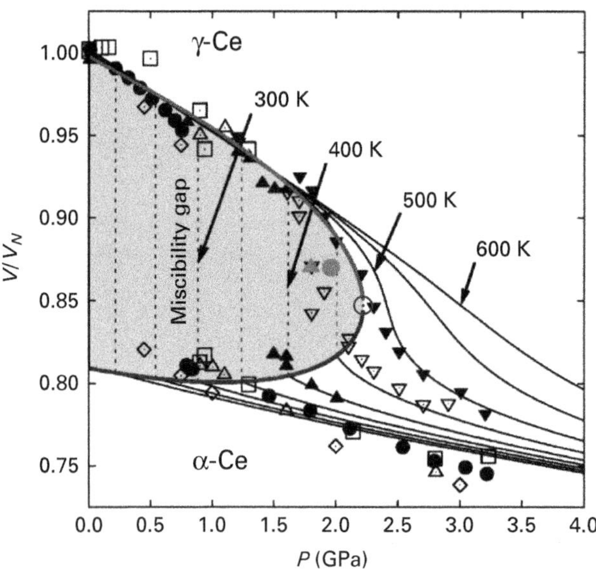

Figure 9.5 *Equation of states for Ce. The black solid lines represent the calculated isotherms from* 200 *to* 600 K *at* $\Delta T = 50$ K *increments, from [69] with permission from American Physical Society.*

calculated isotherms show an anomalous slope change which closely matches the behavior near V/V_N = ~0.85 from experiment.

In a more complex model, one adds the antiferromagnetic microstate. The *E–V* and Helmholtz energy curves obtained at 0 K are shown in Figure 9.6. The equilibrium volume energies reveal that the energy of the antiferromagnetic microstate at the equilibrium volume is close to that of the non-magnetic microstate but substantially lower than that of the ferromagnetic microstate. It should be noted that the magnetic spin disordering in the system is taken into account by the two magnetic microstates, and the contribution denoted by the mean-field theory, i.e. Eq. 9.9, should thus not be added to either magnetic microstate, to avoid double counting. The predicted critical point values are 546 K and 2.05 GPa, closer to the experimental data than the previous prediction for two microstates as shown in the temperature–pressure phase diagram in Figure 9.7 in comparison with experimental data.

The calculated entropy changes are plotted in Figure 9.8a in terms of lattice vibration only (dashed line), lattice vibration plus thermal electron (dot-dashed line), and lattice vibration plus thermal electron and plus configuration coupling (solid line). The square gives the estimated vibrational entropy change at 0.7 GPa of γ-Ce relative to α-Ce, and other open (solid) symbols are from experimental measurements of total entropy. Various contributions to the Helmholtz energy along the γ–α phase boundary, along with experimental data, are plotted in Figure 9.8b; these are $T\Delta S$ (diamonds), ΔE (circles), and $P\Delta V$ (squares). Excellent agreement with experimental data is shown.

The predicted fractions of the three microstates as a function of temperature and heat capacity at the critical pressure of 2.05 Pa are shown in Figure 9.9a and

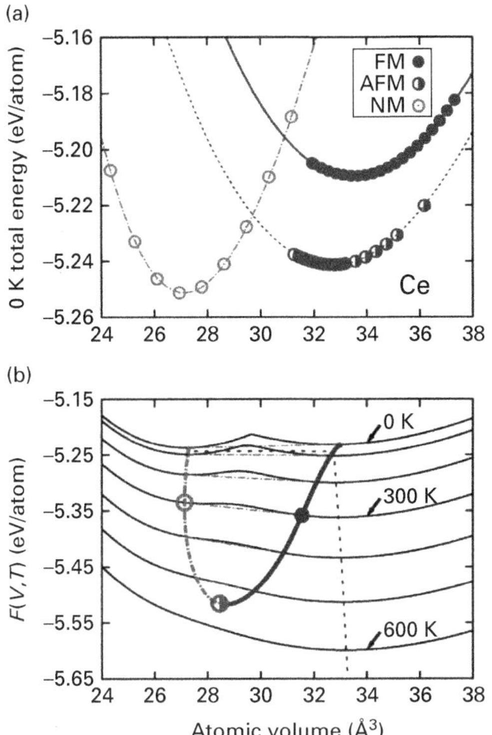

Figure 9.6 *(a) The dot-dashed line with* ○, *dashed line with* ◑, *and solid line with* • *represent the* 0 K *static total energies for non-magnetic (NM), antiferromagnetic (AFM), and ferromagnetic (FM) microstates of Ce, respectively. (b) The solid lines denote Helmholtz energy (per atom) from* 0 *to* 600 K *at* $\Delta T = 100$ K*; the heavier dot-dashed (α-Ce) and solid (γ-Ce) looping curves enclose the two-phase region with the lighter dot-dashed lines connecting the common tangents of each isotherm; the black dashed lines denote zero pressure equilibrium state at given T;* ○ *and* • *emphasize the phase boundary at 300 K while* ◑ *is the critical point, from [70] with permission from IOP Publishing.*

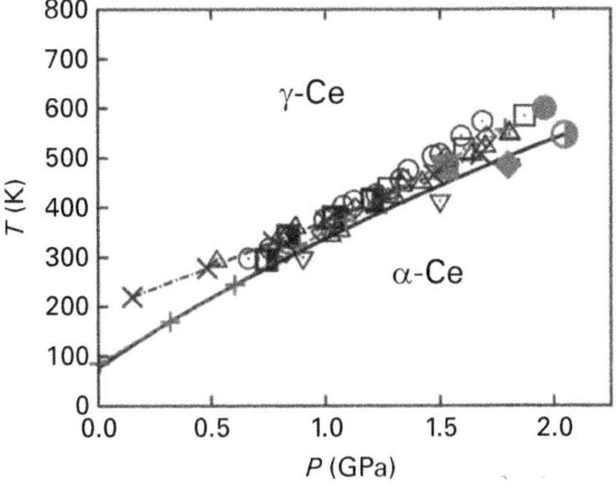

Figure 9.7 *Calculated temperature–pressure phase diagram along with experimental data, from [70] with permission from IOP Publishing.*

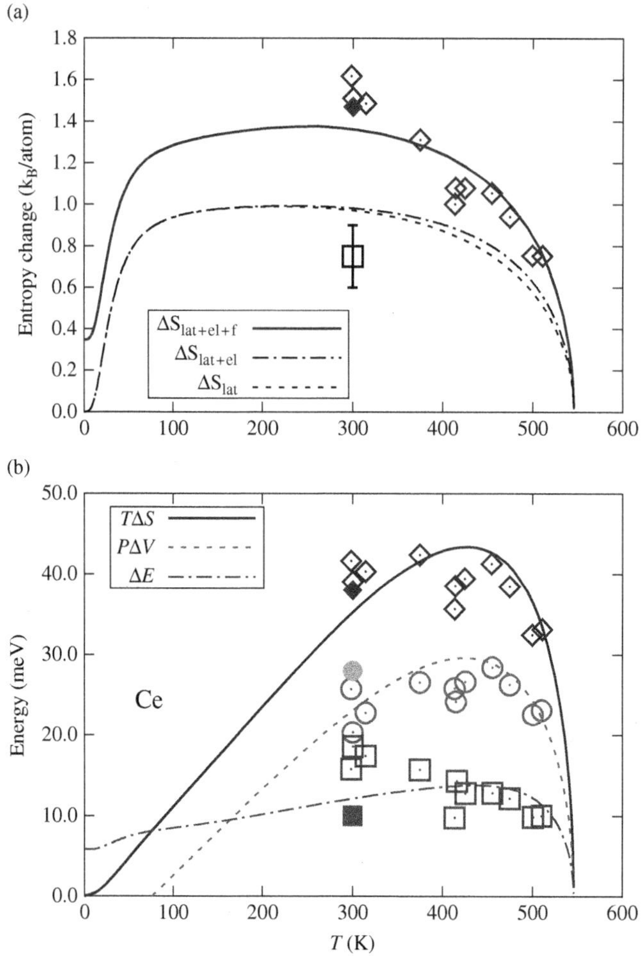

Figure 9.8 *Calculated (a) entropy, (b) various contributions to the Helmholtz energy of Ce, along with experimental data, from [70] with permission from IOP Publishing.*

Figure 9.9b, respectively. Near the critical point, the theory predicts that the system is a mixture of the various microstates. Figure 9.9a depicts that for $T < 300$ K, the system consists mainly of the non-magnetic Ce state, α-Ce. For $T > 300$ K, the thermal populations of the magnetic states increase with increasing temperature. Finally, for $T > 546$ K (the critical point), 70% of the system is composed of the antiferromagnetic Ce state with the remaining 30% consisting of the non-magnetic and ferromagnetic Ce states. This is in agreement with the common belief that γ-Ce is magnetic but with a partially disordered local moment (paramagnetic) and that α-Ce is non-magnetic.

Figure 9.9b shows the predicted temperature evolution of the contributions to the heat capacity: vibrational and magnetic (C_f/T), electronic (C_{el}/T), and their sum (C_{f+el}/T) at 2.05 GPa. The theory suggests the following: (a) below ~500 K, C_{f+el}/T shows an exponential temperature dependence due to statistical fluctuations between the

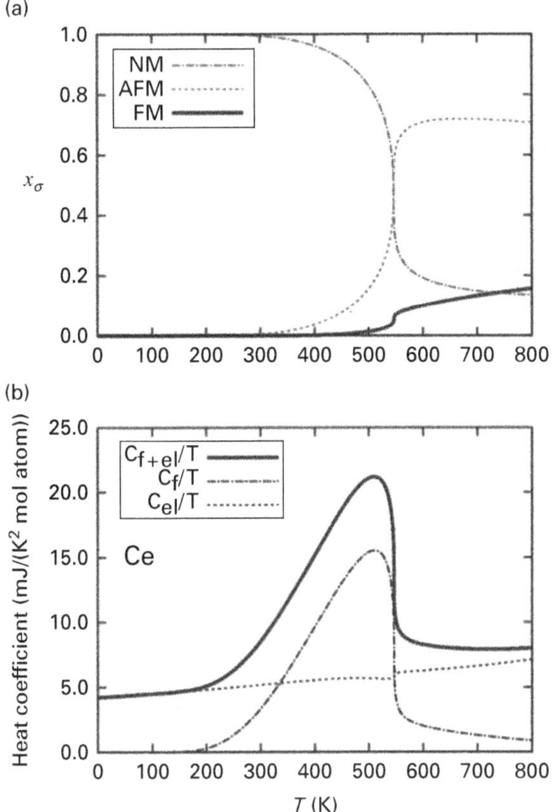

Figure 9.9 *(a) Thermal populations of the non-magnetic (dot-dashed), antiferromagnetic (dashed), and ferromagnetic (solid) states as functions of temperature at the critical pressure of* 2.05 GPa. *(b)* C_{el}/T *(dashed line),* C_f/T *(dot-dashed line), and their sum* C_{f+el}/T *(solid line) at* 2.05 GPa, *from [70] with permission from IOP Publishing.*

non-magnetic, ferromagnetic, and antiferromagnetic states; (b) a peak appears at ~500 K in the C_{f+el}–T curve, which typically suggests the Schottky anomaly; (c) the electronic specific heat coefficient (C_{el}/T) is linear against T; (d) above ~500 K the sum of C_f/T and C_{el}/T renders C_{f+el}/T temperature independent.

9.3 Application to Fe_3Pt

Invar was first discovered in the intermetallic $Fe_{65}Ni_{35}$ alloy and is characterized by "anomalies" in the thermal expansion, equation of state, elastic modulus, heat capacity, magnetization, and Curie temperature. There are a number of theoretical models for Invar such as the Weiss 2-γ model, the non-collinear spin model, and the disordered local moment approach, as reviewed in reference [73]. In this section, the application of the MMS model to ordered $L1_2$ Fe_3Pt is presented to study the Invar anomaly at finite temperatures.

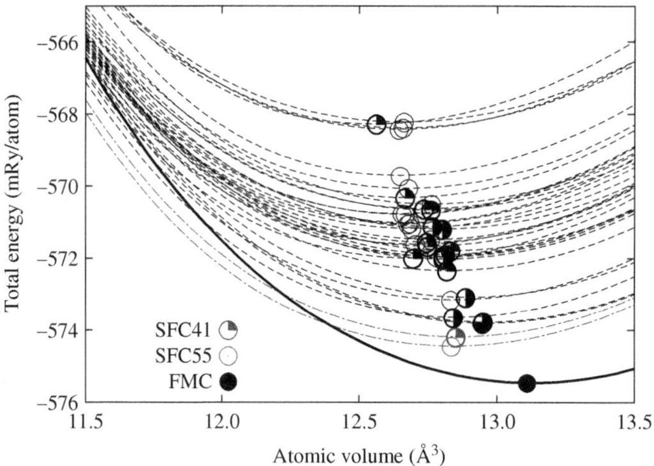

Figure 9.10 *Total energies at* 0 K*. The heavy black line represents the FMC. The symbols* ○, ◔, ◑*, and* ◕ *with dashed lines indicate the minima of the energy–volume curves of the SFCs with spin polarization rates* 1/9, 3/9, 5/9, *and* 7/9*, respectively. The symbols* ○ *and* ◔ *with the dot-dashed lines mark the two SFCs lowest in energy, from [73] with permission from Taylor and Francis Group.*

For a supercell of 12 atoms with nine magnetic Fe atoms in Fe_3Pt, if only the up and down spins are considered, the system contains $2^9 = 512$ spin configurations, which are, by symmetry, reduced to 37 non-equivalent configurations. They are the microstates in the MMS model, and FMC and SFC are used to represent the ferromagnetic and spin flipping microstates, respectively. For the first-principles calculations of each microstate, the VASP package [14] within the projector-augmented wave (PAW) method and the exchange-correlation part of the density functional treated within the GGA of Perdew, Burke, and Ernzerhof (PBE) [32] are employed with details given in reference [73]. For the lattice vibration, the Debye–Grüneisen approach described in Section 5.2.4 is used.

Figure 9.10 presents the first-principles 0 K total energies of 36 non-equivalent SFCs as well as the FMC as functions of atomic volume. It can be seen that there are a number of SFCs, whose energies are within the range of ~1 mRy/atom of that of the FMC. It is noted that all the SFCs studied here have equilibrium averaged atomic volumes at least 1.8 % smaller than that of the FMC, the 0 GPa ground state. In Figure 9.10, the two lowest energy SFCs are labeled as SFC55 and SFC41 with their spin arrangements very similar to the double layer antiferromagnetic state. The non-magnetic configuration has a very small atomic volume of 11.66 Å^3/atom, and much higher energy than both FMC and all SFCs, and is thus not shown here.

The normalized Helmholtz energies of all SFCs are plotted in Figure 9.11 with the Helmholtz energy of FMC as the reference state, showing that the FMC has the lowest Helmholtz energy at all temperatures considered. If only the relative Helmholtz energies of microstates were considered, FMC would be stable at all temperatures. However, the configurational entropy due to the mixing among multiple microstates, i.e. Eq. 5.51, lowers the system free energy by introducing the statistical probability of metastable SFCs.

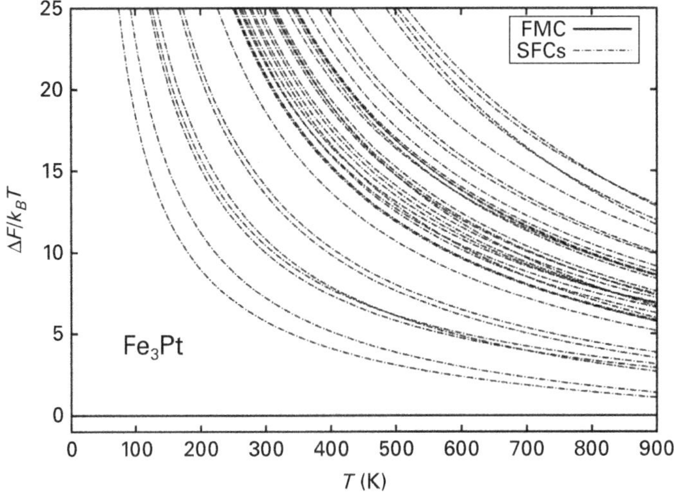

Figure 9.11 *Normalized Helmholtz energy of all SFCs with respect to that of FMC.*

By minimization of the Helmholtz energy of Eq. 5.50, the temperature–pressure and temperature–volume phase diagrams are obtained and are shown in Figure 9.12 [72]. A critical point at 141 K and 5.81 GPa is predicted with $V = 12.61$ Å^3. Below the critical point, there is a two-phase miscibility gap (the shaded area), the dominant microstates being FMC and SFCs, respectively, and the transition between these states is first order. Above the critical point, the macroscopically homogeneous single phase is stable, and phase transitions between the ferromagnetic dominant phase with large volume and the SFC dominant paramagnetic phase with small volume are of second order. The second-order transition pressures and volumes are determined by the condition that the weighted Helmholtz energy counting all SFCs equals the Helmholtz energy counting only FMC.

In Figure 9.12a, the data points give the measured pressure dependence of the Curie temperature, and the agreement between the measurements and predictions is remarkable. It should be pointed out that the classical Weiss 2-γ model predicts only first-order phase transitions while the non-collinear spin model yields only second-order phase transitions at all temperatures. In Figure 9.12b, four isobaric volume curves are also plotted with the predicted NTE regions marked by the open diamonds and the experimental volume data under ambient pressure superimposed, showing excellent agreement. It also depicts the gigantic NTE around the critical point on the isobaric curve at 7 GPa.

Figure 9.11 indicates that the entropies of SFCs are larger than that of FMC, so their Helmholtz energy differences decrease with temperature. This is in line with the origin of NTE in a single phase due to the statistical existence of metastable microstates with lower volumes and higher entropies than the stable state in a temperature range where their probabilities change dramatically. Figure 9.13 plots the calculated thermal populations of the FMC (solid line) and that of the sum over all SFCs (dot-dashed line) under ambient pressure. The two major contributions to the paramagnetic (PM) phase are from SFC55 and SFC41, which are plotted using dashed and long dashed lines, respectively.

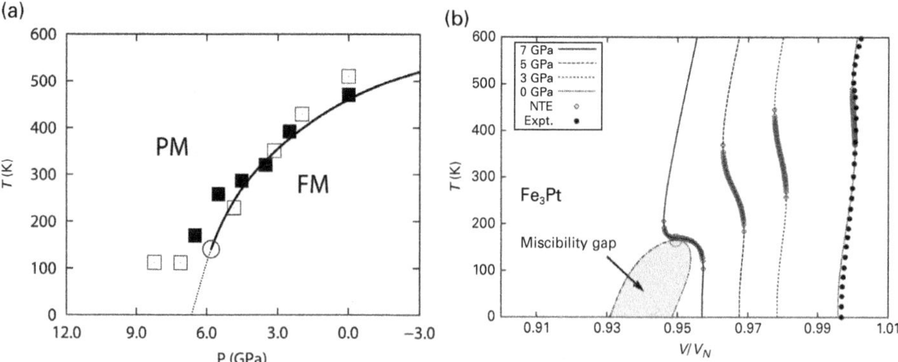

Figure 9.12 *Calculated phase diagrams of Fe_3Pt, (a) temperature-pressure, from [73] with permission from Taylor and Francis, and (b) temperature-volume, from [72] with permission from the Nature Publishing Group.*

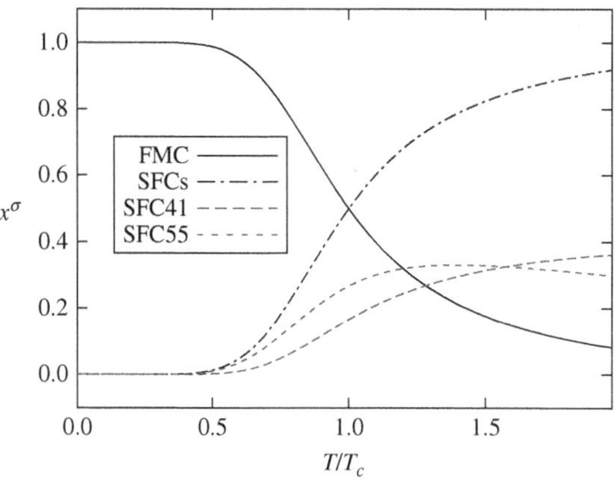

Figure 9.13 *Calculated thermal populations of FMC, SFC41, SFC55, and the sum of all SFCS, respectively, from [73] with permission from Taylor and Francis Group.*

The system is dominated by the FMC at temperatures below half the transition temperature, and the populations of SFCs increase monotonically at temperatures higher than half the transition temperature. As mentioned above, the transition temperature is defined as that when the population of all SFCs is the same as that of FMC due to their equal Helmholtz energies.

The predicted thermal volume expansion and the derived linear thermal expansion coefficient (LTEC) under ambient pressure are plotted in Figure 9.14. A positive thermal expansion is predicted from 100 K to 288 K, followed by a negative thermal expansion in the range 289 to ~ 449 K, and then a positive thermal expansion again at > 450 K, in excellent agreement with experiment. The only disagreement between the predictions and

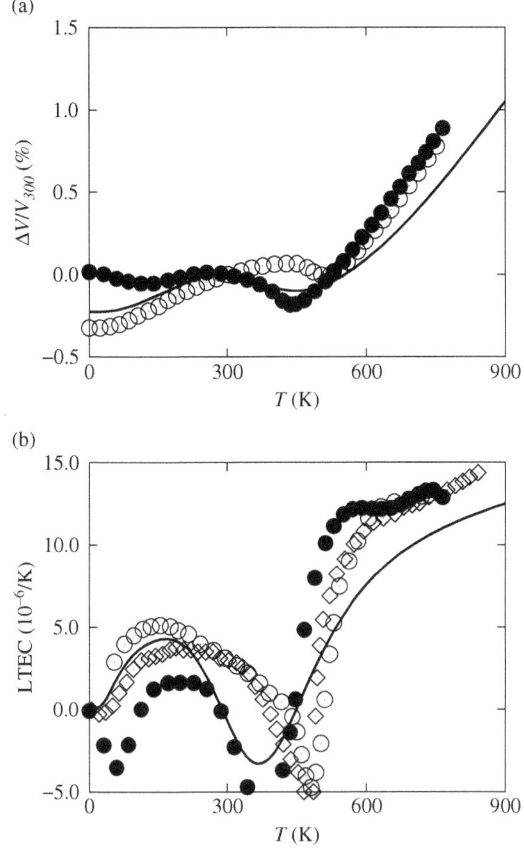

Figure 9.14 *(a) Relative volume increase (V− V_{300})/V_{300} with V_{300} the equilibrium volume at* 300 K *and* 0 GPa *for ordered Fe_3Pt. (b) Linear thermal expansion coefficient (LTEC) along with various experimental data (symbols), with details in reference [73] with permission from Taylor and Francis. Copy from Figure 9.10.*

experiments occurs at $T = 100$ K where the calculations do not reproduce the negative thermal expansion for Fe_3Pt. A larger supercell or more spin configurations may be needed.

9.4 Concept of Materials Genome®

"A genome is a set of information encoded in the language of DNA that serves as a blueprint for an organism's growth and development. The word genome, when applied in nonbiological contexts, connotes a fundamental building block toward a larger purpose" [74]. Materials Genome® (a trademark of Materials Genome, Inc., Pennsylvania, PA) thus concerns the building blocks of materials. Most of this book focuses on the Gibbs and/or Helmholtz energies of individual phases as a function of their natural variables, and the same applies in the CALPHAD modeling of thermodynamics and other properties of individual phases. Multi-component materials systems and their properties are built on the individual phases and their properties. Individual phases are thus naturally considered as building blocks of materials. Consequently, the language of CALPHAD thermodynamics and kinetics contains the genomics of materials by representing experimental and theoretical results in databases to make them applicable to a much wider context than the original experiments or calculations [75]. The variation of individual phases in terms of their properties, amounts, and interactions with other phases with respect to external conditions thus determines the performance of the materials.

On the other hand, at critical points and beyond, phases lose their individuality and form one macroscopically homogeneous system, and the properties of the system change dramatically with respect to external conditions. As shown in this chapter, these dramatic responses can be predicted through the statistical competition of stable and metastable microstates. From the thermodynamic point of view, under any given conditions, one of the individual microstates has the lowest Gibbs/Helmholtz energy and is stable, while all other microstates have higher free energy and are metastable or unstable. These metastable or even unstable microstates are brought into statistical existence in the matrix of the stable microstate due to the entropy of mixing of all the configurations. Those microstates may thus be considered as the building blocks of individual phases [76].

It has been demonstrated in this chapter that the properties of a macroscopically homogeneous system with multiple microstates are not linear combinations of the properties of the constituent microstates and depend significantly on the rate of change of the statistical probability of microstates with respect to external fields. This rate of change is determined by the free energy difference between the stable and metastable microstates and its rate of change with respect to external fields. As shown in Figure 9.3 and Figure 9.12, this rate of change can be correlated to the distance of the system with respect to the critical point in the system. At the critical point, there is a mathematical singularity when the single phase becomes unstable. When the macroscopically homogeneous single-phase system moves away from the critical point, its properties become less and less dramatic, but always retain a certain degree of anomaly. The properties of a system can thus be dramatically altered and designed by changing the position of the critical point through adjustments of chemical compositions and strain energy in thin films.

Appendix A: YPHON

Currently there are essentially two methods in use for the first-principles calculations of phonon frequencies: the linear response theory and the direct approach. The linear response theory evaluates the dynamical matrix through the density functional perturbation theory. In comparison, one advantage of the direct or mixed-space approach over the linear-response method is that it can be applied with the use of any code capable of computing forces. The direct approach is also referred to as the small displacement approach, the supercell method, or the frozen phonon approach.

However, none of the previous implementations of the supercell approach are able to accurately handle long-range dipole–dipole interactions when calculating phonon properties of polar materials. The problem has been solved by the parameter-free mixed-space approach, which makes full use of the accurate force constants from the supercell approach in real space and the dipole–dipole interactions from the linear response theory in reciprocal space. The mixed-space approach is the only existing method that can accurately calculate the phonon properties of polar materials within the framework of the supercell or small displacement approach.

The program YPHON is written in C++ and can be downloaded at http://cpc.cs.qub.ac.uk/summaries/AETS_v1_0.html. The precompiled executable binaries should work for most Linux and Windows systems. Recompiling YPHON requires the GNU Scientific Library (GSL), which is a numerical library for C and C++ programmers. If one is just interested in phonon dispersions, the phonon density-of-states (PDOS), or the neutron scattering cross-section weighted PDOS the so-called generalized phonon density-of-states (GPDOS), this is enough. YPHON also makes it a lot easier to plot phonon dispersions and PDOS. In this case, it is required that Gnuplot be installed.

The static energy and force constants from first-principles calculations need to be formatted to the YPHON input formats (text formats as detailed later). At present, YPHON works closely with VASP.5 or later. The mixed-space approach has built up a unique base of the supercell approach to polar materials and has been adopted in a number of software tools such as CRYSTAL14 by R. Dovesi, ShengBTE (a solver of the Boltzmann transport equation for phonons) by W. Li, J. Carrete, N. A. Katcho, and N. Mingo, the Phonopy package by Atsushi Togo, and the Phonon Transport Simulator (PhonTS) by Chernatynskiy and Phillpot.

A.1 General software requirements

As mentioned above, YPHON is written in C++. The precompiled executable binaries should work for most Linux and Windows systems. Recompiling YPHON requires the GNU Scientific Library (GSL). If one is just interested in the phonon dispersions, phonon density-of-states (PDOS), or the neutron scattering cross-section weighted PDOS, the so-called generalized phonon density-of-states (GPDOS), this is enough. YPHON also makes it a lot easier to plot phonon dispersions and PDOS. In this case, it is required that Gnuplot be installed.

At present, YPHON works closely with VASP.5 or later. The static energy and force constants from other codes for first-principles calculations need to be organized into the YPHON input formats, as discussed below.

A.2 Get and unpack YPHON

Get the zipped file "yphon.tar.gz" from the website, unpack the package, and install YPHON using the following series of Linux commands:

```
tar zxf yphon.tar.gz
cd YPHON; make
```

Notes: Mostly you do not need to recompile the codes with "make." The precompiled executable binaries enclosed with the package work quite well in a number of computational centers and up to the present time we have not seen any exceptions.

For a csh user, depending on the specific management of your Linux system, the Linux command search PATH should be modified by inserting into the .cshrc, .tcshrc, or .cshrc.ext files the following line:

```
set PATH = (. $HOME/YPHON $PATH)
```

assuming that $HOME/YPHON is the path where the YPHON package is unpacked.

For a bsh user, the Linux command search PATH should be modified by inserting the two lines below into the .bash,_profile, or .bashrc files:

```
PATH=.:$HOME/YPHON:$PATH
export PATH
```

A.3 Contents of the YPHON package

Yphon – The main C++ code performing the phonon calculations based on the force constants.

Ycell – A C++ code to build the supercell in the VASP.5 POSCAR format.

vasp_fij – A Linux script for user convenience to collect the force constant matrix from the OUTCAR and CONTCAR files of VASP.5 or the Hessian matrix from the vasprun.xml file of VASP.5. The output file superfij.out is the main input to Yphon.

vaspfijxml – A C++ code to convert the Hessian matrix from the vasprun.xml file into the force constant matrix; vaspfijxml is only called vasp_fij when it sees the vasprun.xml file.

vasp_BE – A Linux script for user convenience to collect the data of Born effective charge and high frequency dielectric tensors from the OUTCAR and CONTCAR files of VASP.5. The output file dielecfij.out forms the input to YPHON when calculating polar materials.

Ydemo – A folder containing several subfolders as exercises of the YPHON package.

A.4 Command line options and files used by YPHON

The usage of YPHON follows the command convention of the Linux operating system. A YPHON command is followed by a series of keywords and parameters. Different keywords and parameters are separated by a space character in the command line. All keywords in YPHON commands are case sensitive.

A.4.1 Ycell

Usage: Ycell [options] <yourprimitiveposcarfile>yoursupercellposcarfile

where yourprimitiveposcarfile is a file from the standard input and yoursupercellposcarfile is a file for the standard output. Both files are in the VASP.5 POSCAR format.

The options are as follows.

-bc2

Make a supercell by a kind of doubling of the primitive unit cell.

-bc3

Make a supercell by a kind of tripling of the primitive unit cell.

-bc4

Make a supercell with a size four times the primitive unit cell.

-mat matrix 3×3

Make a supercell by transforming the primitive unit cell with a 3×3 matrix (nine parameters), for example, -mat 2 -2 0 2 2 0 0 2.

–ss *n*

Make an $n \times n \times n$ supercell of the primitive unit cell.

A.4.2 Yphon

Usage: Yphon [options] <superfij.out

where superfij.out is the name of the file created by vasp_fij containing the crystal structure as well as force constant information. By default, Yphon calculates the PDOS and outputs it into the file vdos.out.

The main options are as follows.

-Born dielecfij.out

This considers the vibration-induced dipole–dipole interaction (called LO-TO splitting in the literature) in the calculation, where dielecfij.out is the name of the file created by vasp_BE containing all information about the Born effective charge and high frequency dielectric tensors.

-pdis *yourdisfile*

This option instructs Yphon to calculate the phonon dispersion instead of the PDOS. *yourdisfile* is a file defining the directions for the dispersion calculation; see the subsection "**File for dispersion calculation**" for instruction on how to prepare the file *yourdisfile*.

-pvdos

This calculates also the generalized PDOS (GPDOS, the neutron scattering cross-section weighted PDOS), following GPDOS $= \sum_i \frac{\sigma_i}{\mu_i} \text{pDOS}_i$ where σ_i and pDOS_i represent respectively the atomic scattering cross-section and the partial phonon density of states projected onto the individual atoms. The results are saved in the file pvdos.out.

-sqs

This is used together with the options -noNA –nof. It calculates the phonon dispersions of a random alloy with respect to the wave vector space of the ideal lattice, by averaging over the force constants calculated using a special quasi-random structure (SQS). The detailed formulism can be found in our publication [19].

-thr2 parameter

This defines the threshold on how to determine the atomic position relation between the high symmetry structure and the low symmetry structure. Care should be taken in this kind of calculation. One should gradually increase the value of the parameter from 0.01 to 0.15.

-Mass *yourmassfile*

This option tells Yphon to redefine the atomic mass, as required for the SQS phonon dispersion calculation. The context in the *yourmassfile* file contains lines like

Cu 96.8975

Au 96.8975

–plot

This instructs Yphon to display the plot in the terminal using gnuplot to check the calculated results.

–expt exp01.dat

This instructs Yphon to plot the experimental data contained in the file "exp01.dat" together with the calculations.

A.5 Files used by YPHON

A.5.1 *superfij.out file*

superfij.out is the main input file of YPHON through the standard input stream stdin in the Linux environment. It contains information about the lattice vectors of the primitive unit cell, the lattice vectors of the supercell, the atomic positions in the supercell, and the matrix of the negative values of the force constants. One does not have to prepare this file by hand if you use VASP.5 since the enclosed script vasp_fij can prepare this file by extracting data from the output files of VASP.5, namely, CONTCAR, OUTCAR, and vasprun.xml. The format of the data contained in superfij.out is illustrated below:

```
          0.0000000000    2.1060000000    2.1060000000
          2.1060000000    0.0000000000    2.1060000000
          2.1060000000    2.1060000000    0.0000000000
          0.000000        4.212000        4.212000
          4.212000        0.000000        4.212000
          4.212000        4.212000        0.000000
16 8
Direct
          0.00000000      0.00000000      0.00000000      Mg
          0.00000000      0.00000000      0.50000000      Mg
          0.00000000      0.50000000      0.00000000      Mg
          0.00000000      0.50000000      0.50000000      Mg
          0.50000000      0.00000000      0.00000000      Mg
          0.50000000      0.00000000      0.50000000      Mg
          0.50000000      0.50000000      0.00000000      Mg
          0.50000000      0.50000000      0.50000000      Mg
          0.25000000      0.25000000      0.25000000      O
          0.25000000      0.25000000      0.75000000      O
          0.25000000      0.75000000      0.25000000      O
          0.25000000      0.75000000      0.75000000      O
          0.75000000      0.25000000      0.25000000      O
          0.75000000      0.25000000      0.75000000      O
          0.75000000      0.75000000      0.25000000      O
          0.75000000      0.75000000      0.75000000      O
        −10.944494        0.000000        0.000000        1.260786    3.229127    0.000000
1.260786  0.000000        3.229127       −0.377911        0.000000    0.000000   −0.377911
0.000000  0.000000        1.260786        0.000000       −3.229127    1.260786   −3.229127
0.000000  1.697668        0.000000        0.000000       −0.151592    0.000000    0.000000
0.788701  0.000000        0.000000        0.788701        0.000000    0.000000    1.065411
0.000000  0.000000        1.065411        0.000000        0.000000    0.788701    0.000000
0.000000  0.788701        0.000000        0.000000       −0.151592    0.000000    0.000000
          0.000000      −10.944494        0.000000        3.229127    1.260786    0.000000
0.000000 −0.377911        0.000000        0.000000        1.260786   −3.229127    0.000000
1.260786  3.229127        0.000000       −0.377911        0.000000   −3.229127    1.260786
0.000000  0.000000        1.697668        0.000000        0.000000   −0.151592    0.000000
0.000000  0.788701        0.000000        0.000000        1.065411    0.000000    0.000000
0.788701  0.000000        0.000000        0.788701        0.000000    0.000000    1.065411
0.000000  0.000000        0.788701        0.000000        0.000000   −0.151592    0.000000

...
```

where

line 1 – line 3 are lattice vectors of the primitive unit cell

line 4 – line 6 are lattice vectors of the supercell

line 7 gives the number of atoms in the supercell, and the number of primitive unit cells in the supercell (serves as an error check) line 8 has the same meaning as VASP

line 9 – line of 8+"Number of atoms in supercell" give the internal atomic positions in the supercell, which again have the same meaning as VASP, and the data after these lines are the negative of the values of the force constant matrix.

A.5.2 *dielecfij.out file*

When one wants to consider the effects of vibration-induced dipole–dipole interaction on phonon frequencies, the dielecfij.out file is required in the command line option in the form of -Born dielecfij.out in running YPHON. The file dielecfij.out contains information about the lattice vectors of the primitive unit cell, the atomic positions, the high frequency dielectric tensor, and the Born effective charge tensor. You do not have to prepare this file by hand if you use VASP.5 since the enclosed script vasp_BE can prepare this file by extracting data from the OUTCAR of VASP.5, if you have the setting "LEPSILON = .T." and "NSW=0" in the INCAR file when running VASP. The format of the data contained in dielecfij.out is illustrated below:

```
0.000000            2.106000            2.106000
2.106000            0.000000            2.106000
2.106000            2.106000            0.000000
0.0000000000000000  0.0000000000000000  0.0000000000000000  Mg
0.5000000000000000  0.5000000000000000  0.5000000000000000  O
3.147               0.000               0.000
0.000               3.147               0.000
0.000               0.000               3.147
ion 1
1                   1.96085             0.00000             0.00000
2                   0.00000             1.96085             0.00000
3                   0.00000             0.00000             1.96085
ion 2
1                   -1.96142            0.00000             0.00000
2                   0.00000             -1.96141            0.00000
3                   0.00000             0.00000             -1.96141
```

where

line 1 – line 3 are lattice vectors of the primitive unit cell

line 4 – line of 3+"Number of atoms in the primitive unit cell" give the internal atomic positions in the primitive unit cell, which again have the same meaning as VASP. The lines following the atomic positions are the dielectric constant tensor and the Born effective charge tensor.

Note: Calculation of the dielectric properties should be performed using the primitive unit cell separately.

A.5.3 *vdos.plt file*

The file vdos.plt is a gnuplot script made by YPHON when no "-pdis" command line option is supplied. Its purpose is to use gnuplot to plot the PDOS, typically using the Linux command "gnuplot –persist vdos.plt."

A.5.4 *vdos.out file*

The file vdos.out is a default output file of YPHON when no "-pdis" command line option is defined. It contains the PDOS data.

Note: The PDOS in vdos.out has been normalized to $3M$ where M is the number of atoms in the primitive unit cell. Also be careful that the unit in the vdos.out file for the frequency is Hz. Therefore, if you want to compare the calculated phonon frequencies with experiment, you need the proper conversion factor.

A characteristic feature of YPHON is concerned with the PDOS calculation. YPHON does not follow the conventional Gaussian smearing approach to calculate the PDOS, since it is difficult to determine the value of the empirical Gaussian broadening parameter. Instead, YPHON uses the following convolution average to calculate the PDOS:

$$D(\omega) = \frac{3M}{N_\omega} \frac{1}{S(S+1)\Delta\omega} \sum_{n=0}^{S} [A(\omega + n\Delta\omega) - A(\omega + n\Delta\omega - S\Delta\omega] \qquad \text{A.1}$$

where $D(\omega)$ is the PDOS, N_ω is the total number of phonon frequencies calculated by a uniform **q** mesh in the wave vector space, $A(\omega)$ is the total number of phonon frequencies below the frequency ω, $\Delta\omega$ is the frequency interval of the numerical expression of the PDOS, and S is an integer playing the role of smoothing the PDOS. By default, YPHON calculates $D(\omega)$ with a mesh containing 10001 equally spaced points together with S = 40. In general, the calculated $D(\omega)$ is reasonable from our experience for a **q** mesh providing N_ω= ~3000000.

Note: Using the option "-nq *nqx nqy nqz*," YPHON is still affordable for an even denser **q** mesh, which can provide ~30000000 frequencies if one wants more accurate PDOS.

A.5.5 *pvdos.out file*

The file pvdos.out is an output file of YPHON when the key "-pvdos" is defined in the command line. It contains the neutron scattering cross-section weighted phonon density of states (GPDOS). The format of the data contained in this file is:

column 1:	phonon frequency (in units of THz)
column 2:	PDOS (in units of THz^{-1}).
column 3:	weight factor due to the different atomic neutron scattering cross-sections, therefore, (column 2)*(column 3) in this file is the GPDOS
column 4:	not useful
column 5-(M +4):	partial PDOSs of the atoms following their orders in the primitive unit cell.

Note: When using gnuplot, one can plot GPDOS by plotting "pvdos.out" using 1:($2*$3) w l.

A.5.6 *vdis.plt file*

The file vdis.plt is a gnuplot script made by YPHON when the "-pdis" command line option is defined. It is used by gnuplot to plot the phonon dispersions. After you run YPHON, you can always display the calculated phonon dispersion using gnuplot –persist vdis.plt

In particular, if you delete the comment sign "#" in the vdis.plt file and then run it with "gnuplot vdis.plt," it can produce a postscript file almost ready for publication.

A.5.7 *vdis.out file*

The file vdis.out is a file made by YPHON when the "-pdis" command line option is defined. It contains phonon dispersion data.

A.6 File for dispersion calculation

To calculate phonon dispersions, one needs to use the command line option "-pdis *yourdisfile*" with YPHON. The file *yourdisfile* is a file defining the direction for the dispersion calculation, and typically its format is as follows, using GaAs as an example:

```
0 0 0 0 0 .5 Gamma X 0 $1 2 0 $1 3
0 .5 .5 0 0 0 X Gamma 1 (1.-$1) 2 1 (1.-$1) 3 1 (1.-$1) 4
0 0 0 .25 .25 .25 Gamma L 2 (2*$1) 2 2 (2*$1) 3
.25 .25 .25 0 0 .5 L X 3 (2*$1) 2 3 (2*$1) 3 3 (2*$1) 4
0 0 .5 0 .25 .5 X W 4 (2*$1) 2 4 (2*$1) 3 4 (2*$1) 4
0 .25 .5 .25 .25 .25 W L 5 (2*$1) 2 5 (2*$1) 3 5 (2*$1) 4
```

Columns 1–3:	the reciprocal reduced coordinate of the starting **q** point along the dispersion path
Columns 4–6:	the reciprocal reduced coordinate of the end **q** point along the dispersion path
Column 7:	the label of the starting **q** point along the dispersion path
Column 8:	the label of the end **q** point along the dispersion path
Column 9:	multi-sets of data, each set containing three columns wherein the first column is the index of the data group (separated by two blank lines following the convention of gnuplot) in the experimental data file, the second column tells gnuplot which column is being considered and how to transform the column into the **q** point following the convention of gnuplot, and the third column tells gnuplot which column of experimental data will be used as the frequency data. This can save a lot of time if you can learn to use it. The calculated phonon dispersions are contained in the file vdis.out.

Note: Care should be taken with the suffix of the *yourdisfile* file. The suffixes .fcc, .bcc, .hcp, and .tet2 are reserved for fcc, bcc, hcp, and tetragonal crystals only. For these crystals, YPHON internally converts the **q** vector of the primitive unit cell into that of the conventional unit cell.

For the cases of bcc and fcc crystals, most neutron scattering data are reported with respect to the cubic conventional cell instead of the primitive unit cell. YPHON does the conversion internally, assuming that you defined the shape of the primitive unit cell of the fcc crystal in the POSCAR file as

```
0   .5  .5
.5  0   .5
.5  .5  0
```

and you defined the shape of the primitive unit cell of the bcc crystal in the POSCAR file as

```
-.5   .5   .5
 .5  -.5   .5
 .5   .5  -.5
```

For the case of an hcp crystal, YPHON assumes that you defined the shape of the primitive unit cell of the crystal in the POSCAR file as (note that c in the third line below is the relative lattice parameter in the *c* direction)

```
0.8660254037844   -.5   0.
0.0000000000000   1.    0.
0.0000000000000   0.    c
```

For the case of a tetragonal crystal, YPHON assumes that you defined the shape of the primitive unit cell of the crystal in the POSCAR file as

```
1.0   0.    0.
0.    1.0   0.
0.5   0.5   c
```

If you do not use the suffixes .fcc, .bcc, .hcp, and .tet2 for the *yourdisfile* file, YPHON will define the direction of the wave vector using the reciprocal lattice vector of the primitive unit cell. The reciprocal lattice vector is printed out as the last three lines, like for example

```
-0.000000000000   -0.362182366065   -0.000000000000
-0.313659560187   -0.181091751042   -0.000000000000
-0.000000000000   -0.000000000000   -0.192811598003
```

You can refer to these (only the direction is important) to define the direction of the phonon dispersion calculation.

For the dispersion calculation, we strongly recommend that one should refer to the web site www.cryst.ehu.es/ for the definition of the KVEC, i.e. the *k*-vector types and Brillouin zone for a specific space group.

A.7 Troubleshooting

Make sure that the Linux command search PATH has been modified correctly to include YPHON.

Make sure Python is installed correctly, otherwise pos2s will not work.

After your VASP.5 job is done, always have a look at the OSZICAR file to make sure the VASP.5 calculation finished normally.

Before running YPHON, always check the superfij.out file for data completeness and correctness. Nowadays the Linux clusters are not that stable, and sometimes very weird results can be observed as a result of certain problems due to RAM, disk space, node sharing by different Linux processes etc. Of course, the most frequent problem is that your job has run over the allowed time limit in your VASP.5 batch jobs.

Appendix B: SQS templates

B.1 16-atom SQS for fcc structure with composition $A_{0.25}B_{0.75}$

```
fccsqs16_25_75
1.00
     2.000000     2.000000    -4.000000
     0.000000    -6.000000    -2.000000
    -8.000000     2.000000    -2.000000
AA BB
12 4
D
     0.500000     0.875000     0.125000 AA
     0.250000     0.187500     0.312500 AA
     0.000000     0.500000     0.500000 AA
     0.500000     0.625000     0.375000 AA
     0.500000     0.375000     0.625000 AA
     0.750000     0.812500     0.687500 AA
     0.000000     0.250000     0.750000 AA
     0.250000     0.937500     0.562500 AA
     0.750000     0.562500     0.937500 AA
     0.250000     0.687500     0.812500 AA
     0.500000     0.125000     0.875000 AA
     0.000000     0.000000     0.000000 AA
     0.750000     0.312500     0.187500 BB
     0.250000     0.437500     0.062500 BB
     0.000000     0.750000     0.250000 BB
     0.750000     0.062500     0.437500 BB
```

B.2 16-atom SQS for fcc structure with composition $A_{0.5}B_{0.5}$

```
sqsfcc16_50
1.00
     4.000000    -2.000000    -2.000000
     0.000000     4.000000    -4.000000
     4.000000     6.000000     6.000000
AA BB
8 8
D
     0.250000     0.250000     0.250000 AA
     0.250000     0.750000     0.250000 AA
```

```
0.750000   0.500000   0.250000 AA
0.250000   0.000000   0.750000 AA
0.000000   0.250000   0.500000 AA
0.500000   0.750000   0.000000 AA
0.000000   0.500000   0.000000 AA
0.000000   0.000000   0.000000 AA
0.500000   0.500000   0.500000 BB
0.750000   0.000000   0.250000 BB
0.500000   0.000000   0.500000 BB
0.250000   0.500000   0.750000 BB
0.000000   0.750000   0.500000 BB
0.750000   0.250000   0.750000 BB
0.750000   0.750000   0.750000 BB
0.500000   0.250000   0.000000 BB
```

B.3 16-atom SQS for bcc structure with composition $A_{0.25}B_{0.75}$

```
bccsqs16_25_75
1.00
    4.000000   -8.000000    0.000000
    0.000000   -8.000000    4.000000
   -8.000000    0.000000   -8.000000
AA BB
12 4
D
    0.750000    0.750000    0.375000 AA
    0.250000    0.250000    0.125000 AA
    0.625000    0.625000    0.562500 AA
    0.875000    0.875000    0.687500 AA
    0.125000    0.125000    0.312500 AA
    0.375000    0.375000    0.437500 AA
    0.750000    0.750000    0.875000 AA
    0.000000    0.000000    0.000000 AA
    0.250000    0.250000    0.625000 AA
    0.500000    0.500000    0.750000 AA
    0.125000    0.125000    0.812500 AA
    0.375000    0.375000    0.937500 AA
    0.000000    0.000000    0.500000 BB
    0.500000    0.500000    0.250000 BB
    0.625000    0.625000    0.062500 BB
    0.875000    0.875000    0.187500 BB
```

B.4 16-atom SQS for bcc structure with composition $A_{0.5}B_{0.5}$

```
sqsbcc16_50
1.00
   -2.000000   -6.000000  -10.000000
   -2.000000   10.000000    6.000000
    6.000000    2.000000   -2.000000
```

```
AA BB
8 8
D
        0.000000    0.500000      0.500000 AA
        0.500000    0.000000      0.500000 AA
        0.250000    0.000000      0.750000 AA
        0.000000    0.250000      0.750000 AA
        0.250000    0.250000      0.500000 AA
        0.750000    0.000000      0.250000 AA
        0.500000    0.750000      0.750000 AA
        0.750000    0.250000      0.000000 AA
        0.500000    0.250000      0.250000 BB
        0.250000    0.500000      0.250000 BB
        0.750000    0.750000      0.500000 BB
        0.000000    0.750000      0.250000 BB
        0.750000    0.500000      0.750000 BB
        0.000000    0.000000      0.000000 BB
        0.500000    0.500000      0.000000 BB
        0.250000    0.750000      0.000000 BB
```

B.5 16-atom SQS for hcp structure with composition $A_{0.25}B_{0.75}$

```
hcpsqs16_25_75
1.00
         1.657150      2.870268    5.220900
        -3.314300      0.000000    5.220900
         6.628600    -11.481071    0.000000
AA BB
12 4
D
        0.000000      0.000000    0.250000 AA
        0.500000      0.500000    0.125000 AA
        0.583333      0.916667    0.062500 AA
        0.000000      0.000000    0.500000 AA
        0.500000      0.500000    0.375000 AA
        0.083333      0.416667    0.437500 AA
        0.083333      0.416667    0.187500 AA
        0.583333      0.916667    0.312500 AA
        0.000000      0.000000    0.750000 AA
        0.500000      0.500000    0.625000 AA
        0.083333      0.416667    0.687500 AA
        0.083333      0.416667    0.937500 AA
        0.583333      0.916667    0.562500 BB
        0.583333      0.916667    0.812500 BB
        0.000000      0.000000    0.000000 BB
        0.500000      0.500000    0.875000 BB
```

B.6 16-atom SQS for hcp structure with composition $A_{0.5}B_{0.5}$

```
hcpsqs16_50
1.00
     1.717000  -2.973931  -5.390800
    -3.434000  -5.947862   0.000000
    -8.585000   2.973931  -5.390800
AA BB
8 8
D
     0.958333   0.625000   0.541667 AA
     0.750000   0.250000   0.250000 AA
     0.250000   0.250000   0.750000 AA
     0.958333   0.125000   0.541667 AA
     0.500000   0.000000   0.500000 AA
     0.708333   0.375000   0.791667 AA
     0.000000   0.000000   0.000000 AA
     0.708333   0.875000   0.791667 AA
     0.750000   0.750000   0.250000 BB
     0.458333   0.625000   0.041667 BB
     0.250000   0.750000   0.750000 BB
     0.500000   0.500000   0.500000 BB
     0.208333   0.375000   0.291667 BB
     0.458333   0.125000   0.041667 BB
     0.208333   0.875000   0.291667 BB
     0.000000   0.500000   0.000000 BB
```

B.7 64-atom SQS for $L1_2$ structure with composition $(A_{0.25}B_{0.75})B_3$

```
sqs l12 (Al,Ni)Ni3 0.25
1.00
      3.580000   3.580000  -7.160000
      3.580000  -3.580000  -7.160000
    -14.320000   0.000000   0.000000
AA BB
4 60
D
      0.250000   0.250000   0.875000 AA
      0.250000   0.250000   0.125000 AA
      0.000000   0.000000   0.250000 AA
      0.000000   0.000000   0.000000 AA
      0.750000   0.750000   0.875000 BB
      0.750000   0.750000   0.625000 BB
      0.750000   0.750000   0.375000 BB
      0.750000   0.750000   0.125000 BB
      0.500000   0.500000   0.750000 BB
      0.500000   0.500000   0.500000 BB
      0.500000   0.500000   0.250000 BB
      0.500000   0.500000   0.000000 BB
      0.250000   0.250000   0.625000 BB
      0.250000   0.250000   0.375000 BB
```

0.000000	0.000000	0.750000 BB
0.000000	0.000000	0.500000 BB
0.000000	0.500000	0.750000 BB
0.625000	0.625000	0.687500 BB
0.875000	0.375000	0.812500 BB
0.000000	0.500000	0.500000 BB
0.625000	0.625000	0.437500 BB
0.875000	0.375000	0.562500 BB
0.000000	0.500000	0.250000 BB
0.625000	0.625000	0.187500 BB
0.875000	0.375000	0.312500 BB
0.000000	0.500000	0.000000 BB
0.625000	0.625000	0.937500 BB
0.875000	0.375000	0.062500 BB
0.750000	0.250000	0.625000 BB
0.375000	0.375000	0.562500 BB
0.625000	0.125000	0.687500 BB
0.750000	0.250000	0.375000 BB
0.375000	0.375000	0.312500 BB
0.625000	0.125000	0.437500 BB
0.750000	0.250000	0.125000 BB
0.375000	0.375000	0.062500 BB
0.625000	0.125000	0.187500 BB
0.750000	0.250000	0.875000 BB
0.375000	0.375000	0.812500 BB
0.625000	0.125000	0.937500 BB
0.500000	0.000000	0.750000 BB
0.125000	0.125000	0.687500 BB
0.375000	0.875000	0.812500 BB
0.500000	0.000000	0.500000 BB
0.125000	0.125000	0.437500 BB
0.375000	0.875000	0.562500 BB
0.500000	0.000000	0.250000 BB
0.125000	0.125000	0.187500 BB
0.375000	0.875000	0.312500 BB
0.500000	0.000000	0.000000 BB
0.125000	0.125000	0.937500 BB
0.375000	0.875000	0.062500 BB
0.250000	0.750000	0.625000 BB
0.875000	0.875000	0.562500 BB
0.125000	0.625000	0.687500 BB
0.250000	0.750000	0.375000 BB
0.875000	0.875000	0.312500 BB
0.125000	0.625000	0.437500 BB
0.250000	0.750000	0.125000 BB
0.875000	0.875000	0.062500 BB
0.125000	0.625000	0.187500 BB
0.250000	0.750000	0.875000 BB
0.875000	0.875000	0.812500 BB
0.125000	0.625000	0.937500 BB

B.8 64-atom SQS for $L1_2$ structure with composition $(A_{0.5}B_{0.5})B_3$

```
sqs l12 (Al,Ni)Ni3 0.50
1.00
      3.580000   7.160000     3.580000
      3.580000   0.000000    -3.580000
    -10.740000   7.160000   -10.740000
AA BB
8 56
D
      0.125000   0.500000     0.875000 AA
      0.750000   0.500000     0.750000 AA
      0.125000   0.000000     0.375000 AA
      0.875000   0.000000     0.625000 AA
      0.000000   0.000000     0.000000 AA
      0.500000   0.500000     0.000000 AA
      0.750000   0.000000     0.250000 AA
      0.875000   0.500000     0.125000 AA
      0.000000   0.500000     0.500000 BB
      0.250000   0.000000     0.750000 BB
      0.375000   0.500000     0.625000 BB
      0.625000   0.000000     0.875000 BB
      0.250000   0.500000     0.250000 BB
      0.500000   0.000000     0.500000 BB
      0.625000   0.500000     0.375000 BB
      0.375000   0.000000     0.125000 BB
      0.375000   0.750000     0.875000 BB
      0.250000   0.500000     0.750000 BB
      0.375000   0.250000     0.875000 BB
      0.250000   0.750000     0.500000 BB
      0.125000   0.500000     0.375000 BB
      0.250000   0.250000     0.500000 BB
      0.500000   0.250000     0.750000 BB
      0.375000   0.000000     0.625000 BB
      0.500000   0.750000     0.750000 BB
      0.625000   0.750000     0.625000 BB
      0.500000   0.500000     0.500000 BB
      0.625000   0.250000     0.625000 BB
      0.875000   0.250000     0.875000 BB
      0.750000   0.000000     0.750000 BB
      0.875000   0.750000     0.875000 BB
      0.000000   0.750000     0.750000 BB
      0.875000   0.500000     0.625000 BB
      0.000000   0.250000     0.750000 BB
      0.375000   0.250000     0.375000 BB
      0.250000   0.000000     0.250000 BB
      0.375000   0.750000     0.375000 BB
      0.500000   0.750000     0.250000 BB
      0.375000   0.500000     0.125000 BB
      0.500000   0.250000     0.250000 BB
      0.750000   0.250000     0.500000 BB
```

```
          0.625000   0.000000    0.375000 BB
          0.750000   0.750000    0.500000 BB
          0.875000   0.750000    0.375000 BB
          0.750000   0.500000    0.250000 BB
          0.875000   0.250000    0.375000 BB
          0.125000   0.250000    0.625000 BB
          0.000000   0.000000    0.500000 BB
          0.125000   0.750000    0.625000 BB
          0.250000   0.250000    0.000000 BB
          0.125000   0.000000    0.875000 BB
          0.250000   0.750000    0.000000 BB
          0.625000   0.250000    0.125000 BB
          0.500000   0.000000    0.000000 BB
          0.625000   0.750000    0.125000 BB
          0.750000   0.750000    0.000000 BB
          0.625000   0.500000    0.875000 BB
          0.750000   0.250000    0.000000 BB
          0.000000   0.250000    0.250000 BB
          0.875000   0.000000    0.125000 BB
          0.000000   0.750000    0.250000 BB
          0.125000   0.750000    0.125000 BB
          0.000000   0.500000    0.000000 BB
          0.125000   0.250000    0.125000 BB
```

B.9 64-atom SQS for $L1_2$ structure with composition $A(A_{0.25}B_{0.75})B_2$

```
sqs l12 Al(Al,Ni)Ni2 0.25
1.00
          3.580000   7.160000    3.580000
          3.580000   0.000000   -3.580000
        -10.740000   7.160000  -10.740000
AA BB
8 56
D
          0.125000   0.500000    0.875000 AA
          0.750000   0.500000    0.750000 AA
          0.125000   0.000000    0.375000 AA
          0.875000   0.000000    0.625000 AA
          0.000000   0.000000    0.000000 AA
          0.500000   0.500000    0.000000 AA
          0.750000   0.000000    0.250000 AA
          0.875000   0.500000    0.125000 AA
          0.000000   0.500000    0.500000 BB
          0.250000   0.000000    0.750000 BB
          0.375000   0.500000    0.625000 BB
          0.625000   0.000000    0.875000 BB
          0.250000   0.500000    0.250000 BB
          0.500000   0.000000    0.500000 BB
          0.625000   0.500000    0.375000 BB
```

```
0.375000  0.000000    0.125000 BB
0.375000  0.750000    0.875000 BB
0.250000  0.500000    0.750000 BB
0.375000  0.250000    0.875000 BB
0.250000  0.750000    0.500000 BB
0.125000  0.500000    0.375000 BB
0.250000  0.250000    0.500000 BB
0.500000  0.250000    0.750000 BB
0.375000  0.000000    0.625000 BB
0.500000  0.750000    0.750000 BB
0.625000  0.750000    0.625000 BB
0.500000  0.500000    0.500000 BB
0.625000  0.250000    0.625000 BB
0.875000  0.250000    0.875000 BB
0.750000  0.000000    0.750000 BB
0.875000  0.750000    0.875000 BB
0.000000  0.750000    0.750000 BB
0.875000  0.500000    0.625000 BB
0.000000  0.250000    0.750000 BB
0.375000  0.250000    0.375000 BB
0.250000  0.000000    0.250000 BB
0.375000  0.750000    0.375000 BB
0.500000  0.750000    0.250000 BB
0.375000  0.500000    0.125000 BB
0.500000  0.250000    0.250000 BB
0.750000  0.250000    0.500000 BB
0.625000  0.000000    0.375000 BB
0.750000  0.750000    0.500000 BB
0.875000  0.750000    0.375000 BB
0.750000  0.500000    0.250000 BB
0.875000  0.250000    0.375000 BB
0.125000  0.250000    0.625000 BB
0.000000  0.000000    0.500000 BB
0.125000  0.750000    0.625000 BB
0.250000  0.250000    0.000000 BB
0.125000  0.000000    0.875000 BB
0.250000  0.750000    0.000000 BB
0.625000  0.250000    0.125000 BB
0.500000  0.000000    0.000000 BB
0.625000  0.750000    0.125000 BB
0.750000  0.750000    0.000000 BB
0.625000  0.500000    0.875000 BB
0.750000  0.250000    0.000000 BB
0.000000  0.250000    0.250000 BB
0.875000  0.000000    0.125000 BB
0.000000  0.750000    0.250000 BB
0.125000  0.750000    0.125000 BB
0.000000  0.500000    0.000000 BB
0.125000  0.250000    0.125000 BB
```

B.10 64-atom SQS for $L1_2$ structure with composition $A(A_{0.5}B_{0.5})B_2$

```
sqs l12 Al(Al,Ni)Ni2 0.50
1.00
    0.000000    3.580000    7.160000
    7.160000    0.000000    0.000000
    0.000000   10.740000   -7.160000
AA BB
24 40
D
    0.125000    0.000000    0.625000 AA
    0.125000    0.500000    0.625000 AA
    0.375000    0.000000    0.875000 AA
    0.375000    0.500000    0.875000 AA
    0.000000    0.000000    0.000000 AA
    0.000000    0.500000    0.000000 AA
    0.250000    0.000000    0.250000 AA
    0.250000    0.500000    0.250000 AA
    0.500000    0.000000    0.500000 AA
    0.500000    0.500000    0.500000 AA
    0.750000    0.000000    0.750000 AA
    0.750000    0.500000    0.750000 AA
    0.625000    0.000000    0.125000 AA
    0.625000    0.500000    0.125000 AA
    0.875000    0.000000    0.375000 AA
    0.875000    0.500000    0.375000 AA
    0.125000    0.250000    0.125000 AA
    0.375000    0.250000    0.375000 AA
    0.625000    0.750000    0.625000 AA
    0.875000    0.750000    0.875000 AA
    0.750000    0.250000    0.250000 AA
    0.750000    0.750000    0.250000 AA
    0.000000    0.250000    0.500000 AA
    0.000000    0.750000    0.500000 AA
    0.250000    0.250000    0.750000 BB
    0.250000    0.750000    0.750000 BB
    0.500000    0.250000    0.000000 BB
    0.500000    0.750000    0.000000 BB
    0.125000    0.750000    0.125000 BB
    0.375000    0.750000    0.375000 BB
    0.625000    0.250000    0.625000 BB
    0.875000    0.250000    0.875000 BB
    0.312500    0.250000    0.562500 BB
    0.437500    0.000000    0.687500 BB
    0.312500    0.750000    0.562500 BB
    0.437500    0.500000    0.687500 BB
    0.562500    0.250000    0.812500 BB
    0.687500    0.000000    0.937500 BB
    0.562500    0.750000    0.812500 BB
    0.687500    0.500000    0.937500 BB
    0.187500    0.250000    0.937500 BB
```

```
0.312500   0.000000   0.062500 BB
0.187500   0.750000   0.937500 BB
0.312500   0.500000   0.062500 BB
0.437500   0.250000   0.187500 BB
0.562500   0.000000   0.312500 BB
0.437500   0.750000   0.187500 BB
0.562500   0.500000   0.312500 BB
0.687500   0.250000   0.437500 BB
0.812500   0.000000   0.562500 BB
0.687500   0.750000   0.437500 BB
0.812500   0.500000   0.562500 BB
0.937500   0.250000   0.687500 BB
0.062500   0.000000   0.812500 BB
0.937500   0.750000   0.687500 BB
0.062500   0.500000   0.812500 BB
0.812500   0.250000   0.062500 BB
0.937500   0.000000   0.187500 BB
0.812500   0.750000   0.062500 BB
0.937500   0.500000   0.187500 BB
0.062500   0.250000   0.312500 BB
0.187500   0.000000   0.437500 BB
0.062500   0.750000   0.312500 BB
0.187500   0.500000   0.437500 BB
```

B.11 24-atom SQS for fcc structure with composition $A_{1/3}B_{1/3}C_{1/3}$

```
ternary fcc sqs ABC
1.00
     12.000000   4.000000   -4.000000
    -12.000000   4.000000   -4.000000
      0.000000   2.000000    2.000000
AA BB CC
8 8 8
D
      0.166667   0.083333    0.250000 AA
      0.239583   0.197917    0.125000 AA
      0.125000   0.125000    0.250000 AA
      0.052083   0.010417    0.125000 AA
      0.145833   0.229167    0.250000 AA
      0.083333   0.166667    0.250000 AA
      0.072917   0.114583    0.125000 AA
      0.031250   0.156250    0.125000 AA
      0.229167   0.145833    0.250000 BB
      0.218750   0.093750    0.125000 BB
      0.156250   0.031250    0.125000 BB
      0.187500   0.187500    0.250000 BB
      0.114583   0.072917    0.125000 BB
      0.197917   0.239583    0.125000 BB
      0.020833   0.104167    0.250000 BB
```

```
0.093750 0.218750 0.125000 BB
0.208333 0.041667 0.250000 CC
0.104167 0.020833 0.250000 CC
0.177083 0.135417 0.125000 CC
0.062500 0.062500 0.250000 CC
0.250000 0.250000 0.250000 CC
0.135417 0.177083 0.125000 CC
0.010417 0.052083 0.125000 CC
0.041667 0.208333 0.250000 CC
```

B.12 32-atom SQS for fcc structure with composition $A_{0.5}B_{0.25}C_{0.25}$

```
ternary fcc sqs A2BC
1.00
 4.000000 4.000000  8.000000
 4.000000 4.000000 -8.000000
-4.000000 4.000000  0.000000
AA BB CC
16 8 8
D
0.250000 0.187500 0.062500 AA
0.062500 0.187500 0.250000 AA
0.062500 0.187500 0.125000 AA
0.250000 0.062500 0.062500 AA
0.062500 0.125000 0.062500 AA
0.125000 0.187500 0.062500 AA
0.250000 0.250000 0.250000 AA
0.062500 0.062500 0.250000 AA
0.125000 0.125000 0.250000 AA
0.062500 0.125000 0.187500 AA
0.125000 0.187500 0.187500 AA
0.125000 0.250000 0.250000 AA
0.187500 0.062500 0.250000 AA
0.062500 0.250000 0.187500 AA
0.187500 0.125000 0.187500 AA
0.125000 0.250000 0.125000 AA
0.250000 0.125000 0.250000 BB
0.250000 0.187500 0.187500 BB
0.250000 0.062500 0.187500 BB
0.250000 0.250000 0.125000 BB
0.187500 0.187500 0.125000 BB
0.062500 0.250000 0.062500 BB
0.187500 0.062500 0.125000 BB
0.187500 0.250000 0.187500 BB
0.250000 0.125000 0.125000 CC
0.187500 0.187500 0.250000 CC
0.062500 0.062500 0.125000 CC
0.125000 0.125000 0.125000 CC
0.125000 0.062500 0.062500 CC
```

```
0.187500   0.125000    0.062500 CC
0.125000   0.062500    0.187500 CC
0.187500   0.250000    0.062500 CC
```

B.13 80-atom SQS for ABO_3 structure with composition $A_{0.5}B_{0.25}C_{0.25}$

```
Title: by James
1.00
    0.000000    3.904000    7.808000
    7.808000    0.000000    0.000000
    0.000000   11.712000   -7.808000
AA BB O Ti
8 8 48 16
D
    0.750000    0.500000    0.750000 AA
    0.625000    0.500000    0.125000 AA
    0.500000    0.500000    0.500000 AA
    0.875000    0.500000    0.375000 AA
    0.000000    0.000000    0.000000 AA
    0.250000    0.000000    0.250000 AA
    0.625000    0.000000    0.125000 AA
    0.875000    0.000000    0.375000 AA
    0.375000    0.500000    0.875000 BB
    0.000000    0.500000    0.000000 BB
    0.250000    0.500000    0.250000 BB
    0.125000    0.500000    0.625000 BB
    0.750000    0.000000    0.750000 BB
    0.375000    0.000000    0.875000 BB
    0.500000    0.000000    0.500000 BB
    0.125000    0.000000    0.625000 BB
    0.562500    0.250000    0.812500 O
    0.937500    0.250000    0.687500 O
    0.187500    0.250000    0.937500 O
    0.875000    0.250000    0.875000 O
    0.500000    0.250000    0.000000 O
    0.125000    0.250000    0.125000 O
    0.062500    0.250000    0.312500 O
    0.437500    0.250000    0.187500 O
    0.812500    0.250000    0.062500 O
    0.375000    0.250000    0.375000 O
    0.750000    0.250000    0.250000 O
    0.312500    0.250000    0.562500 O
    0.687500    0.250000    0.437500 O
    0.625000    0.250000    0.625000 O
    0.250000    0.250000    0.750000 O
    0.000000    0.250000    0.500000 O
    0.687500    0.500000    0.937500 O
    0.312500    0.500000    0.062500 O
    0.187500    0.500000    0.437500 O
```

0.562500	0.500000	0.312500 O
0.937500	0.500000	0.187500 O
0.437500	0.500000	0.687500 O
0.812500	0.500000	0.562500 O
0.062500	0.500000	0.812500 O
0.562500	0.750000	0.812500 O
0.937500	0.750000	0.687500 O
0.187500	0.750000	0.937500 O
0.875000	0.750000	0.875000 O
0.500000	0.750000	0.000000 O
0.125000	0.750000	0.125000 O
0.062500	0.750000	0.312500 O
0.437500	0.750000	0.187500 O
0.812500	0.750000	0.062500 O
0.375000	0.750000	0.375000 O
0.750000	0.750000	0.250000 O
0.312500	0.750000	0.562500 O
0.687500	0.750000	0.437500 O
0.625000	0.750000	0.625000 O
0.250000	0.750000	0.750000 O
0.000000	0.750000	0.500000 O
0.687500	0.000000	0.937500 O
0.312500	0.000000	0.062500 O
0.187500	0.000000	0.437500 O
0.562500	0.000000	0.312500 O
0.937500	0.000000	0.187500 O
0.437500	0.000000	0.687500 O
0.812500	0.000000	0.562500 O
0.062500	0.000000	0.812500 O
0.687500	0.250000	0.937500 Ti
0.312500	0.250000	0.062500 Ti
0.187500	0.250000	0.437500 Ti
0.562500	0.250000	0.312500 Ti
0.937500	0.250000	0.187500 Ti
0.437500	0.250000	0.687500 Ti
0.812500	0.250000	0.562500 Ti
0.062500	0.250000	0.812500 Ti
0.687500	0.750000	0.937500 Ti
0.312500	0.750000	0.062500 Ti
0.187500	0.750000	0.437500 Ti
0.562500	0.750000	0.312500 Ti
0.937500	0.750000	0.187500 Ti
0.437500	0.750000	0.687500 Ti
0.812500	0.750000	0.562500 Ti
0.062500	0.750000	0.812500 Ti

References

1. M. Hillert, *Phase Equilibria, Phase Diagrams and Phase Transformations*, Cambridge University Press, Cambridge, 2nd edn., 2007.
2. E. A. Guggenheim, *Mixtures*, Clarendon Press, Oxford, 1952.
3. J. F. Nye, *Physical Properties of Crystals: Their Representation by Tensors and Matrices*, Clarendon Press, Oxford, 1985.
4. L. S. Palatnik and A. I. Landau, *Phase Equilibria in Multicomponent Systems*, Holt, Rinehart and Winston, London, 1964.
5. M. Hillert, "Principles of phase diagrams", *Int. Met. Rev.* 30 (1985) 45–67.
6. Z. K. Liu and Y. A. Chang, "Thermodynamic assessment of the Al–Fe–Si system", *Metall. Mater. Trans. A, Phys. Metall. Mater. Sci.* 30 (1999) 1081–1095.
7. Z. K. Liu, "Design magnesium alloys: how computational thermodynamics can help". In H. I. Kaplan, J. N. Hryn, and B. B. Clow, Eds., *Magnesium Technology 2000, Nashville, TN*, TMS, Warrendale, PA, 2000, pp. 191–198.
8. W. Kohn and L. J. Sham, "Self-consistent equations including exchange and correlation effects", *Phys. Rev.* 140 (1965) A1133–A1138.
9. A. R. H. Goodwin, K. N. Marsh, and W. A. Wakeham, Eds., *Measurement of the Thermodynamic Properties of Single Phases*, Elsevier, Amsterdam, 2003.
10. R. D. Weir and T. W. de Loos, Eds., *Measurement of the Thermodynamic Properties of Multiple Phases*, Elsevier, Amsterdam, 2005.
11. J.-C. Zhao, Ed., *Methods for Phase Diagram Determination*, Elsevier, Amsterdam, 2007.
12. K. N. Marsh and P. A. G. O'Hare, Eds., *Solution Calorimetry*, Blackwell Scientific, Oxford, 1994.
13. G. Kresse and D. Joubert, "From ultrasoft pseudopotentials to the projector augmented-wave method", *Phys. Rev. B* 59 (1999) 1758–1775.
14. G. Kresse and J. Furthmuller, "Efficiency of ab-initio total energy calculations for metals and semiconductors using a plane-wave basis set", *Comput. Mater. Sci.* 6 (1996) 15–50.
15. Y. Wang, L.-Q. Chen, and Z.-K. Liu, "YPHON: a package for calculating phonons of polar materials", *Commun. Comput. Phys.* 185 (2014) 2950–2968.
16. Y. Wang, Z. K. Liu, and L. Q. Chen, "Thermodynamic properties of Al, Ni, NiAl, and Ni_3Al from first-principles calculations", *Acta Mater.* 52 (2004) 2665–2671.
17. D. M. Teter, G. V. Gibbs, M. B. Boisen, D. C. Allan, and M. P. Teter, "First-principles study of several hypothetical silica framework structures", *Phys. Rev. B* 52 (1995) 8064–8073.
18. S. L. Shang, Y. Wang, D. Kim, and Z. K. Liu, "First-principles thermodynamics from phonon and Debye model: application to Ni and Ni_3Al", *Comput. Mater. Sci.* 47 (2010) 1040–1048.
19. J. J. Xie, S. de Gironcoli, S. Baroni, and M. Scheffler, "First-principles calculation of the thermal properties of silver", *Phys. Rev. B* 59 (1999) 965–969.

20. A. van de Walle, M. Asta, and G. Ceder, "The alloy theoretic automated toolkit: a user guide", *CALPHAD* 26 (2002) 539–553.
21. D. Alfe, "PHON: A program to calculate phonons using the small displacement method", *Comput. Phys. Commun.* 180 (2009) 2622–2633.
22. M. Kresch, O. Delaire, R. Stevens, J. Y. Y. Lin, and B. Fultz, "Neutron scattering measurements of phonons in nickel at elevated temperatures", *Phys. Rev. B* 75 (2007) 104301.
23. M. Born and K. Huang, *Dynamical Theory of Crystal Lattices*, Clarendon Press, Oxford, 1954.
24. B. Fultz, L. Anthony, L. J. Nagel, R. M. Nicklow, and S. Spooner, "Phonon densities of states and vibrational entropies of ordered and disordered Ni_3Al", *Phys. Rev. B* 52 (1995) 3315–3321.
25. M. Mostoller, R. M. Nicklow, D. M. Zehner, S. C. Lui, J. M. Mundenar, and E. W. Plummer, "Bulk and surface vibrational-modes in NiAl", *Phys. Rev. B* 40 (1989) 2856–2872.
26. C. Statassis, F. X. Kayser, C.-K. Loong, and D. Rach, "Lattice dynamics of Ni_3Al", *Phys. Rev. B* 24 (1981) 3048–3054.
27. M. E. Manley, G. H. Lander, H. Sinn, A. Alatas, W. L. Hults, R. J. McQueeney, J. L. Smith, and J. Willit, "Phonon dispersion in uranium measured using inelastic x-ray scattering", *Phys. Rev. B* 67 (2003) 052302.
28. Y. Wang, J. J. Wang, H. Zhang, V. R. Manga, S. L. Shang, L. Q. Chen, and Z. K. Liu, "A first-principles approach to elasticity at finite temperatures", *J. Phys. Condens. Matter* 22 (2010) 225404.
29. J. C. Slater, "A simplification of the Hartree–Fock method", *Phys. Rev.* 81 (1951) 385–390.
30. J. P. Perdew and A. Zunger, "Self-interaction correction to density-functional approximations for many-electron systems", *Phys. Rev. B* 23 (1981) 5048–5079.
31. J. P. Perdew and Y. Wang, "Accurate and simple analytic representation of the electron-gas correlation-energy", *Phys. Rev. B* 45 (1992) 13244–13249.
32. J. P. Perdew, K. Burke, and M. Ernzerhof, "Generalized gradient approximation made simple", *Phys. Rev. Lett.* 77 (1996) 3865–3868.
33. D. C. Wallace, *Thermodynamics of Crystals*, John Wiley & Sons, New York, 1972.
34. S. Baroni, S. de Gironcoli, A. Dal Corso, and P. Giannozzi, "Phonons and related crystal properties from density-functional perturbation theory", *Rev. Mod. Phys.* 73 (2001) 515–562.
35. A. van de Walle and G. Ceder, "The effect of lattice vibrations on substitutional alloy thermodynamics", *Rev. Mod. Phys.* 74 (2002) 11–45.
36. S. Baroni, P. Giannozzi, and A. Testa, "Elastic-constants of crystals from linear-response theory", *Phys. Rev. Lett.* 59 (1987) 2662–2665.
37. G. Kern, G. Kresse, and J. Hafner, "Ab initio calculation of the lattice dynamics and phase diagram of boron nitride", *Phys. Rev. B* 59 (1999) 8551–8559.
38. Y. Wang, J. J. Wang, W. Y. Wang, Z. G. Mei, S. L. Shang, L. Q. Chen *et al.*, "A mixed-space approach to first-principles calculations of phonon frequencies for polar materials", *J. Phys. Condens. Matter* 11 (2010) 202201.
39. C. Jiang, Ph.D. thesis, Theoretical studies of aluminum and aluminide alloys using CALPHAD and first-principles approach, Pennsylvania State University, Philadelphia, PA,2004.
40. C. Jiang, C. Wolverton, J. Sofo, L. Q. Chen, and Z. K. Liu, "First-principles study of binary bcc alloys using special quasirandom structures", *Phys. Rev. B* 69 (2004) 214202.
41. C. Sigli, M. Kosugi, and J. Sanchez, "Calculation of thermodynamic properties and phase diagrams of binary transition-metal alloys", *Phys. Rev. Lett.* 57 (1986) 253–256.

42. C. Wolverton and A. Zunger, "Ising-like description of structurally relaxed ordered and disordered alloys", *Phys. Rev. Lett.* 75 (1995) 3162–3165.
43. A. Zunger, S. H. Wei, L. G. Ferreira, and J. E. Bernard, "Special quasirandom structures", *Phys. Rev. Lett.* 65 (1990) 353.
44. A. van de Walle, P. Tiwary, M. de Jong, D. L. Olmsted, M. Asta, A. Dick *et al.*, "Efficient stochastic generation of special quasirandom structures", *CALPHAD* 42 (2013) 13–18.
45. Y. Wang, C. L. Zacherl, S. L. Shang, L. Q. Chen, and Z. K. Liu, "Phonon dispersions in random alloys: a method based on special quasi-random structure force constants", *J. Phys. Condens. Matter* 23 (2011) 485403.
46. B. Dutta, K. Bisht, and S. Ghosh, "Ab initio calculation of phonon dispersions in size-mismatched disordered alloys", *Phys. Rev. B* 82 (2010) 134207.
47. L. Kaufman and H. Bernstein, *Computer Calculation of Phase Diagrams*, Academic Press, New York, 1970.
48. N. Saunders and A. P. Miodownik, *CALPHAD (Calculation of Phase Diagrams): A Comprehensive Guide*, Pergamon, Oxford, 1998.
49. H. L. Lukas, S. G. Fries, and B. Sundman, *Computational Thermodynamics: The CALPHAD Method*, Cambridge University Press, Cambridge, 2007.
50. "Software for CALPHAD modeling", *CALPHAD* 26(2) (2002).
51. "Software for CALPHAD modeling", *CALPHAD* 33(2) (2009).
52. A. T. Dinsdale, "SGTE data for pure elements", *CALPHAD* 15 (1991) 317–425.
53. Y. Wang, S. Curtarolo, C. Jiang, R. Arroyave, T. Wang, G. Ceder *et al.*, "Ab initio lattice stability in comparison with CALPHAD lattice stability", *CALPHAD* 28 (2004) 79–90.
54. V. Ozolins, "First-principles calculations of free energies of unstable phases: the case of fcc W", *Phys. Rev. Lett.* 102 (2009) 065702.
55. M. Hillert, "The Compound energy fermalism". *J. Alloys Compd.* 320 (2001) 161–176.
56. B. Sundman, I. Ohnuma, N. Dupin, U. R. Kattner, and S. G. Fries, "An assessment of the entire Al–Fe system including D0(3) ordering", *Acta Mater.* 57 (2009) 2896–2908.
57. A. Kusoffsky, N. Dupin, and B. Sundman, "On the compound energy formalism applied to fcc ordering", *CALPHAD* 25 (2001) 549–565.
58. T. Abe and B. Sundman, "A description of the effect of short range ordering in the compound energy formalism", *CALPHAD* 27 (2003) 403–408.
59. Z. K. Liu, H. Zhang, S. Ganeshan, Y. Wang, and S. N. Mathaudhu, "Computational modeling of effects of alloying elements on elastic coefficients", *Scr. Matter* 63 (2010) 686–691.
60. M. Hillert, and M. A. Jarl, "Model for alloying effects in ferromagnatic metals", *CALPHAD* 2 (1978) 227–238.
61. W. Xiong, Q. Chen, P. A. Korzhavyi, and M. Selleby, "An improved magnetic model for thermodynamic modelling", *CALPHAD* 39 (2012) 11–20.
62. Haun, M. J., Furman, E., Jang, S. J., McKinstry, H. A. & Cross, L. E. "Thermodynamic theory of PbTiO 3", *J. Appl. Phys.* 62 (1987) 3331–8.
63. Scientific Group Thermodata Europe (SGTE), *Thermodynamic Properties of Inorganic Materials*. Lehrstuhl für Theoretische Hüttenkunde, Ed. Landolt-Boernstein New Series, Group IV, Springer, Berlin, 1999, vol. 19.
64. J. O. Andersson, T. Helander, L. H. Hoglund, P. F. Shi, and B. Sundman, "Thermo-Calc and DICTRA, computational tools for materials science", *CALPHAD* 26 (2002) 273–312.
65. M. Yang, Y. Zhong, and Z. K. Liu, "Defect analysis and thermodynamic modeling of $LaCoO_{3-\delta}$", *Solid State Ionics* 178 (2007) 1027–1032.

66. D. D. Macdonald, "Passivity – the key to our metals-based civilization", *Pure Appl. Chem.* 71 (1999) 951–978.
67. J. Larcin, W. C. Maskell, and F. L. Tye, "Leclanché cell investigations. 1. $Zn(NH_3)_2Cl_2$ solubility and the formation of $ZnCl_2{\cdot}4Zn(OH)_2{\cdot}H_2O$", *Electrochim. Acta* 42 (1997) 2649–2658.
68. Z. K. Liu, Y. Wang, and S. L. Shang, "Origin of negative thermal expansion phenomenon in solids", *Scripta Mater.* 65 (2011) 664–667.
69. Y. Wang, L. G. Hector, H. Zhang, S. L. Shang, L. Q. Chen, and Z. K. Liu, "Thermodynamics of the Ce gamma-alpha transition: density-functional study", *Phys. Rev. B* 78 (2008) 104113.
70. Y. Wang, L. G. Hector, H. Zhang, S. L. Shang, L. Q. Chen, and Z. K. Liu, "A thermodynamic framework for a system with itinerant-electron magnetism", *J. Phys. Condens. Matter* 21 (2009) 326003.
71. S. L. Dudarev, G. A. Botton, S. Y. Savrasov, C. J. Humphreys, and A. P. Sutton, "Electron-energy-loss spectra and the structural stability of nickel oxide: an LSDA+U study", *Phys. Rev. B* 57 (1998) 1505–1509.
72. Z.-K. Liu, Y. Wang, and S. Shang, "Thermal expansion anomaly regulated by entropy", *Sci. Rep.* 4 (2014) 7043.
73. Y. Wang, S. L. Shang, H. Zhang, L. Q. Chen, and Z. K. Liu, "Thermodynamic fluctuations in magnetic states: Fe_3Pt as a prototype", *Philos. Mag. Lett.* 90 (2010) 851–859.
74. National Science and Technology Council, "Materials Genome Initiative for Global Competitiveness", www.whitehouse.gov/sites/default/files/microsites/ostp/materials_genome_initiative-final.pdf, Office of Science and Technology Policy, Washington DC, June 2011.
75. L. Kaufman and J. Agren, "CALPHAD, first and second generation – birth of the materials genome", 70 (2014) 3–6.
76. Z. K. Liu, "Perspective on Materials Genome®", *Chin. Sci. Bull.* 59 (2014) 1619–1623.

Index